Game Theory, Alive

Game Theory, Alive

Anna R. Karlin
Yuval Peres

AMERICAN MATHEMATICAL SOCIETY
Providence, Rhode Island

2010 *Mathematics Subject Classification.* Primary 91A10, 91A12, 91A18, 91B12, 91A24, 91A43, 91A26, 91A28, 91A46, 91B26.

For additional information and updates on this book, visit
www.ams.org/bookpages/mbk-101

Library of Congress Cataloging-in-Publication Data

Names: Karlin, Anna R. | Peres, Y. (Yuval)
Title: Game theory, alive / Anna Karlin, Yuval Peres.
Description: Providence, Rhode Island : American Mathematical Society, [2016] | Includes bibliographical references and index.
Identifiers: LCCN 2016038151 | ISBN 9781470419820 (alk. paper)
Subjects: LCSH: Game theory. | AMS: Game theory, economics, social and behavioral sciences – Game theory – Noncooperative games. msc | Game theory, economics, social and behavioral sciences – Game theory – Cooperative games. msc | Game theory, economics, social and behavioral sciences – Game theory – Games in extensive form. msc | Game theory, economics, social and behavioral sciences – Mathematical economics – Voting theory. msc | Game theory, economics, social and behavioral sciences – Game theory – Positional games (pursuit and evasion, etc.). msc | Game theory, economics, social and behavioral sciences – Game theory – Games involving graphs. msc | Game theory, economics, social and behavioral sciences – Game theory – Rationality, learning. msc | Game theory, economics, social and behavioral sciences – Game theory – Signaling, communication. msc | Game theory, economics, social and behavioral sciences – Game theory – Combinatorial games. msc | Game theory, economics, social and behavioral sciences – Mathematical economics – Market models (auctions, bargaining, bidding, selling, etc.). msc
Classification: LCC QA269 .K3684 2016 | DDC 519.3–dc23 LC record available at https://lccn.loc.gov/2016038151

Contents

Acknowledgements

We are grateful to Alan Hammond, Yun Long, Gábor Pete, and Peter Ralph for scribing early drafts of some of the chapters in this book from lectures by Yuval Peres. These drafts were edited by Liat Kessler, Asaf Nachmias, Sara Robinson, Yelena Shvets, and David Wilson. We are especially indebted to Yelena Shvets for her contributions to the chapters on combinatorial games and voting, and to David Wilson for numerous insightful edits. Their efforts greatly improved the book. David was also a major force in writing the paper [PSSW07], which is largely reproduced in Chapter 9. The expository style of that paper also inspired our treatment of other topics in the book.

We thank TJ Gilbrough, Christine Hill, Isaac Kuek, Davis Shepherd, and Yelena Shvets for creating figures for the book. We also thank Ranjit Samra for the lemon figure and Barry Sinervo for the picture in Figure 7.2. The artistic drawings in the book were created by Gabrielle Cohn, Nate Jensen, and Yelena Shvets. We are extremely grateful to all three of them.

Peter Bartlett, Allan Borodin, Sourav Chatterjee, Amos Fiat, Elchanan Mossel, Asaf Nachmias, Allan Sly, Shobhana Stoyanov, and Markus Vasquez taught from drafts of the book and provided valuable suggestions and feedback. Thanks also to Omer Adelman, John Buscher, Gabriel Carroll, Jiechen Chen, Wenbo Cui, Varsha Dani, Kira Goldner, Nick Gravin, Brian Gu, Kieran Kishore, Elias Koutsoupias, Itamar Landau, Shawn Lee, Eric Lei, Bryan Lu, Andrea McCool, Mallory Monasterio, Andrew Morgan, Katia Nepom, Uuganbaatar Ninjbat, Miki Racz, Colleen Ross, Zhuohong Shen, Davis Shepherd, Stephanie Somersille, Kuai Yu, Sithparran Vanniasegaram, and Olga Zamaraeva for their helpful comments and corrections.

The support of the NSF VIGRE grant to the Department of Statistics at the University of California, Berkeley, and NSF grants DMS-0244479, DMS-0104073, CCF-1016509, and CCF-1420381 is gratefully acknowledged.

Credits

The American Mathematical Society gratefully acknowledges the kindness of the following institutions and individuals in granting permissions:

Cover Art

    Gabrielle Cohn: Front cover

    Nate Jensen: Back cover

Back Cover Photo

    Courtesy of Gireeja Ranade

Preface

    TJ Gilbrough: Figures 1 and 10

    Nate Jensen: Figures 4, 5, 6, 7, 8, and 9

    Yelena Shvets: Figure 2

    David B. Wilson: Figure 3

Chapter 1: Combinatorial games

    TJ Gilbrough: Figures 1.4, 1.14, 1.15, and 1.18

    Yelena Shvets: Figures 1.1, 1.2, 1.3, 1.5, 1.6, 1.7, 1.8, 1.9, 1.10, 1.11, 1.12, 1.13, 1.16, 1.17, 1.19, 1.20, 1.21, 1.22, 1.23, and the figure in Exercise 1.7

Chapter 2: Two-person zero-sum games

    Christine Hill: Figure 2.7

    Nate Jensen: Figures 2.1 and 2.6

    Davis Shepherd: Figure 2.2

    Yelena Shvets: Figures 2.3, 2.4, and 2.5

Chapter 3: Zero-sum games on graphs

    Gabrielle Cohn: Figures 3.1, 3.2, and 3.9

    TJ Gilbrough: Figure 3.15

    Christine Hill: Figures 3.11 and 3.13

    Isaac Kuek: Figures 3.7, 3.8, 3.17, and 3.18

Davis Shepherd: Figure 3.5

Yelena Shvets: Figures 3.3, 3.4, 3.6, and 3.16

Peter Winkler: Figures 3.10, 3.12, and 3.14

Chapter 4: General-sum games

Gabrielle Cohn: Figure 4.12

Christine Hill: Figures 4.10, and 4.13

Nate Jensen: Figures 4.2, 4.6, 4.7, and 4.8

Isaac Kuek: Figures 4.4 and 4.11

Ranjit Samra: Figure 4.14 (www.rojaysoriginalart.com)

Yelena Shvets: Figures 4.1, 4.3, 4.5, and 4.9

Chapter 5: Existence of Nash equilibria and fixed points

TJ Gilbrough: Figure 5.4

TJ Gilbrough and **Yelena Shvets**: Figure 5.8

Christine Hill: Figures 5.3, 5.9, 5.10, 5.11, 5.12, and 5.13

Yelena Shvets: Figures 5.1, 5.2, 5.5, 5.6, and 5.7

Chapter 6: Games in extensive form

Gabrielle Cohn: Figure 6.13

TJ Gilbrough and **Christine Hill**: Figures 6.1, 6.2, 6.5, 6.10, 6.11, and 6.12

Christine Hill: Figures 6.3, 6.4, 6.6, and 6.7

Nate Jensen: Figure 6.8

Isaac Kuek: Figures 6.14, 6.15, 6.16, and 6.17

Yelena Shvets: Figure 6.9

Chapter 7: Evolutionary and correlated equilibria

Gabrielle Cohn: Figure 7.3

TJ Gilbrough: Figures 7.4 and 7.5

Yelena Shvets: Figures 7.1 and 7.6

Barry Sinervo: Figure 7.2

Chapter 8: The price of anarchy

TJ Gilbrough: Figure 8.10

TJ Gilbrough and **Christine Hill**: Figure 8.11

TJ Gilbrough and **Isaac Kuek**: Figures 8.1, 8.2, 8.3, 8.4, 8.5, 8.6, 8.7, 8.8, 8.9, and 8.12

Chapter 9: Random-turn games

Yelena Shvets: Figures 9.2 and 9.3

David B. Wilson: Figures 9.1 and 9.4

Chapter 10: Stable matching and allocation

TJ Gilbrough: Figure 10.8

TJ Gilbrough and **Christine Hill**: Figures 10.5 and 10.6

Alexander E. Holroyd: Figures 10.7. Alexander E. Holroyd, Robin Pemantle, Yuval Peres, and Oded Schramm, Poisson matching, Ann. Inst. Henri Poincaré Probab. Stat. **45**(1):266–287, 2009.

Alexander E. Holroyd: Figure 10.10. Christopher Hoffman, Alexander E. Holroyd, and Yuval Peres, A stable marriage of Poisson and Lebesgue, Ann. Probab. **34**(4):1241–1272, 2006.

Nate Jensen: Figure 10.1

Isaac Kuek: Figure 10.9

Yelena Shvets: Figures 10.2, 10.3, and 10.4

Chapter 11: Fair division

TJ Gilbrough: Figures 11.10 and 11.11

TJ Gilbrough and **Christine Hill**: Figure 11.6

Christine Hill: Figures 11.2, 11.3, 11.4, 11.5, and 11.8

Nate Jensen: Figures 11.1, 11.7, and 11.9

Chapter 12: Cooperative games

TJ Gilbrough: Figures 12.2 and 12.3

Yelena Shvets: Figure 12.1

Chapter 13: Social choice and voting

TJ Gilbrough: Figures 13.14, 13.15, 13.16, and 13.17

TJ Gilbrough and **Yelena Shvets**: Figures 13.1, 13.2, 13.3, 13.4, 13.5, 13.6, 13.7, 13.8, 13.9, 13.10, 13.11, and 13.13

Chapter 14: Auctions

Gabrielle Cohn: Figures 14.1 and 14.4

TJ Gilbrough and **Isaac Kuek**: Figures 14.2, 14.3, 14.6, 14.7

Nate Jensen: Figures 14.5 and 14.8

Chapter 15: Truthful auctions in win/lose settings

TJ Gilbrough and **Isaac Kuek**: Figures 15.2, 15.3, 15.4, 15.6, 15.7, 15.8, 15.9, and 15.10

Nate Jensen: Figures 15.1 and 15.12

Davis Shepherd: Figure 15.11

Chapter 16: VCG and scoring rules

Gabrielle Cohn: Figure 16.1

TJ Gilbrough and **Christine Hill**: Figure 16.3

Isaac Kuek: Figure 16.2

Chapter 17: Matching markets

TJ Gilbrough: Figure 17.2

TJ Gilbrough and **Isaac Kuek**: Figure 17.1

Nate Jensen: Figure 17.3

Chapter 18: Adaptive decision making

Gabrielle Cohn: Figure 18.2

TJ Gilbrough: Figure 18.1

TJ Gilbrough and **Isaac Kuek**: Figure 18.3

Appendix A: Linear programming

Davis Shepherd: Figures A.1 and A.2

Appendix C: Convex functions

TJ Gilbrough: Figures C.1, C.2, C.3, C.4, and C.5

Appendix D: Solution sketches for selected exercises

Davis Shepherd: Figure D.1

Yelena Shvets: Figure D.2

Photos (in alphabetical order)

George Akerlof: Source: Wikimedia Commons, credit: Yan Chi Vinci Chow (http://www.flickr.com/photos/ticoneva/444805175). The work is licensed under Creative Commons Attribution 3.0 Unported.

Kenneth Arrow: Source: Wikimedia Commons, credit: Linda A. Cicero/ Stanford News Service, 2008. The work is licensed under Creative Commons Attribution 3.0 Unported.

Robert John Aumann: Source: Wikimedia Commons, credit: DMY

David Blackwell: Courtesy of Archives of the Mathematisches Forschungsinstitut Oberwolfach, © George Bergman

Gabrielle Demange: © Aurore Bagarry–2015

David Gale: Courtesy of Archives of the Mathematisches Forschungsinstitut Oberwolfach, © George Bergman

James Hannan: Courtesy of Bettie Hannan

Anna Karlin and **Yuval Peres** (back cover): Photo by Gireeja Ranade

Elias Koutsoupias: Courtesy of Elias Koutsoupias

Oskar Morgenstern: © Archiv der Universität Wien

Roger Myerson: Source: Wikimedia Commons, credit: MS!Mit8rd. The work is licensed under Creative Commons Attribution-Share Alike 3.0 Unported.

John Nash: Picture credit: Princeton University Library. Reference citation: John F. Nash, Jr.; circa 1948–1950, 1994, 2002; Graduate Alumni Records, Princeton University Archives, Department of Rare Books and Special Collections, Princeton University Library.

Noam Nisan: Courtesy of Noam Nisan

Christos Papadimitriou: Credit: Eirene Markenscoff-Papadimitriou

Julia Robinson: Courtesy of Archives of the Mathematisches Forschungsinstitut Oberwolfach, © George Bergman

Alvin E. Roth: Source: Wikimedia Commons, credit: Bengt Nyman (http://www.flickr.com/people/97469566@N00). The work is licensed under Creative Commons Attribution 2.0 Generic.

Tim Roughgarden: Courtesy of Tim Roughgarden

Donald Saari: © Donald Saari

Thomas Schelling: Source: Wikimedia Commons, credit: Hessam Armandehi. The work is licensed under GNU Free Documentation License and Creative Commons Attribution 3.0 Unported.

L. S. Shapley: © Archives of the Mathematisches Forschungsinstitut Oberwolfach

John Maynard Smith: Source: Wikimedia Commons, credit: Web of Stories. The work is licensed under
Creative Commons Attribution-Share Alike 3.0 Unported.

Marilda Sotomayor: Courtesy of Gustavo Morita

Eva Tardos: Courtesy of Eva Tardos

William Vickery: Courtesy of University Archives, Columbia University in the City of New York

John von Neumann: American Mathematical Society

Preface

We live in a highly connected world, with multiple self-interested agents interacting, leading to myriad opportunities for conflict and cooperation. Understanding these is the goal of game theory. It finds application in fields such as economics, business, political science, biology, psychology, sociology, computer science, and engineering. Conversely, ideas from the social sciences (e.g., fairness), from biology (evolutionary stability), from statistics (adaptive learning), and from computer science (complexity of finding equilibria) have greatly enriched game theory. In this book, we present an introduction to this field. We will see applications from a variety of disciplines and delve into some of the fascinating mathematics that underlies game theory.

An overview of the book

Part I: Analyzing games: Strategies and equilibria. We begin in **Chapter 1** with **combinatorial games**, in which two players take turns making moves until a winning position for one of them is reached.

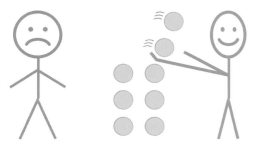

FIGURE 1. Two people playing Nim.

A classic example of a combinatorial game is **Nim**. In this game, there are several piles of chips, and players take turns removing one or more chips from a single pile. The player who takes the last chip wins. We will describe a winning strategy for Nim and show that a large class of combinatorial games can be reduced to it.

Other well-known combinatorial games are Chess, Go, and Hex. The youngest of these is **Hex**, which was invented by Piet Hein in 1942 and independently by John Nash in 1947. Hex is played on a rhombus-shaped board tiled with small hexagons (see Figure 2). Two players, Blue and Yellow, alternate coloring in hexagons in their assigned color, blue or yellow, one hexagon per turn. Blue wins if she produces

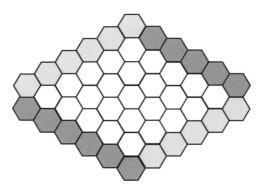

FIGURE 2. The board for the game of Hex.

a blue chain crossing between her two sides of the board and Yellow wins if he produces a yellow chain connecting the other two sides.

We will show that the player who moves first has a winning strategy; finding this strategy remains an unsolved problem, except when the board is small.

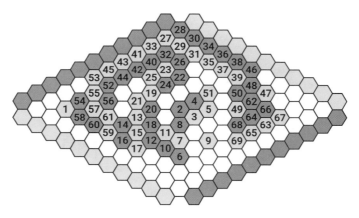

FIGURE 3. The board position near the end of the match between Queenbee and Hexy at the 5th Computer Olympiad. Each hexagon is labeled by the time at which it was placed on the board. Blue moves next, but Yellow has a winning strategy. Can you see why?

In an interesting variant of the game, the players, instead of alternating turns, toss a coin to determine who moves next. In this case, we can describe optimal strategies for the players. Such **random-turn combinatorial games** are the subject of **Chapter 9**.

In **Chapters 2–5**, we consider games in which the players simultaneously select from a set of possible actions. Their selections are then revealed, resulting in a payoff to each player. For two players, these payoffs are represented using the matrices $A = (a_{ij})$ and $B = (b_{ij})$. When player I selects action i and player II selects action j, the payoffs to these players are a_{ij} and b_{ij}, respectively. Two-person games where one player's gain is the other player's loss, that is, $a_{ij} + b_{ij} = 0$ for all i, j, are called **zero-sum games**. Such games are the topic of **Chapter 2**.

We show that every zero-sum game has a **value** V such that player I can ensure her expected payoff is at least V (no matter how II plays) and player II can ensure he pays I at most V (in expectation) no matter how I plays.

For example, in **Penalty Kicks**, a zero-sum game inspired by soccer, one player, the kicker, chooses to kick the ball either to the left or to the right of the other player, the goalie. At the same instant as the kick, the goalie guesses whether to dive left or right.

FIGURE 4. The game of Penalty Kicks.

The goalie has a chance of saving the goal if he dives in the same direction as the kick. The kicker, who we assume is right-footed, has a greater likelihood of success if she kicks right. The probabilities that the penalty kick scores are displayed in the table below:

		goalie	
		L	R
kicker	L	0.5	1
	R	1	0.8

For this set of scoring probabilities, the optimal strategy for the kicker is to kick left with probability $2/7$ and kick right with probability $5/7$ — then regardless of what the goalie does, the probability of scoring is $6/7$. Similarly, the optimal strategy for the goalie is to dive left with probability $2/7$ and dive right with probability $5/7$.

Chapter 3 goes on to analyze a number of interesting zero-sum games on graphs. For example, we consider a game between a **Troll and a Traveler**. Each of them chooses a route (a sequence of roads) from Syracuse to Troy, and then they simultaneously disclose their routes. Each road has an associated toll. For each road chosen by both players, the traveler pays the toll to the troll. We find optimal strategies by developing a connection with electrical networks.

In **Chapter 4** we turn to **general-sum games**. In these games, players no longer have optimal strategies. Instead, we focus on situations where each player's strategy is a *best response* to the strategies of the opponents: a **Nash equilibrium** is an assignment of (possibly randomized) strategies to the players, with the property that no player can gain by unilaterally changing his strategy. It turns out that every general-sum game has at least one Nash equilibrium. The proof of this fact requires an important geometric tool, the **Brouwer fixed-point theorem**, which is covered in **Chapter 5**.

FIGURE 5. Prisoner's Dilemma: the prisoners considering the possible consequences of confessing or remaining silent.

The most famous general-sum game is the **Prisoner's Dilemma**. If one prisoner confesses and the other remains silent, then the first goes free and the second receives a ten-year sentence. They will be sentenced to eight years each if they both confess and to one year each if they both remain silent. The only equilibrium in this game is for both to confess, but the game becomes more interesting when it is repeated, as we discuss in **Chapter 6**. More generally, in Chapter 6 we consider games where players alternate moves as in Chapter 1, but the payoffs are general as in Chapter 4. These are called **extensive-form** games. Often these games involve **imperfect information**, where players do not know all actions that have been taken by their opponents. For instance, in the 1962 Cuban Missile Crisis, the U.S. did not know whether the U.S.S.R. had installed nuclear missiles in Cuba and had to decide whether to bomb the missile sites in Cuba without knowing whether or not they were fitted with nuclear warheads. (The U.S. used a naval blockade instead.) We also consider games of **incomplete information** where the players do not even know exactly what game they are playing. For instance, in poker, the potential payoffs to a player depend on the cards dealt to his opponents.

One criticism of optimal strategies and equilibria in game theory is that finding them requires hyperrational players that can analyze complicated strategies. However, it was observed that populations of termites, spiders, and lizards can arrive at a Nash equilibrium just via natural selection. The equilibria that arise in such populations have an additional property called **evolutionary stability**, which is discussed in **Chapter 7**.

In the same chapter, we also introduce **correlated equilibria**. When two drivers approach an intersection, there is no good Nash equilibrium. For example, the convention of yielding to a driver on your right is problematic in a four-way intersection. A traffic light serves as a correlating device that ensures each driver is incentivized to follow the indications of the light. Correlated equilibria generalize this idea.

In **Chapter 8**, we compare outcomes in Nash equilibrium to outcomes that could be achieved by a central planner optimizing a global objective function. For example, in Prisoner's Dilemma, the total loss (combined jail time) in the unique Nash equilibrium is 16 years; the minimum total loss is 2 years (if both stay silent). Thus, the ratio, known as the **price of anarchy** of the game, is 8. Another example compares the average driving time in a road network when the drivers are selfish (i.e., in a Nash equilibrium) to the average driving time in an optimal routing.

FIGURE 6. An unstable pair.

Part II: Designing games and mechanisms. So far, we have considered predefined games, and our goal was to understand the outcomes that we can expect from rational players. In the second part of the book, we also consider **mechanism design** where we start with desired properties of the outcome (e.g., high profit or fairness) and attempt to *design* a game (or market or scheme) that incentivizes players to reach an outcome that meets our goals. Applications of mechanism design include voting systems, auctions, school choice, environmental regulation, and organ donation.

For example, suppose that there are n men and n women, where each man has a preference ordering of the women and vice versa. A matching between them is **stable** if there is no *unstable pair*, i.e., a man and woman who prefer each other to their partners in the matching. In **Chapter 10**, we introduce the Gale-Shapley algorithm for finding a stable matching. A generalization of stable matching is used by the National Resident Matching Program, which matches about 20,000 new doctors to residency programs at hospitals every year.

Chapter 11 considers the design of mechanisms for **fair division**. Consider the problem of dividing a cake with several different toppings among several people. Each topping is distributed over some portion of the cake, and each person prefers some toppings to others. If there are just two people, there is a well-known mechanism for dividing the cake: One cuts it in two, and the other chooses which piece to take. Under this system, each person is at least as happy with what he receives as he would be with the other person's share. What if there are three or more people? We also consider a 2000-year-old problem: how to divide an estate between several creditors whose claims exceed the value of the estate.

The topic of **Chapter 12** is **cooperative game theory**, in which players form coalitions in order to maximize their utility. As an example, suppose that three people have gloves to sell. Two are each selling a single, left-handed glove, while the third is selling a right-handed one. A wealthy tourist enters the store in dire need of a pair of gloves. She refuses to deal with the glove-bearers individually, so at least two of them must form a coalition to sell a left-handed and a right-handed glove to her. The third player has an advantage because his commodity is in scarcer supply. Thus, he should receive a higher fraction of the price the tourist pays. However, if he holds out for too high a fraction of the payment, the other players may agree between themselves that he must pay both of them in order to obtain a left glove. A related topic discussed in the chapter is **bargaining**, where the classical solution is again due to Nash.

FIGURE 7. Voting in Florida during the 2000 U.S. presidential election.

In **Chapter 13** we turn to **social choice**: designing mechanisms that aggregate the preferences of a collection of individuals. The most basic example is the design of **voting schemes**. We prove Arrow's Impossibility Theorem, which implies that all voting systems are strategically vulnerable. However, some systems are better than others. For example, the widely used system of runoff elections is not even monotone; i.e., transferring votes from one candidate to another might lead the second candidate to lose an election he would otherwise win. In contrast, Borda count and approval voting are monotone and more resistant to manipulation.

Chapter 14 studies **auctions** for a single item. We compare different auction formats such as **first-price** (selling the item to the highest bidder at a price equal to his bid) and **second-price** (selling the item to the highest bidder at a price

FIGURE 8. An auction for a painting.

equal to the second highest bid). In first-price auctions, bidders must bid below their value if they are to make any profit; in contrast, in a second-price auction, it is optimal for bidders to simply bid their value. Nevertheless, the **Revenue Equivalence Theorem** shows that, in equilibrium, if the bidders' values are independent and identically distributed, then the expected auctioneer revenue in the first-price and second-price auctions is the same. We also show how to design optimal (i.e., revenue-maximizing) auctions under the assumption that the auctioneer has good prior information about the bidders' values for the item he is selling.

    Chapters 15 and 16 discuss truthful mechanisms that go beyond the second-price auction, in particular, the **Vickrey-Clarke-Groves (VCG)** mechanism for maximizing **social surplus**, the total utility of all participants in the mechanism. A key application is to **sponsored search auctions**, the auctions that search engines like Google and Bing run every time you perform a search. In these auctions, the bidders are companies that wish to place their advertisements in one of the slots you see when you get the results of your search. In Chapter 16, we also discuss **scoring rules**. For instance, how can we incentivize a meteorologist to give the most accurate prediction he can?

    Chapter 17 considers **matching markets**. A certain housing market has n homeowners and n potential buyers. Buyer i has a value v_{ij} for house j. The goal is to find an allocation of houses to buyers and corresponding prices that are stable; i.e., there is no pair of buyer and homeowner that can strike a better deal. A related problem is allocating rooms to renters in a shared rental house. See Figure 9.

    Finally, **Chapter 18** concerns **adaptive decision making**. Suppose that each day several experts suggest actions for you to take; each possible action has a reward (or penalty) that varies between days and is revealed only after you choose.

FIGURE 9. Three roommates need to decide who will get each room and how much of the rent each person will pay.

Surprisingly, there is an algorithm that ensures your average reward over many days (almost) matches that of the best expert. If two players in a repeated zero-sum game employ such an algorithm, the empirical distribution of play for each of them will converge to an optimal strategy.

For the reader and instructor

Prerequisites. Readers should have taken basic courses in probability and linear algebra. Starred sections and subsections are more difficult; some require familiarity with mathematical analysis that can be acquired, e.g., in [Rud76].

Courses. This book can be used for different kinds of courses. For instance, an undergraduate game theory course could include Chapter 1 (combinatorial games), Chapter 2 and most of Chapter 3 on zero-sum games, Chapters 4 and 7 on general-sum games and different types of equilibria, Chapter 10 (stable matching), parts of Chapters 11 (fair division), 13 (social choice), and possibly 12 (especially the Shapley value). Indeed, this book started from lecture notes to such a course that was given at Berkeley for several years by the second author.

A course for computer science students might skip some of the above chapters (e.g., combinatorial games) and instead emphasize Chapter 9 on price of anarchy, Chapters 14–16 on auctions and VCG, and possibly parts of Chapters 17 (matching markets) and 18 (adaptive decision making). The topic of stable matching (Chapter 10) is a gem that requires no background and could fit in any course. The logical dependencies between the chapters are shown in Figure 10.

There are solution outlines to some problems in Appendix D. Such solutions are labeled with an "S" in the text. More difficult problems are labeled with a *. Additional exercises and material can be found at:

http://homes.cs.washington.edu/~karlin/GameTheoryAlive

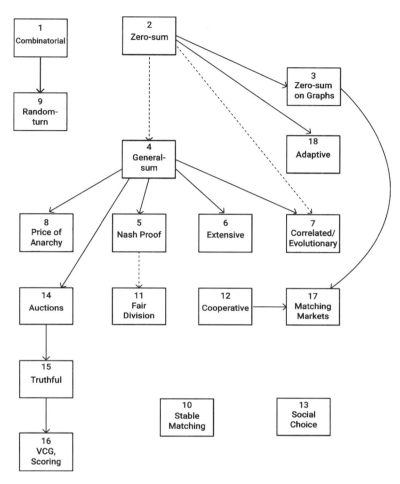

FIGURE 10. Chapter dependencies.

Notes

There are many excellent books on game theory. In particular, in writing this book, we consulted Ferguson [Fer08], Gintis [Gin00], González-Díaz et al. [GDGJFJ10], Luce and Raiffa [LR57], Maschler, Solan, and Zamir [MSZ13], Osborne and Rubinstein [OR94], Owen [Owe95], the survey book on algorithmic game theory [NRTV07], and the handbooks of game theory, Volumes 1–4 (see, e.g., [AH92]).

The entries in the payoff matrices for zero-sum games represent the **utility** of the players, and throughout the book we assume that the goal of each agent is maximizing his expected utility. Justifying this assumption is the domain of utility theory, which is discussed in most game theory books.

The Penalty Kicks matrix we gave was idealized for simplicity. Actual data on 1,417 penalty kicks from professional games in Europe was collected and analyzed by Palacios-Huerta [PH03]. The resulting matrix is

		goalie	
		L	R
kicker	L	0.58	0.95
	R	0.93	0.70

Here 'R' represents the dominant (natural) side for the kicker. Given these probabilities, the optimal strategy for the kicker is $(0.38, 0.62)$ and the optimal strategy for the goalie is $(0.42, 0.58)$. The observed frequencies were $(0.40, 0.60)$ for the kicker and $(0.423, 0.577)$ for the goalie.

The early history of the theory of strategic games from Waldegrave to Borel is discussed in [DD92].

Part I: Analyzing games: Strategies and equilibria

CHAPTER 1

Combinatorial games

In a combinatorial game, there are two players, a set of positions, and a set of
legal moves between positions. The players take turns moving from one position
to another. Some of the positions are terminal. Each terminal position is labelled
as winning for either player I or player II. We will concentrate on combinatorial
games that terminate in a finite number of steps.

EXAMPLE 1.0.1 (**Chomp**). In Chomp, two players take turns biting off a chunk
of a rectangular bar of chocolate that is divided into squares. The bottom left corner
of the bar has been removed and replaced with a broccoli floret. Each player, in
his turn, chooses an uneaten chocolate square and removes it along with all the
squares that lie above and to the right of it. The person who bites off the last piece
of chocolate wins and the loser has to eat the broccoli (i.e., the terminal position
is when all the chocolate is gone.) See Figure 1.1. We will return to Chomp in
Example 1.1.6.

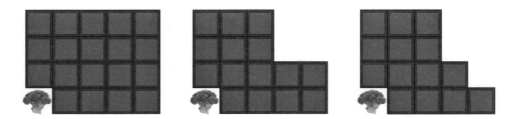

FIGURE 1.1. Two moves in a game of Chomp.

DEFINITION 1.0.2. A combinatorial game with a position set X is said to be
progressively bounded if, for every starting position $x \in X$, there is a finite
bound on the number of moves until the game terminates. Let $B(x)$ be the maximum number of moves from x to a terminal position.

Combinatorial games generally fall into two categories: Those for which the
winning positions and the available moves are the same for both players (e.g., Nim),
are called **impartial**. The player who first reaches one of the terminal positions
wins the game. All other games are called **partisan**. In such games (e.g., Hex),
either the players have different sets of winning positions, or from some position
their available moves differ.[1]

---

[1] In addition, some partisan games (e.g., Chess) may terminate in a draw (or tie), but we will
not consider those here.

For a given combinatorial game, our goal will be to find out whether one of the players can always force a win and, if so, to determine the *winning strategy* – the moves this player should make under every contingency. We will show that, in a progressively bounded combinatorial game with no ties, one of the players has a winning strategy.

1.1. Impartial games

EXAMPLE 1.1.1 (**Subtraction**). Starting with a pile of $x \in \mathbb{N}$ chips, two players alternate taking 1 to 4 chips. The player who removes the last chip wins.

Observe that starting from any $x \in \mathbb{N}$, this game is progressively bounded with $B(x) = x$. If the game starts with 4 or fewer chips, the first player has a winning move: she just removes them all. If there are 5 chips to start with, however, the second player will be left with between 1 and 4 chips, regardless of what the first player does.

What about 6 chips? This is again a winning position for the first player because if he removes 1 chip, the second player is left in the losing position of 5 chips. The same is true for 7, 8, or 9 chips. With 10 chips, however, the second player again can guarantee that he will win. Define:

$$\mathbf{N} = \left\{ x \in \mathbb{N} : \begin{array}{l} \text{the first (``\underline{n}ext'') player can ensure a win} \\ \text{if there are } x \text{ chips at the start} \end{array} \right\},$$

$$\mathbf{P} = \left\{ x \in \mathbb{N} : \begin{array}{l} \text{the second (``\underline{p}revious'') player can ensure a win} \\ \text{if there are } x \text{ chips at the start} \end{array} \right\}.$$

So far, we have seen that $\{1, 2, 3, 4, 6, 7, 8, 9\} \subseteq \mathbf{N}$ and $\{0, 5\} \subseteq \mathbf{P}$. Continuing this line of reasoning, we find that $\mathbf{P} = \{x \in \mathbb{N} : x \text{ is divisible by } 5\}$ and $\mathbf{N} = \mathbb{N} \setminus \mathbf{P}$.

The approach that we used to analyze the Subtraction game can be extended to other impartial games.

DEFINITION 1.1.2. An **impartial combinatorial game** has two players and a set of possible positions. To make a **move** is to take the game from one position to another. More formally, a move is an ordered pair of positions. A terminal position is one from which there are no legal moves. For every nonterminal position, there is a set of legal moves, the same for both players. Under **normal play**, the player who moves to a terminal position wins.

We can think of the game positions as **nodes** and the moves as directed **links**. Such a collection of nodes (vertices) and links (edges) between them is called a (directed) **graph**. At the start of the game, a token is placed at the node corresponding to the initial position. Subsequently, players take turns moving the token along directed edges until one of them reaches a terminal node and is declared the winner.

With this definition, it is clear that the Subtraction Game is an impartial game under normal play. The only terminal position is $x = 0$. Figure 1.2 gives a directed graph corresponding to the Subtraction Game with initial position $x = 14$.

We saw that starting from a position $x \in \mathbf{N}$, the next player to move can force a win by moving to one of the elements in $\mathbf{P} = \{5n : n \in \mathbb{N}\}$, namely $5\lfloor x/5 \rfloor$.

DEFINITION 1.1.3. A **strategy** for a player is a function that assigns a legal move to each nonterminal position. A **winning strategy** from a position x is a

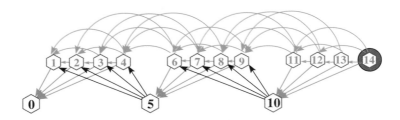

FIGURE 1.2. Moves in the Subtraction Game. Positions in **N** are marked in red and those in **P** are marked in black.

strategy that, starting from x, is guaranteed to result in a win for that player in a finite number of steps.

We can extend the notions of **N** and **P** to any impartial game.

DEFINITION 1.1.4. For any impartial combinatorial game, define **N** (for "next") to be the set of positions such that the first player to move can guarantee a win. The set of positions for which every move leads to an **N**-position is denoted by **P** (for "previous"), since the player who can force a **P**-position can guarantee a win.

Let $\mathbf{N}_i$ (respectively, $\mathbf{P}_i$) be the set of positions from which the next player (respectively, the previous player) can guarantee a win within at most i moves (of either player). Note that $\mathbf{P}_0 \subseteq \mathbf{P}_1 \subseteq \mathbf{P}_2 \subseteq \cdots$ and $\mathbf{N}_1 \subseteq \mathbf{N}_2 \subseteq \cdots$. Clearly

$$\mathbf{N} = \bigcup_{i \geq 1} \mathbf{N}_i, \quad \mathbf{P} = \bigcup_{i \geq 0} \mathbf{P}_i.$$

The sets $\mathbf{N}_i$ and $\mathbf{P}_i$ can be determined recursively:

$$\mathbf{P}_1 := \mathbf{P}_0 := \{\text{ terminal positions }\},$$
$$\mathbf{N}_{i+1} := \{\text{ positions } x \text{ for which there is a move leading to } \mathbf{P}_i\},$$
$$\mathbf{P}_{i+1} := \{\text{ positions } y \text{ such that each move leads to } \mathbf{N}_i\}.$$

In the Subtraction Game, we have

$$\mathbf{P}_1 = \mathbf{P}_0 = \{0\},$$
$$\mathbf{N}_2 = \mathbf{N}_1 = \{1, 2, 3, 4\}, \qquad \mathbf{P}_3 = \mathbf{P}_2 = \{0, 5\},$$
$$\mathbf{N}_4 = \mathbf{N}_3 = \{1, 2, 3, 4, 6, 7, 8, 9\}, \qquad \mathbf{P}_5 = \mathbf{P}_4 = \{0, 5, 10\},$$

$$\vdots \qquad\qquad\qquad \vdots$$

$$\mathbf{N} = \mathbb{N} \smallsetminus 5\mathbb{N}. \qquad\qquad \mathbf{P} = 5\mathbb{N}.$$

THEOREM 1.1.5. *In a progressively bounded impartial combinatorial game under normal play*[2], *all positions lie in* $\mathbf{N} \cup \mathbf{P}$. *Thus, from any initial position, one of the players has a winning strategy.*

PROOF. Recall that $B(x)$ is the maximum number of moves from x to a terminal position. We prove by induction on n, that all positions x with $B(x) \leq n$ are in $\mathbf{N}_n \cup \mathbf{P}_n$.

---

[2]Recall that normal play means that the player who moves to a terminal position wins.

Certainly, for all x such that $B(x) = 0$, we have that $x \in \mathbf{P}_0 \subseteq \mathbf{P}$. Now consider any position z for which $B(z) = n + 1$. Then every move from z leads to a position w with $B(w) \leq n$. There are two cases:

Case 1: Each move from z leads to a position in $\mathbf{N}_n$. Then $z \in \mathbf{P}_{n+1}$.

Case 2: There is a move from z to a position $w \notin \mathbf{N}_n$. Since $B(w) \leq n$, the inductive hypothesis implies that $w \in \mathbf{P}_n$. Thus, $z \in \mathbf{N}_{n+1}$.

Hence, all positions lie in $\mathbf{N} \cup \mathbf{P}$. If the starting position is in $\mathbf{N}$, then the first player has a winning strategy, otherwise, the second player does. $\qquad\square$

EXAMPLE 1.1.6 (**Chomp Revisited**). Recall the game of Chomp from Example 1.0.1. Since Chomp is progressively bounded, Theorem 1.1.5 implies that one of the players must have a winning strategy. We will show that it is the first player.

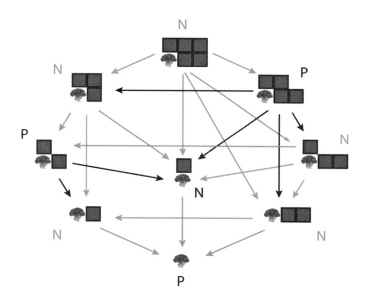

FIGURE 1.3. The graph representation of a 2×3 game of Chomp: Every move from a **P**-position leads to an **N**-position (bold black links); from every **N**-position there is at least one move to a **P**-position (red links).

THEOREM 1.1.7. *Starting from a position in which the remaining chocolate bar is rectangular of size greater than 1×1, the next player to move has a winning strategy.*

PROOF. Given a rectangular bar of chocolate R of size greater than 1×1, let R^- be the result of chomping off the upper-right 1×1 corner of R.

If $R^- \in \mathbf{P}$, then $R \in \mathbf{N}$, and a winning move is to chomp off the upper-right corner.

If $R^- \in \mathbf{N}$, then there is a move from R^- to some position x in $\mathbf{P}$. But if we can chomp R^- to get x, then chomping R in the same way will also give x, since the upper-right corner will be removed by any such chomp. Since there is a move from R to the position x in $\mathbf{P}$, it follows that $R \in \mathbf{N}$. $\qquad\square$

The technique used in this proof is called *strategy-stealing*. Note that the proof does *not* show that chomping the upper-right corner is a winning move. In the 2×3 case, chomping the upper-right corner happens to be a winning move (since this leads to a move in **P**; see Figure 1.3), but for the 3×3 case, chomping the upper-right corner is *not* a winning move. The strategy-stealing argument merely shows that a winning strategy for the first player must exist; it does not help us identify the strategy.

1.1.1. Nim. Next we analyze the game of Nim, a particularly important progressively bounded impartial game.

EXAMPLE 1.1.8 (**Nim**). In Nim, there are several piles, each containing finitely many chips. A legal move is to remove a positive number of chips from a single pile. Two players alternate turns with the aim of removing the last chip. Thus, the terminal position is the one where there are no chips left.

Because Nim is progressively bounded, all the positions are in **N** or **P**, and one of the players has a winning strategy. We will describe the winning strategy explicitly in the next section.

As usual, we will analyze the game by working backwards from the terminal positions. We denote a position in the game by $(n_1, n_2, \ldots, n_k)$, meaning that there are k piles of chips and that the first has n_1 chips in it, the second has n_2, and so on.

Certainly $(0, 1)$ and $(1, 0)$ are in **N**. On the other hand, $(1, 1) \in$ **P** because either of the two available moves leads to $(0, 1)$ or $(1, 0)$. We see that $(1, 2), (2, 1) \in$ **N** because the next player can create the position $(1, 1) \in$ **P**. More generally, $(n, n) \in$ **P** for $n \in \mathbb{N}$ and $(n, m) \in$ **N** if $n, m \in \mathbb{N}$ are not equal.

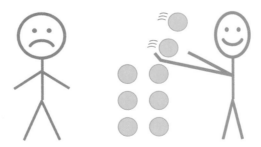

FIGURE 1.4. This figure shows why (n, m) with $n < m$ is in **N**: The next player's winning strategy is to remove $m - n$ chips from the bigger pile.

Moving to three piles, we see that $(1, 2, 3) \in$ **P**, because whichever move the first player makes, the second can force two piles of equal size. It follows that $(1, 2, 3, 4) \in$ **N** because the next player to move can remove the fourth pile.

To analyze $(1, 2, 3, 4, 5)$, we will need the following lemma:

LEMMA 1.1.9. *For two Nim positions* $X = (x_1, \ldots, x_k)$ *and* $Y = (y_1, \ldots, y_\ell)$, *we denote the position* $(x_1, \ldots, x_k, y_1, \ldots, y_\ell)$ *by* (X, Y).

  (1) *If* X *and* Y *are in* **P**, *then* $(X, Y) \in$ **P**.
  (2) *If* $X \in$ **P** *and* $Y \in$ **N** *(or vice versa), then* $(X, Y) \in$ **N**.

PROOF. If (X, Y) has 0 chips, then X, Y, and (X, Y) are all **P**-positions, so the lemma is true in this case.

Next, we suppose by induction that whenever (X, Y) has n or fewer chips,

$$X \in \mathbf{P} \text{ and } Y \in \mathbf{P} \text{ implies } (X, Y) \in \mathbf{P}$$

and

$$X \in \mathbf{P} \text{ and } Y \in \mathbf{N} \text{ implies } (X, Y) \in \mathbf{N}.$$

Suppose (X, Y) has at most $n + 1$ chips. If $X \in \mathbf{P}$ and $Y \in \mathbf{N}$, then the next player to move can reduce Y to a position in **P**, creating a $(\mathbf{P}, \mathbf{P})$ configuration with at most n chips, so by the inductive hypothesis it must be in **P**. Thus, $(X, Y) \in \mathbf{N}$.

If $X \in \mathbf{P}$ and $Y \in \mathbf{P}$, then the next player must move to an $(\mathbf{N}, \mathbf{P})$ or $(\mathbf{P}, \mathbf{N})$ position with at most n chips, which by the inductive hypothesis is an **N** position. Thus, $(X, Y) \in \mathbf{P}$. $\square$

Going back to our example, $(1, 2, 3, 4, 5)$ can be divided into two subgames: $(1, 2, 3) \in \mathbf{P}$ and $(4, 5) \in \mathbf{N}$. By the lemma, $(1, 2, 3, 4, 5) \in \mathbf{N}$.

REMARK 1.1.10. Note that if $X, Y \in \mathbf{N}$, then (X, Y) can be either in **P** or in **N**. E.g., $(1, 1) \in \mathbf{P}$ but $(1, 2) \in \mathbf{N}$. Thus, the divide-and-sum method (that is, using Lemma 1.1.9) for analyzing a position is limited. For instance, it does not classify any configuration of three piles of different sizes, since every nonempty subset of piles is in **N**.

1.1.2. Bouton's solution of Nim. We next describe a simple way of determining if a state is in **P** or **N**: We explicitly describe a set Z of configurations (containing the terminal position) such that, from every position in Z, all moves lead to Z^c, and from every position in Z^c, there is a move to Z. It will then follow by induction that $Z = \mathbf{P}$.

Such a set Z can be defined using the notion of *Nim-sum*. Given integers $x_1, x_2, \ldots, x_k$, the Nim-sum $x_1 \oplus x_2 \oplus \cdots \oplus x_k$ is obtained by writing each x_i in binary and then adding the digits in each column mod 2. For example:

	decimal	binary
x_1	3	0 0 1 1
x_2	9	1 0 0 1
x_3	13	1 1 0 1
$x_1 \oplus x_2 \oplus x_3$	7	0 1 1 1

DEFINITION 1.1.11. The **Nim-sum** $x_1 \oplus x_2 \oplus \cdots \oplus x_k$ of a configuration $(x_1, x_2, \ldots, x_k)$ is defined as follows: Write each pile size x_i in binary; i.e., $x_i = \sum_{j \geq 0} x_{ij} 2^j$ where $x_{ij} \in \{0, 1\}$. Then

$$x_1 \oplus x_2 \oplus \cdots \oplus x_k = \sum_{j \geq 0} (x_{1j} \oplus \cdots \oplus x_{kj}) 2^j$$

where for bits,

$$x_{1j} \oplus x_{2j} \oplus \cdots \oplus x_{kj} = \left(\sum_{i=1}^{k} x_{ij} \right) \bmod 2.$$

THEOREM 1.1.12 (**Bouton's Theorem**). *A Nim position* $x = (x_1, x_2, \ldots, x_k)$ *is in* **P** *if and only if the Nim-sum of its components is 0.*

To illustrate the theorem, consider the starting position $(1, 2, 3)$:

decimal	binary
1	0 1
2	1 0
3	1 1
0	0 0

Adding the two columns of the binary expansions modulo two, we obtain 00. The theorem affirms that $(1, 2, 3) \in \mathbf{P}$. Now, we prove Bouton's Theorem.

PROOF OF THEOREM 1.1.12. Define Z to be those positions with Nim-sum zero. We will show that:

(a) From every position in Z, all moves lead to Z^c.

(b) From every position in Z^c, there is a move to Z.

(a) Let $x = (x_1, \ldots, x_k) \in Z \setminus \mathbf{0}$. Suppose that we reduce pile ℓ, leaving $x'_\ell < x_\ell$ chips. This must result in some bit in the binary representation of x_ℓ, say the jth, changing from 1 to 0. The number of 1's in the jth column was even, so after the reduction it is odd.

(b) Next, suppose that $x = (x_1, x_2, \ldots, x_k) \notin Z$. Let $s = x_1 \oplus \cdots \oplus x_k \neq 0$. Let j be the position of the leftmost 1 in the expression for s. There are an odd number of values of $i \in \{1, \ldots, k\}$ with a 1 in position j. Choose one such i. Now $x_i \oplus s$ has a 0 in position j and agrees with x_i in positions $j+1, j+2, \ldots$ to the left of j, so $x_i \oplus s < x_i$. Consider the move which reduces the ith pile size from x_i to $x_i \oplus s$. The Nim-sum of the resulting position $(x_1, \ldots, x_{i-1}, x_i \oplus s, x_{i+1}, \ldots, x_k)$ is 0, so this new position lies in Z. Here is an example with $i = 1$ and $x_1 \oplus s = 3$.

	decimal	binary		decimal	binary
x_1	6	0 1 1 0	$x_1 \oplus s$	3	0 0 1 1
x_2	12	1 1 0 0	x_2	12	1 1 0 0
x_3	15	1 1 1 1	x_3	15	1 1 1 1
$s = x_1 \oplus x_2 \oplus x_3$	5	0 1 0 1	$(x_1 \oplus s) \oplus x_2 \oplus x_3$	0	0 0 0 0

This verifies (b).

It follows by induction on n that Z and $\mathbf{P}$ coincide on configurations with at most n chips. We also obtain the winning strategy: For any Nim-position that is not in Z, the next player should move to a position in Z, as described in the proof of (b). $\square$

1.1.3. Other impartial games. We next consider two other games that are just Nims in disguise.

EXAMPLE 1.1.13 (**Rims**). A starting position consists of a finite number of dots in the plane and a finite number of continuous loops that do not intersect. Each loop must pass through at least one dot. Each player, in his turn, draws a new loop that does not intersect any other loop. The goal is to draw the last such loop.

Next, we analyze the game. For a given position of Rims, we say that two uncovered dots are *equivalent* if there is a continuous path between them that does not intersect any loops. This partitions the uncovered dots into equivalence classes.

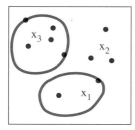
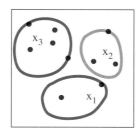
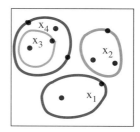

FIGURE 1.5. Two moves in a game of Rims.

Classical plane topology ensures that for any equivalence class of, say, k dots, and any integers $w, u, v \geq 0$ such that $w + u + v = k$ and $w \geq 1$, a loop can be drawn though w dots so that u dots are inside the loop (forming one equivalence class) and v dots are outside (forming another).

To see the connection to Nim, think of each class of dots as a pile of chips. A loop, because it passes through at least one dot, in effect, removes at least one chip from a pile and splits the remaining chips into two new piles. This last part is not consistent with the rules of Nim unless the player draws the loop so as to leave the remaining chips in a single pile.

FIGURE 1.6. Equivalent sequence of moves in Nim with splittings allowed.

Thus, Rims is equivalent to a variant of Nim where players have the option of splitting a pile into two piles after removing chips from it. As the following theorem shows, the fact that players have the option of splitting piles has no impact on the analysis of the game.

THEOREM 1.1.14. *The sets* **N** *and* **P** *coincide for Nim and Rims.*

PROOF. Let $x = (x_1, \ldots, x_k)$ be a position in Rims, represented by the number of dots in each equivalence class. Let Z be the collection of Rims positions with Nim-sum 0.

From any position $x \notin Z$, there is a move in Nim, which is also legal in Rims, to a position in Z.

Given a Rims position $x \in Z \setminus \mathbf{0}$, we must verify that every Rims move leads to Z^c. We already know this for Nim moves, so it suffices to consider a move where some equivalence class x_ℓ is reduced to two new equivalence classes of sizes u and v, where $u + v < x_\ell$. Since $u \oplus v \leq u + v < x_\ell$, it follows that $u \oplus v$ and x_ℓ must disagree in some binary digit, so replacing x_ℓ by (u, v) must change the Nim-sum. $\square$

EXERCISE 1.a (**Staircase Nim**). This game is played on a staircase of n steps. On each step j for $j = 1, \ldots, n$ is a stack of coins of size $x_j \geq 0$.

Each player, in his turn, picks j and moves one or more coins from step j to step $j - 1$. Coins reaching the ground (step 0) are removed from play. The game ends when all coins are on the ground, and the last player to move wins. See Figure 1.7.

Show that the **P**-positions in Staircase Nim are the positions such that the stacks of coins on the odd-numbered steps have Nim-sum 0.

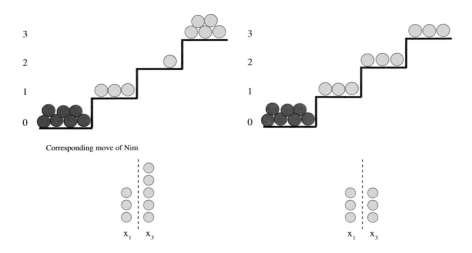

FIGURE 1.7. A move in Staircase Nim, in which two coins are moved from step 3 to step 2. Considering the odd stairs only, the above move is equivalent to the move in regular Nim from $(3, 5)$ to $(3, 3)$.

1.2. Partisan games

A combinatorial game in which the legal moves in some positions, or the sets of winning positions, differ for the two players, are called **partisan**.

While an impartial combinatorial game can be represented as a graph with a single edge-set, a partisan game is most often given by a set of nodes X representing the positions of the game and two sets of directed edges that represent the legal moves available to either player. Let $E_\mathrm{I}, E_\mathrm{II}$ be the two edge-sets for players I and II, respectively. If (x, y) is a legal move for player $i \in \{\mathrm{I}, \mathrm{II}\}$, then $(x, y) \in E_i$, and we say that y is a **successor** of x. We write $S_i(x) = \{y : (x, y) \in E_i\}$.

We start with a simple example:

EXAMPLE 1.2.1 (**A partisan subtraction game**). Starting with a pile of $x \in \mathbb{N}$ chips, two players, I and II, alternate taking a certain number of chips. Player I can remove 1 or 4 chips. Player II can remove 2 or 3 chips. The last player who removes chips wins the game.

This is a progressively bounded partisan game where both the terminal nodes and the moves are different for the two players. See Figure 1.8.

From this example we see that the number of steps it takes to complete the game from a given position now depends on the **state of the game**, $s = (x, i)$,

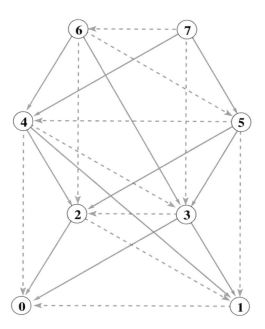

FIGURE 1.8. The partisan Subtraction game: The red dotted (respectively, green solid) edges represent moves of player I (respectively, II). Node 0 is terminal for either player, and node 1 is also terminal if the last move was by player I.

where x denotes the position and $i \in \{\text{I}, \text{II}\}$ denotes the player that moves next. We let $B(x, i)$ denote the maximum number of moves to complete the game from state (x, i).

The following theorem is analogous to Theorem 1.1.5.

THEOREM 1.2.2. *In any progressively bounded combinatorial game with no ties allowed, one of the players has a winning strategy which depends only upon the current state of the game.*

EXERCISE 1.b. Prove Theorem 1.2.2.

Theorem 1.2.2 relies essentially on the game being progressively bounded. Next we show that many games have this property.

LEMMA 1.2.3. *In a game with a finite position set, if the players cannot move to repeat a previous game state, then the game is progressively bounded.*

PROOF. If there are n positions x in the game, there are $2n$ possible game states (x, i), where i is one of the players. When the players play from position (x, i), the game can last at most $2n$ steps, since otherwise a state would be repeated. $\square$

The games of Chess and Go both have special rules to ensure that the game is progressively bounded. In Chess, whenever the board position (together with whose turn it is) is repeated for a third time, the game is declared a draw. (Thus the real game state effectively has built into it all previous board positions.) In Go, it is not legal to repeat a board position (together with whose turn it is), and this has a big effect on how the game is played.

1.2.1. The game of Hex. Recall the description of Hex from the preface.

EXAMPLE 1.2.4 (**Hex**). Hex is played on a rhombus-shaped board tiled with hexagons. Each player is assigned a color, either blue or yellow, and two opposing sides of the board. The players take turns coloring in empty hexagons. The goal for each player is to link his two sides of the board with a chain of hexagons in his color. Thus, the terminal positions of Hex are the full or partial colorings of the board that have a chain crossing.

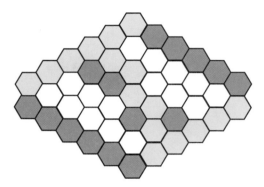

FIGURE 1.9. A completed game of Hex with a yellow chain crossing.

Note that Hex is a partisan, progressively bounded game where both the terminal positions and the legal moves are different for the two players. In Theorem 1.2.6 below, we will prove that any fully colored, standard Hex board contains either a blue crossing or a yellow crossing but not both. This topological fact guarantees that ties are not possible, so one of the players must have a winning strategy. We will now prove, again using a strategy-stealing argument, that the first player can always win.

THEOREM 1.2.5. *On a standard, symmetric Hex board of arbitrary size, the first player has a winning strategy.*

PROOF. We know that one of the players has a winning strategy. Suppose that the second player has a winning strategy S. The first player, on his first move, just colors in an arbitrarily chosen hexagon. Subsequently, the first player ignores his first move and plays S rotated by 90°. If S requires that the first player move in the spot that he chose in his first turn and there are empty hexagons left, he just picks another arbitrary spot and moves there instead.

Having an extra hexagon on the board can never hurt the first player — it can only help him. In this way, the first player, too, is guaranteed to win, implying that both players have winning strategies, a contradiction.   □

1.2.2. Topology and Hex: A path of arrows*. We now present two proofs that any colored standard Hex board contains a monochromatic crossing (and all such crossings have the same color). The proof in this section is quite general and can be applied to nonstandard boards. The proof in the next section has the advantage of showing that there can be no more than one crossing, a statement that seems obvious but is quite difficult to prove.

In the following discussion, precolored hexagons are referred to as **boundary**. Uncolored hexagons are called **interior**. Without loss of generality, we may assume that the edges of the board are made up of precolored hexagons (see Figure 1.10). Thus, the interior hexagons are surrounded by hexagons on all sides.

THEOREM 1.2.6. *For a filled-in standard Hex board with nonempty interior and with the boundary divided into two disjoint yellow and two disjoint blue segments, there is always at least one crossing between a pair of segments of like color.*

PROOF. Along every edge separating a blue hexagon and a yellow one, insert an arrow so that the blue hexagon is to the arrow's left and the yellow one to its right. In the initial position, there will be four such arrows, two directed toward the interior of the board (call these entry arrows) and two directed away from the interior (call these exit arrows). See the left side of Figure 1.10.

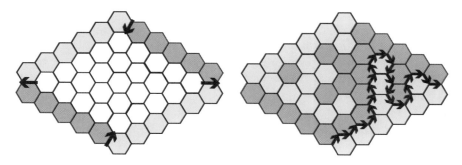

FIGURE 1.10. The left figure shows the entry and exit arrows on an empty board. The right figure shows a filled-in board with a blue crossing on the left side of the directed path.

Now, suppose the board has been arbitrarily filled with blue and yellow hexagons. Starting with one of the entry arrows, we will show that it is possible to construct a continuous path by adding arrows tail-to-head always keeping a blue hexagon on the left and a yellow on the right.

In the interior of the board, when two hexagons share an edge with an arrow, there is always a third hexagon which meets them at the vertex toward which the arrow is pointing. If that third hexagon is blue, the next arrow will turn to the right. If the third hexagon is yellow, the arrow will turn to the left. See (a) and (b) of Figure 1.11. Thus, every arrow (except exit arrows) has a unique successor. Similarly, every arrow (except entry arrows) has a unique predecessor.

Because we started our path at the boundary, where yellow and blue meet, our path will never contain a loop. If it did, the first arrow in the loop would have two predecessors. See (c) of Figure 1.11. Since there are finitely many available edges on the board and our path has no loops, it must eventually exit the board via one of the exit arrows.

All the hexagons on the left of such a path are blue, while those on the right are yellow. If the exit arrow touches the same yellow segment of the boundary as the entry arrow, there is a blue crossing (see Figure 1.10). If it touches the same blue segment, there is a yellow crossing.   □

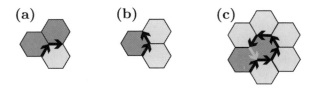

FIGURE 1.11. In (a) the third hexagon is blue and the next arrow turns to the right; in (b) the next arrow turns to the left; in (c) we see that in order to close the loop an arrow would have to pass between two hexagons of the same color.

1.2.3. Hex and Y. That there cannot be more than one crossing in the game of Hex seems obvious until you actually try to prove it carefully. To do this directly, we would need a discrete analog of the Jordan curve theorem, which says that a continuous closed curve in the plane divides the plane into two connected components. The discrete version of the theorem is considerably easier than the continuous one, but it is still quite challenging to prove.

Thus, rather than attacking this claim directly, we will resort to a trick: We will instead prove a similar result for a related, more general game — the game of Y, also known as Tripod.

EXAMPLE 1.2.7. The **Game of Y** is played on a triangular board tiled with hexagons. As in Hex, the two players take turns coloring in hexagons, each using his assigned color. A player wins when he establishes a Y, a monochromatic connected region in his color that meets all three sides of the triangle.

Playing Hex is equivalent to playing Y with some of the hexagons precolored, as shown in Figure 1.12.

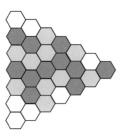
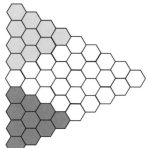

Blue has a winning Y here.　　　　　Reduction of Hex to Y

FIGURE 1.12. Hex is a special case of Y.

We will first show below that a filled-in Y board always contains a single Y. Because Hex is equivalent to Y with certain hexagons precolored, the existence and uniqueness of the chain crossing is inherited by Hex from Y.

THEOREM 1.2.8. *Any blue/yellow coloring of the triangular board contains either a blue Y or a yellow Y, but not both.*

PROOF. We can reduce a colored board with sides of length n to one with sides of length $n-1$ as follows: Think of the board as an arrow pointing right. Except for the leftmost column of hexagons, each hexagon is the right tip of a small arrow-shaped cluster of three adjacent hexagons pointing the same way as the board. We call such a triple a *triangle*. See Figure 1.13. Starting from the right, recolor each hexagon the majority color of the triangle that it tips, removing the leftmost column of hexagons altogether.

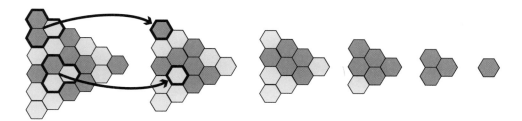

FIGURE 1.13. A step-by-step reduction of a colored Y board.

We claim that a board of side-length n contains a monochromatic Y if and only if the resulting board of size $n-1$ does: Suppose the board of size n contains, say, a blue Y. Let $(h_1, h_2, \ldots, h_L)$ be a path in this Y from one side of the board to another, where each h_i is a blue hexagon. We claim that there is a corresponding path $T_1, \ldots, T_{L-1}$ of triangles where T_i is the unique triangle containing (h_i, h_{i+1}). The set of rightmost hexagons in each of these triangles yields the desired blue path in the reduced graph. Similarly, there is a blue path from h_1 to the third side of the original board that becomes a blue path in the reduced board, creating the desired Y.

FIGURE 1.14. The construction of a blue path in the reduced board.

For the converse, a blue path between two sides A and B in the smaller board induces a path $T_1, \ldots, T_\ell$ of overlapping, majority blue triangles between A and B on the larger board. Suppose that we have shown that there is a path $h_1, \ldots, h_\ell$ (possibly with repetitions) of blue hexagons that starts at A and ends at $h_\ell \in T_k$ in the larger board. If $T_k \cap T_{k+1}$ is blue, take $h_{\ell+1} = T_k \cap T_{k+1}$. Otherwise, the four hexagons in the symmetric difference $T_k \triangle T_{k+1}$ are all blue and form a connected set. Extending the path by at most two of these hexagons, $h_{\ell+1}, h_{\ell+2}$ will reach T_{k+1}. See Figure 1.15.

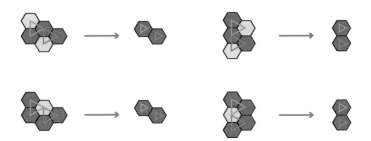

FIGURE 1.15. An illustration of the four possibilities for how a blue path going through T_k and T_{k+1} reduces to a blue path in the smaller board.

Thus, we can inductively reduce the board of size n to a board of size one, a single, colored cell. By the argument above, the color of this last cell is the color of a winning Y on the original board.   □

Because any colored Y board contains one and only one winning Y, it follows that any colored Hex board contains one and only one crossing.

REMARK 1.2.9. Why did we introduce Y instead of carrying out the proof directly for Hex? Hex corresponds to a subclass of Y boards, but this subclass is not preserved under the reduction we applied in the proof.

1.2.4. More general boards*. The statement that any colored Hex board contains exactly one crossing is stronger than the statement that every sequence of moves in a Hex game always leads to a crossing, i.e., a terminal position. To see why it's stronger, consider the following variant of Hex.

EXAMPLE 1.2.10. **Six-sided Hex** is similar to ordinary Hex, but the board is hexagonal, rather than square. Each player is assigned three nonadjacent sides and the goal for each player is to create a crossing in his color between two of his assigned sides. Thus, the terminal positions are those that contain one and only one monochromatic crossing between two like-colored sides.

In Six-sided Hex, there can be crossings of both colors in a completed board, but the game ends when the first crossing is created.

THEOREM 1.2.11. *Consider an arbitrarily shaped simply-connected[3] filled-in Hex board with nonempty interior with its boundary partitioned into n blue and n yellow segments, where $n \geq 2$. Then there is at least one crossing between some pair of segments of like color.*

EXERCISE 1.c. Adapt the proof of Theorem 1.2.6 to prove Theorem 1.2.11. (As in Hex, each entry and exit arrow lies on the boundary between a yellow and blue segment. Unlike in Hex, in shapes with with six or more sides, these four segments can be distinct. In this case there is both a blue and a yellow crossing. See Figure 1.16.)

---

[3] "Simply-connected" means that the board has no holes. Formally, it requires that every continuous closed curve on the board can be continuously shrunk to a point.

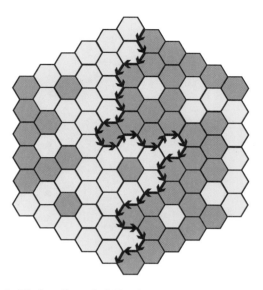

FIGURE 1.16. A filled-in Six-sided Hex board can have both blue and yellow crossings. In a game when players take turns to move, one of the crossings will occur first, and that player will be the winner.

1.2.5. Other partisan games played on graphs. We now discuss several other partisan games which are played on graphs. For each of our examples, we can explicitly describe a winning strategy for the first player.

EXAMPLE 1.2.12. The **Shannon Switching Game** is a variant of Hex played by two players, Cut and Short, on a connected graph with two distinguished nodes, A and B. Short, in his turn, reinforces an edge of the graph, making it immune to being cut. Cut, in her turn, deletes an edge that has not been reinforced. Cut wins if she manages to disconnect A from B. Short wins if he manages to link A to B with a reinforced path.

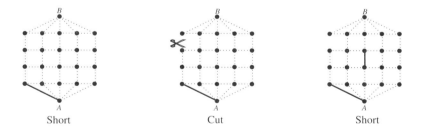

FIGURE 1.17. Shannon Switching Game played on a 5×6 grid (the top and bottom rows have been merged to the points A and B). Shown are the first three moves of the game, with Short moving first. Available edges are indicated by dotted lines, and reinforced edges by thick lines. Scissors mark the edge that Cut deleted.

We focus here on the case where the graph is an $L \times (L + 1)$ grid with the vertices of the bottom side merged into a single vertex, A, and the vertices on the top side merged into another node, B. In this case, the roles of the two players are symmetric, due to planar duality. See Figure 1.18.

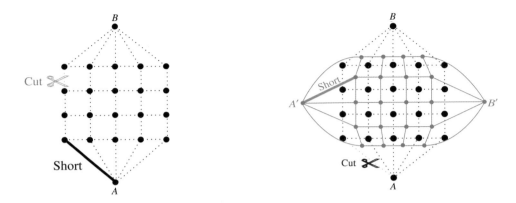

FIGURE 1.18. The dual graph $G^\dagger$ of planar graph G is defined as follows: Associated with each face of G is a vertex in $G^\dagger$. Two faces of G are adjacent if and only if there is an edge between the corresponding vertices of $G^\dagger$. Cutting in G is shorting in $G^\dagger$ and vice versa.

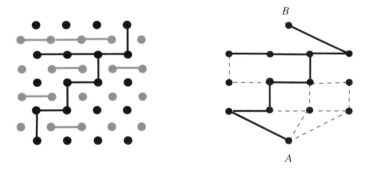

FIGURE 1.19. The figure shows corresponding positions in the Shannon Switching Game and an equivalent game known as Bridg-It. In Bridg-It, Black, in his turn, chooses two adjacent black dots and connects them with a edge. Green tries to block Black's progress by connecting an adjacent pair of green dots. Black and green edges cannot cross. Black's goal is to construct a path from top to bottom, while Green's goal is to block him by building a left-to-right path. The black dots are on the square lattice, and the green dots are on the dual square lattice.

EXERCISE 1.d. Use a strategy-stealing argument to show that the first player in the Shannon switching game has a winning strategy.

Next we will describe a winning strategy for the first player, which we will assume is Short. We will need some definitions from graph theory.

DEFINITION 1.2.13. A **tree** is a connected undirected graph without cycles.

(1) Every tree must have a *leaf*, a vertex of degree 1.
(2) A tree on n vertices has $n-1$ edges.
(3) A connected graph with n vertices and $n-1$ edges is a tree.
(4) A graph with no cycles, n vertices, and $n-1$ edges is a tree.

The proofs of these properties of trees are left as an exercise (Exercise 1.10).

THEOREM 1.2.14. *In Shannon's Switching Game on an $L \times (L+1)$ board, Short has a winning strategy if he moves first.*

PROOF. Short begins by reinforcing an edge of the graph G, connecting A to an adjacent dot, a. We identify A and a by "fusing" them into a single new A. On the resulting graph there is a pair of edge-disjoint trees such that each tree *spans* (contains all the nodes of) G. (Indeed, there are many such pairs.)

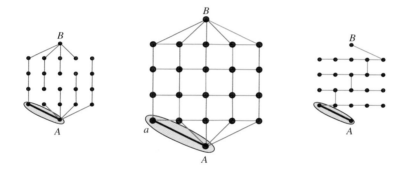

FIGURE 1.20. Two spanning trees — the blue one is constructed by first joining top and bottom using the leftmost vertical edges and then adding other vertical edges, omitting exactly one edge in each row along a diagonal; the red tree contains the remaining edges. The two circled nodes are identified.

For example, the blue and red subgraphs in the 4×5 grid in Figure 1.20 are such a pair of spanning trees: Each of them is connected and has the right number of edges. The same construction can be repeated on an arbitrary $L \times (L+1)$ grid.

Using these two spanning trees, which necessarily connect A to B, we can define a strategy for Short.

The first move by Cut disconnects one of the spanning trees into two components. Short can repair the tree as follows: Because the other tree is also a spanning tree, it must have an edge, e, that connects the two components. Short reinforces e. See Figure 1.21.

If we think of a reinforced edge e as being both red and blue, then the resulting red and blue subgraphs will still be spanning trees for G. To see this, note that both subgraphs will be connected and they will still have n edges and $n-1$ vertices. Thus, by property (3), they will be trees that span every vertex of G.

Continuing in this way, Short can repair the spanning trees with a reinforced edge each time Cut disconnects them. Thus, Cut will never succeed in disconnecting A from B, and Short will win. $\square$

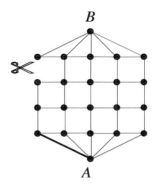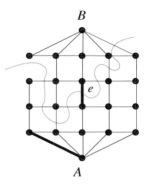

FIGURE 1.21. The left side of the figure shows how Cut separates the blue tree into two components. The right side shows how Short reinforces a red edge to reconnect the two components.

EXAMPLE 1.2.15. **Recursive Majority** is a game played on a complete ternary tree of height h (see Figure 1.22). The players take turns marking the leaves, player I with a "+" and player II with a "−". A parent node acquires the majority sign of its children. Because each interior (non-leaf) vertex has three children, its sign is determined unambiguously. The player whose mark is assigned to the root wins.

This game always ends in a win for one of the players, so one of them has a winning strategy.

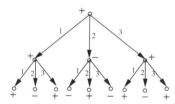

FIGURE 1.22. A ternary tree of height 2; the leftmost leaf is denoted by 11. Here player I wins the Recursive Majority game.

To analyze the game, label each of the three edges emanating downward from a single node 1, 2 or 3 from left to right. (See Figure 1.22.) Using these labels, we can identify each node below the root with the label sequence on the path from the root that leads to it. For instance, the leftmost leaf is denoted by $11\ldots1$, a word of length h consisting entirely of ones. A strategy-stealing argument implies that the first player to move has the advantage.

We can describe his winning strategy explicitly: On his first move, player I marks the leaf $11\ldots1$ with a "+". To determine his moves for the remaining even number of leaves, he first pairs up the leaves as follows: Letting 1^k be shorthand for a string of ones of fixed length $k \geq 0$ and letting w stand for an arbitrary fixed word of length $h - k - 1$, player I pairs the leaves by the following map: $1^k 2w \mapsto 1^k 3w$. (See Figure 1.23).

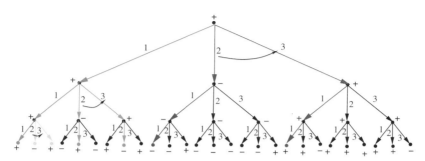

FIGURE 1.23. Player I marks the leftmost leaf in the first step. Some matched leaves are marked with the same shade of green or blue.

Once the pairs have been identified, whenever player II marks a leaf with a "−", player I marks its mate with a "+".

THEOREM 1.2.16. *The player I strategy described above is a winning strategy.*

PROOF. The proof is by induction on the height h of the tree. The base case of $h = 1$ is immediate. By the induction hypothesis, we know that player I wins in the left subtree of depth $h - 1$.

As for the remaining two subtrees of depth $h - 1$, whenever player II wins in one, player I wins in the other because each leaf in the middle subtree is paired with the corresponding leaf in the right subtree. Hence, player I is guaranteed to win two of the three subtrees, thus determining the sign of the root. □

Notes

The birth of combinatorial game theory as a mathematical subject can be traced to Bouton's 1902 characterization of the winning positions in Nim [Bou02]. In this chapter, we saw that many impartial combinatorial games can be reduced to Nim. This has been formalized in the Sprague-Grundy theory [Spr36, Spr37, Gru39] for analyzing all progressively bounded impartial combinatorial games.

The game of Chomp was invented by David Gale [BCG82a]. It is an open research problem to describe a general winning strategy for Chomp.

The game of Hex was invented by Piet Hein and reinvented by John Nash. Nash proved that Hex cannot end in a tie and that the first player has a winning strategy [Mil95, Gal79]. Shimon Even and Robert Tarjan [ET76] showed that determining whether a position in the game of Hex is a winning position is PSPACE-complete. This result was further generalized by Stefan Reisch [Rei81]. These results mean that an efficient algorithm for solving Hex on boards of arbitrary size is unlikely to exist. For small boards, however, an Internet-based community of Hex enthusiasts has made substantial progress (much of it unpublished). Jing Yang [Yan], a member of this community, has announced the solution of Hex (and provided associated computer programs) for boards of size up to 9×9. Usually, Hex is played on an 11×11 board, for which a winning strategy for player I is not yet known.

The game of Y and the Shannon Switching Game were introduced by Claude Shannon [Gar88]. The Shannon Switching Game can be played on any graph. The special case where the graph is a rectangular grid was invented independently by David Gale under the name Bridg-It (see Figure 1.19). Oliver Gross proved that the player who moves first in Bridg-It has a winning strategy. Several years later, Alfred B. Lehman [Leh64] (see also [Man96]) devised a solution to the general Shannon Switching Game. For more on the Shannon Switching Game, see [BCG82b].

For a more complete account of combinatorial games, see the books [Sie13, BCG82a, BCG82b, BCG04, Now96, Now02] and [YZ15, Chapter 15].

Exercises

1.1.   Consider a game of Nim with four piles, of sizes $9, 10, 11, 12$.
(a) Is this position a win for the *next* player or the *previous* player (assuming optimal play)? Describe the winning first move.
(b) Consider the same initial position, but suppose that each player is allowed to remove at most 9 chips in a single move (other rules of Nim remain in force). Is this an **N**- or **P**-position?

1.2.   Consider a game where there are two piles of chips. On a player's turn, he may remove between 1 and 4 chips from the first pile or else remove between 1 and 5 chips from the second pile. The person who takes the last chip wins. Determine for which $m, n \in \mathbb{N}$ it is the case that $(m, n) \in \mathbf{P}$.

1.3.   In the game of **Nimble**, there are n slots arranged from left to right, and a finite number of coins in each slot. In each turn, a player moves one of the coins to the left, by any number of places. The first player who can't move (since all coins are in the leftmost slot) loses. Determine which of the starting positions are **P**-positions.

1.4.   Given combinatorial games $G_1, \ldots, G_k$, let $G_1 + G_2 + \cdots + G_k$ be the following game: A state in the game is a tuple $(x_1, x_2, \ldots, x_k)$, where x_i is a state in G_i. In each move, a player chooses one G_i and takes a step in that game. A player who is unable to move, because all games are in a terminal state, loses.

Let G_1 be the Subtraction game with subtraction set $S_1 = \{1, 3, 4\}$, G_2 be the Subtraction game with $S_2 = \{2, 4, 6\}$, and G_3 be the Subtraction game with $S_3 = \{1, 2, \ldots, 20\}$. Who has a winning strategy from the starting position $(100, 100, 100)$ in $G_1 + G_2 + G_3$?

1.5.   Consider two arbitrary progressively bounded combinatorial games G_1 and G_2 with positions x_1 and x_2. If for any third such game G_3 and position x_3, the outcome of (x_1, x_3) in $G_1 + G_3$ (i.e., whether it's an **N**- or **P**-position) is the same as the outcome of (x_2, x_3) in $G_2 + G_3$, then we say that the game-position pairs (G_1, x_1) and (G_2, x_2) are *equivalent*.

Prove that equivalence for game-position pairs is transitive, reflexive, and symmetric. (Thus, it is indeed an equivalence relation.)

1.6.   Let G_i, $i = 1, 2$, be progressively bounded impartial combinatorial games. Prove that the position (x_1, x_2) in $G_1 + G_2$ is a **P**-position if and only if (G_1, x_1) and (G_2, x_2) are equivalent.

1.7.   Consider the game of Up-and-Down Rooks played on a standard chess-board. Player I has a set of white rooks initially located in the bottom row, while player II has a set of black rooks in the top row. In each turn, a player moves one of its rooks up or down a column, without skipping over the other rook or occupying its position. The first player who cannot move loses. This game is not progressively bounded, yet an optimal strategy exists. Find such a strategy by relating this game to a Nim position with 8 piles.

 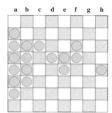 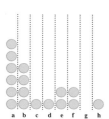

1.8.   Two players take turns placing dominos on an $n \times 1$ board of squares, where each domino covers two squares and dominos cannot overlap. The last player to play wins.
    (a) Where would you place the first domino when $n = 11$?
    (b) Show that for n even and positive, the first player can guarantee a win.

1.9.   Recall the game of Y shown in Figure 1.12. Prove that the first player has a winning strategy.

1.10.   Prove the following statements. Hint: Use induction.
    (a) Every tree on $n > 1$ vertices must have a **leaf** — a vertex of degree 1. (Indeed, it must have at least two leaves.)
    (b) A tree on n vertices has $n - 1$ edges.
    (c) A connected graph with n vertices and $n - 1$ edges is a *tree*.
    (d) A graph with no cycles, n vertices, and $n - 1$ edges is a *tree*.

CHAPTER 2

Two-person zero-sum games

We begin with the theory of **two-person zero-sum games**, developed in a seminal paper by John von Neumann [vN28]. In these games, one player's loss is the other player's gain. The central theorem for two-person zero-sum games is that even if each player's strategy is known to the other, there is an amount that one player can guarantee as her expected gain, and the other, as his maximum expected loss. This amount is known as the **value** of the game.

2.1. Examples

FIGURE 2.1. Two people playing Pick-a-Hand.

Consider the following game:

EXAMPLE 2.1.1 (**Pick a Hand, a betting game**). There are two players, Chooser (player I) and Hider (player II). Hider has two gold coins in his back pocket. At the beginning of a turn, he [1] puts his hands behind his back and either

---

[1] In almost all two-person games, we adopt the convention that player I is female and player II is male.

takes out one coin and holds it in his left hand (strategy $L1$) or takes out both and holds them in his right hand (strategy $R2$). Chooser picks a hand and wins any coins the hider has hidden there. She may get nothing (if the hand is empty), or she might win one coin, or two. How much should Chooser be willing to pay in order to play this game?

The following matrix summarizes the payoffs to Chooser in each of the cases:

<table>
<tr><td></td><td></td><td colspan="2">Hider</td></tr>
<tr><td></td><td></td><td>$L1$</td><td>$R2$</td></tr>
<tr><td rowspan="2">Chooser</td><td>L</td><td>1</td><td>0</td></tr>
<tr><td>R</td><td>0</td><td>2</td></tr>
</table>

How should Hider and Chooser play? Imagine that they are conservative and want to optimize for the worst-case scenario. Hider can guarantee himself a loss of at most 1 by selecting action $L1$, whereas if he selects $R2$, he has the potential to lose 2. Chooser cannot guarantee herself any positive gain since, if she selects L, in the worst-case, Hider selects $R2$, whereas if she selects R, in the worst case, Hider selects $L1$.

Now consider expanding the possibilities available to the players by incorporating randomness. Suppose that Hider selects $L1$ with probability y_1 and $R2$ with probability $y_2 = 1 - y_1$. Hider's expected loss is y_1 if Chooser plays L, and $2(1 - y_1)$ if Chooser plays R. Thus Hider's worst-case expected loss is $\max(y_1, 2(1 - y_1))$. To minimize this, Hider will choose $y_1 = 2/3$. Thus, **no matter how Chooser plays, Hider can guarantee himself an expected loss of at most 2/3.** See Figure 2.2.

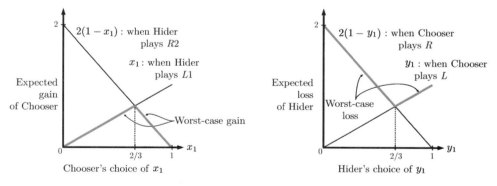

FIGURE 2.2. The left side of the figure shows the worst-case expected gain of Chooser as a function of x_1, the probability with which she plays L. The right side of the figure shows the worst-case expected loss of Hider as a function of y_1, the probability with which he plays $L1$. (In this example, the two graphs "look" the same because the payoff matrix is symmetric. See Exercise 2.a for a game where the two graphs are different.)

Similarly, suppose that Chooser selects L with probability x_1 and R with probability $x_2 = 1 - x_1$. Then Chooser's worst-case expected gain is $\min(x_1, 2(1 - x_1))$. To maximize this, she will choose $x_1 = 2/3$. Thus, **no matter how Hider plays, Chooser can guarantee herself an expected gain of at least 2/3.**

Observe that without some extra incentive, it is not in Hider's interest to play Pick a Hand, because he can only lose by playing. To be enticed into joining the game, Hider will need to be paid at least 2/3. Conversely, Chooser should be willing to pay any sum below 2/3 to play the game. Thus, we say that the **value** of this game is 2/3.

S

EXERCISE 2.a. Consider the betting game with the following payoff matrix:

player II

		L	R
player I	T	0	2
	B	5	1

Draw graphs for this game analogous to those shown in Figure 2.2, and determine the value of the game.

2.2. Definitions

A two-person zero-sum game can be represented by an $m \times n$ **payoff matrix** $A = (a_{ij})$, whose rows are indexed by the m possible actions of player I and whose columns are indexed by the n possible actions of player II. Player I selects an action i and player II selects an action j, each unaware of the other's selection. Their selections are then revealed and player II pays player I the amount a_{ij}.

If player I selects action i, in the worst case her gain will be $\min_j a_{ij}$, and thus the largest gain she can guarantee is $\max_i \min_j a_{ij}$. Similarly, if II selects action j, in the worst case his loss will be $\max_i a_{ij}$, and thus the smallest loss he can guarantee is $\min_j \max_i a_{ij}$. It follows that

$$\max_i \min_j a_{ij} \leq \min_j \max_i a_{ij} \qquad (2.1)$$

since player I can guarantee gaining the left-hand side and player II can guarantee losing no more than the right-hand side. (For a formal proof, see Lemma 2.6.3.) As in Example 2.1.1, without randomness, the inequality is usually strict.

A strategy in which each action is selected with some probability is a **mixed strategy**. A mixed strategy for player I is determined by a vector $(x_1, \ldots, x_m)^T$ where x_i represents the probability of playing action i. The set of mixed strategies for player I is denoted by

$$\Delta_m = \left\{ \mathbf{x} \in \mathbb{R}^m : x_i \geq 0, \sum_{i=1}^{m} x_i = 1 \right\}.$$

Similarly, the set of mixed strategies for player II is denoted by

$$\Delta_n = \left\{ \mathbf{y} \in \mathbb{R}^n : y_j \geq 0, \sum_{j=1}^{n} y_j = 1 \right\}.$$

A mixed strategy in which a particular action is played with probability 1 is called a **pure strategy**. Observe that in this vector notation, pure strategies are represented by the standard basis vectors, though we often identify the pure strategy $\mathbf{e}_i$ with the corresponding action i.

If player I employs strategy $\mathbf{x}$ and player II employs strategy $\mathbf{y}$, the expected gain of player I (which is the same as the expected loss of player II) is

$$\mathbf{x}^T A \mathbf{y} = \sum_i \sum_j x_i a_{ij} y_j.$$

Thus, if player I employs strategy $\mathbf{x}$, she can guarantee herself an expected gain of

$$\min_{\mathbf{y} \in \Delta_n} \mathbf{x}^T A \mathbf{y} = \min_j (\mathbf{x}^T A)_j \tag{2.2}$$

since for any $\mathbf{z} \in \mathbb{R}^n$, we have $\min_{\mathbf{y} \in \Delta_n} \mathbf{z}^T \mathbf{y} = \min_j z_j$.

A conservative player will choose $\mathbf{x}$ to maximize (2.2), that is, to maximize her worst-case expected gain. This is a safety strategy.

DEFINITION 2.2.1. A mixed strategy $\mathbf{x}^* \in \Delta_m$ is **a safety strategy for player I** if the maximum over $\mathbf{x} \in \Delta_m$ of the function

$$\mathbf{x} \mapsto \min_{\mathbf{y} \in \Delta_n} \mathbf{x}^T A \mathbf{y}$$

is attained at $\mathbf{x}^*$. The value of this function at $\mathbf{x}^*$ is the **safety value for player I**. Similarly, a mixed strategy $\mathbf{y}^* \in \Delta_n$ is **a safety strategy for player II** if the minimum over $\mathbf{y} \in \Delta_n$ of the function

$$\mathbf{y} \mapsto \max_{\mathbf{x} \in \Delta_m} \mathbf{x}^T A \mathbf{y}$$

is attained at $\mathbf{y}^*$. The value of this function at $\mathbf{y}^*$ is the **safety value for player II**.

REMARK 2.2.2. For the existence of safety strategies see Lemma 2.6.3.

2.3. The Minimax Theorem and its meaning

Safety strategies might appear conservative, but the following celebrated theorem shows that the two players' safety values coincide.

THEOREM 2.3.1 (**Von Neumann's Minimax Theorem**). *For any two-person zero-sum game with $m \times n$ payoff matrix A, there is a number V, called the value of the game, satisfying*

$$\max_{\mathbf{x} \in \Delta_m} \min_{\mathbf{y} \in \Delta_n} \mathbf{x}^T A \mathbf{y} = V = \min_{\mathbf{y} \in \Delta_n} \max_{\mathbf{x} \in \Delta_m} \mathbf{x}^T A \mathbf{y}. \tag{2.3}$$

We will prove the Minimax Theorem in §2.6.

REMARKS 2.3.2.

(1) It is easy to check that the left-hand side of equation (2.3) is upper bounded by the right-hand side, i.e.

$$\max_{\mathbf{x}\in\Delta_m} \min_{\mathbf{y}\in\Delta_n} \mathbf{x}^T A \mathbf{y} \le \min_{\mathbf{y}\in\Delta_n} \max_{\mathbf{x}\in\Delta_m} \mathbf{x}^T A \mathbf{y}. \tag{2.4}$$

(See the argument for (2.1) and Lemma 2.6.3.) The magic of zero-sum games is that, in mixed strategies, this inequality becomes an equality.

(2) If $\mathbf{x}^*$ is a safety strategy for player I and $\mathbf{y}^*$ is a safety strategy for player II, then it follows from Theorem 2.3.1 that

$$\min_{\mathbf{y}\in\Delta_n} (\mathbf{x}^*)^T A \mathbf{y} = V = \max_{\mathbf{x}\in\Delta_m} \mathbf{x}^T A \mathbf{y}^*. \tag{2.5}$$

In words, this means that the mixed strategy $\mathbf{x}^*$ yields player I an expected gain of at least V, no matter how II plays, and the mixed strategy $\mathbf{y}^*$ yields player II an expected loss of at most V, no matter how I plays. Therefore, from now on, we will refer to the safety strategies in zero-sum games as **optimal strategies**.

(3) The Minimax Theorem has the following interpretation: If, for every strategy $\mathbf{y} \in \Delta_m$ of player II, player I has a counterstrategy $\mathbf{x} = \mathbf{x}(\mathbf{y})$ that yields her expected payoff at least V, then player I has **one** strategy $\mathbf{x}^*$ that yields her expected payoff at least V against **all** strategies of player II.

2.4. Simplifying and solving zero-sum games

In this section, we will discuss techniques that help us understand zero-sum games and solve them (that is, find their value and determine optimal strategies for the two players).

2.4.1. Pure optimal strategies: Saddle points. Given a zero-sum game, the first thing to check is whether or not there is a pair of optimal strategies that is pure.

For example, in the following game, by playing action 1, player I guarantees herself a payoff at least 2 (since that is the smallest entry in the row). Similarly, by playing action 1, player II guarantees himself a loss of at most 2. Thus, the value of the game is 2.

		player II	
		action 1	action 2
player I	action 1	2	3
	action 2	1	0

DEFINITION 2.4.1. A **saddle point**[2] of a payoff matrix A is a pair (i^*, j^*) such that

$$\max_i a_{ij^*} = a_{i^*j^*} = \min_j a_{i^*j} \tag{2.6}$$

---

[2]The term saddle point comes from the continuous setting where a function $f(x, y)$ of two variables has a point (x^*, y^*) at which locally $\max_x f(x, y^*) = f(x^*, y^*) = \min_y f(x^*, y)$. Thus, the surface resembles a saddle that curves up in the y-direction and curves down in the x-direction.

If (i^*, j^*) is a saddle point, then $a_{i^* j^*}$ is the value of the game. A saddle point is also called a **pure Nash equilibrium**: Given the action pair (i^*, j^*), neither player has an incentive to deviate. See §2.5 for a more detailed discussion of Nash equilibria.

2.4.2. Equalizing payoffs. Most zero-sum games do not have pure optimal strategies. At the other extreme, some games have a pair $(\mathbf{x}^*, \mathbf{y}^*)$ of optimal strategies that are fully mixed, that is, where each action is assigned positive probability. In this case, it must be that against $\mathbf{y}^*$, player I obtains the same payoff from each action. If not, say $(A\mathbf{y}^*)_1 > (A\mathbf{y}^*)_2$, then player I could increase her gain by moving probability from action 2 to action 1: This contradicts the optimality of $\mathbf{x}^*$. Applying this observation to both players enables us to solve for optimal strategies by equalizing payoffs. Consider, for example, the following payoff matrix, where each row and column is labeled with the probability that the corresponding action is played in the optimal strategy:

		player II	
		y_1	$1 - y_1$
player I	x_1	3	0
	$1 - x_1$	1	4

Equalizing the gains for player I's actions, we obtain

$$3y_1 = y_1 + 4(1 - y_1),$$

i.e., $y_1 = 2/3$. Thus, if player II plays $(2/3, 1/3)$, his loss will not depend on player I's actions; it will be 2 no matter what I does.

Similarly, equalizing the losses for player II's actions, we obtain

$$3x_1 + (1 - x_1) = 4(1 - x_1),$$

i.e., $x_1 = 1/2$. So if player I plays $(1/2, 1/2)$, her gain will not depend on player II's action; again, it will be 2 no matter what II does. We conclude that the value of the game is 2.

See Proposition 2.5.3 for a general version of the **equalization principle.**

EXERCISE 2.b. Show that any 2×2 game (i.e., a game in which each player has exactly two strategies) has a pair of optimal strategies that are both pure or both fully mixed. Show that this can fail for 3×3 games.

2.4.3. The technique of domination. Domination is a technique for reducing the size of a game's payoff matrix, enabling it to be more easily analyzed. Consider the following example.

EXAMPLE 2.4.2 (**Plus One**). Each player chooses a number from $\{1, 2, \ldots, n\}$ and writes it down; then the players compare the two numbers. If the numbers differ by one, the player with the higher number wins \$1 from the other player. If the players' choices differ by two or more, the player with the higher number pays \$2 to the other player. In the event of a tie, no money changes hands.

The payoff matrix for the game is

player II

player I	1	2	3	4	5	6	$\cdots$			n
1	0	−1	2	2	2	2	$\cdots$			2
2	1	0	−1	2	2	2	$\cdots$			2
3	−2	1	0	−1	2	2	$\cdots$			2
4	−2	−2	1	0	−1	2	$\cdots$			2
5	−2	−2	−2	1	0	−1	2			2
6	−2	−2	−2	−2	1	0	2			2
$n-1$	−2	−2	$\cdots$						0	−1
n	−2	−2	$\cdots$					-2	1	0

In this payoff matrix, every entry in row 4 is at most the corresponding entry in row 1. Thus player I has no incentive to play 4 since it is **dominated** by row 1. In fact, rows 4 through n are all dominated by row 1, and hence player I can ignore these rows.

By symmetry, we see that player II need never play any of actions 4 through n. Thus, in Plus One we can search for optimal strategies in the reduced payoff matrix:

player II

player I	1	2	3
1	0	−1	2
2	1	0	−1
3	−2	1	0

To analyze the reduced game, let $\mathbf{x}^T = (x_1, x_2, x_3)$ be player I's mixed strategy. For $\mathbf{x}$ to be optimal, each component of

$$\mathbf{x}^T A = (x_2 - 2x_3,\ -x_1 + x_3,\ 2x_1 - x_2) \tag{2.7}$$

must be at least the value of the game. In this game, there is complete symmetry between the players. This implies that the payoff matrix is **antisymmetric:** the game matrix is square, and $a_{ij} = -a_{ji}$ for every i and j.

CLAIM 2.4.3. *If the payoff matrix of a zero-sum game is antisymmetric, then the game has value* 0.

PROOF. This is intuitively clear by symmetry. Formally, suppose that the value of the game is V. Then there is a vector $\mathbf{x} \in \Delta_n$ such that for all $\mathbf{y} \in \Delta_n$, $\mathbf{x}^T A\mathbf{y} \geq V$. In particular

$$\mathbf{x}^T A\mathbf{x} \geq V. \tag{2.8}$$

Taking the transpose of both sides yields $\mathbf{x}^T A^T \mathbf{x} = -\mathbf{x}^T A\mathbf{x} \geq V$. Adding this latter inequality to (2.8) yields $V \leq 0$. Similarly, there is a $\mathbf{y} \in \Delta_n$ such that for all $\tilde{\mathbf{x}} \in \Delta_n$ we have $\tilde{\mathbf{x}}^T A\mathbf{y} \leq V$. Taking $\tilde{\mathbf{x}} = \mathbf{y}$ yields in the same way that $0 \leq V$. □

We conclude that for any optimal strategy $\mathbf{x}$ in Plus One

$$
\begin{aligned}
x_2 - 2x_3 &\geq 0, \\
-x_1 + x_3 &\geq 0, \\
2x_1 - x_2 &\geq 0.
\end{aligned}
$$

If one of these inequalities were strict, then adding the first, twice the second, and the third, we could deduce $0 > 0$, so in fact each of them must be an equality. Solving the resulting system, with the constraint $x_1 + x_2 + x_3 = 1$, we find that the optimal strategy for each player is $(1/4, 1/2, 1/4)$.

Summary of domination. We say a row ℓ of a two-person zero-sum game dominates row i if $a_{\ell j} \geq a_{ij}$ for all j. When row i is dominated, then there is no loss to player I if she never plays it. More generally, we say that subset I of rows dominates row i if some convex combination of the rows in I dominates row i; i.e., there is a probability vector $(\beta_\ell)_{\ell \in I}$ such that for every j

$$
\sum_{\ell \in I} \beta_\ell a_{\ell j} \geq a_{ij}. \tag{2.9}
$$

Similar definitions hold for columns.

S   EXERCISE 2.c. Prove that if equation (2.9) holds, then player I can safely ignore row i.

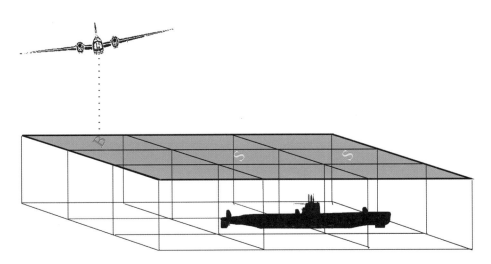

FIGURE 2.3. The bomber chooses one of the nine squares to bomb. She cannot see which squares represent the location of the submarine.

2.4.4. Using symmetry.

EXAMPLE 2.4.4 (**Submarine Salvo**). A submarine is located on two adjacent squares of a 3×3 grid. A bomber (player I), who cannot see the submerged craft, hovers overhead and drops a bomb on one of the nine squares. She wins \$1 if she hits the submarine and \$0 if she misses it. (See Figure 2.3.) There are nine pure strategies for the bomber and twelve for the submarine, so the payoff matrix for the game is quite large. To determine some, but not all, optimal strategies, we can use symmetry arguments to simplify the analysis.

There are three types of moves that the bomber can make: She can drop a bomb in the center, in the middle of one of the sides, or in a corner. Similarly, there are three types of positions that the submarine can assume: taking up the center square, taking up a corner square and the adjacent square clockwise, or taking up a corner square and the adjacent square counter-clockwise. It is intuitive (and true) that both players have optimal strategies that assign equal probability to actions of the same type (e.g., *corner-clockwise*). To see this, observe that in Submarine Salvo a 90° rotation describes a permutation π of the possible submarine positions and a permutation σ of the possible bomber actions. Clearly π^4 (rotating by 90° four times) is the identity and so is σ^4. For any bomber strategy $\mathbf{x}$, let $\pi\mathbf{x}$ be the rotated row strategy. (Formally $(\pi\mathbf{x})_i = x_{\pi(i)}$). Clearly, the probability that the bomber will hit the submarine if they play $\pi\mathbf{x}$ and $\sigma\mathbf{y}$ is the same as it is when they play $\mathbf{x}$ and $\mathbf{y}$, and therefore

$$\min_{\mathbf{y}} \mathbf{x}^T A \mathbf{y} = \min_{\mathbf{y}} (\pi\mathbf{x})^T A (\sigma\mathbf{y}) = \min_{\mathbf{y}} (\pi\mathbf{x})^T A \mathbf{y}.$$

Thus, if V is the value of the game and $\mathbf{x}$ is optimal, then $\pi^k\mathbf{x}$ is also optimal for all k.

Fix any submarine strategy $\mathbf{y}$. Then $\pi^k\mathbf{x}$ gains at least V against $\mathbf{y}$; hence so does

$$\mathbf{x}^* = \frac{1}{4}(\mathbf{x} + \pi\mathbf{x} + \pi^2\mathbf{x} + \pi^3\mathbf{x}).$$

Therefore $\mathbf{x}^*$ is a rotation-invariant optimal strategy.

Using these equivalences, we may write down a more manageable payoff matrix:

		submarine		
		center	*corner-clockwise*	*corner-counterclockwise*
bomber	*corner*	0	1/4	1/4
	midside	1/4	1/4	1/4
	middle	1	0	0

Note that the values for the new payoff matrix are different from those in the standard payoff matrix. They incorporate the fact that when, say, the bomber is playing *corner* and the submarine is playing *corner-clockwise*, there is only a one-in-four chance that there will be a hit. In fact, the pure strategy of corner for the bomber in this reduced game corresponds to the mixed strategy of bombing each corner with probability 1/4 in the original game. Similar reasoning applies to each of the pure strategies in the reduced game.

Since the rightmost two columns yield the same payoff to the submarine, it's natural for the submarine to give them the same weight. This yields the mixed strategy of choosing uniformly one of the eight positions containing a corner. We can use domination to simplify the matrix even further. This is because for the bomber, the strategy *midside* dominates that of *corner* (because the submarine, when touching a corner, must also be touching a midside). This observation reduces the matrix to

	submarine	
bomber	*center*	*corner*
midside	1/4	1/4
middle	1	0

Now note that for the submarine, *corner* dominates *center*, and thus we obtain the reduced matrix:

	submarine
bomber	*corner*
midside	1/4
middle	0

The bomber picks the better alternative — technically, another application of domination — and picks *midside* over *middle*. The value of the game is 1/4; an optimal strategy for the bomber is to hit one of the four midsides with probability 1/4 each, and an optimal submarine strategy is to hide with probability 1/8 each in one of the eight possible pairs of adjacent squares that exclude the center. The symmetry argument is generalized in Exercise 2.21.

REMARK 2.4.5. It is perhaps surprising that in Submarine Salvo there also exist optimal strategies that do not assign equal probability to all actions of the same type. (See Exercise 2.15.)

2.5. Nash equilibria, equalizing payoffs, and optimal strategies

A notion of great importance in game theory is **Nash equilibrium**. In §2.4.1, we introduced pure Nash equilibria. In this section, we introduce mixed Nash equilibria.

DEFINITION 2.5.1. A pair of strategies $(\mathbf{x}^*, \mathbf{y}^*)$ is a **Nash equilibrium** in a zero-sum game with payoff matrix A if

$$\min_{\mathbf{y} \in \Delta_n} (\mathbf{x}^*)^T A \mathbf{y} = (\mathbf{x}^*)^T A \mathbf{y}^* = \max_{\mathbf{x} \in \Delta_m} \mathbf{x}^T A \mathbf{y}^*. \tag{2.10}$$

Thus, $\mathbf{x}^*$ is a **best response** to $\mathbf{y}^*$ and vice versa.

REMARK 2.5.2. If $\mathbf{x}^* = \mathbf{e}_{i^*}$ and $\mathbf{y}^* = \mathbf{e}_{j^*}$, then by (2.2), this definition coincides with Definition 2.4.1.

PROPOSITION 2.5.3. *Let* $\mathbf{x} \in \Delta_m$ *and* $\mathbf{y} \in \Delta_n$ *be a pair of mixed strategies. The following are equivalent:*

   (i) *The vectors* $\mathbf{x}$ *and* $\mathbf{y}$ *are in Nash equilibrium.*
   (ii) *There are* V_1, V_2 *such that*

$$\sum_i x_i a_{ij} \begin{cases} = V_1 & \textit{for every } j \textit{ such that } y_j > 0, \\ \geq V_1 & \textit{for every } j \textit{ such that } y_j = 0. \end{cases} \tag{2.11}$$

   and

$$\sum_j a_{ij} y_j \begin{cases} = V_2 & \textit{for every } i \textit{ such that } x_i > 0, \\ \leq V_2 & \textit{for every } i \textit{ such that } x_i = 0. \end{cases} \tag{2.12}$$

   (iii) *The vectors* $\mathbf{x}$ *and* $\mathbf{y}$ *are optimal.*

REMARK 2.5.4. If (2.11) and (2.12) hold, then

$$V_1 = \sum_j y_j \sum_i x_i a_{ij} = \sum_i x_i \sum_j a_{ij} y_j = V_2.$$

PROOF. (i) implies (ii): Clearly, $\mathbf{y}$ is a best response to $\mathbf{x}$ if and only if $\mathbf{y}$ assigns positive probability only to actions that yield II the minimum loss given $\mathbf{x}$; this is precisely (2.11). The argument for (2.12) is identical. Thus (i) and (ii) are equivalent.

(ii) implies (iii): Player I guarantees herself a gain of at least V_1 and player II guarantees himself a loss of at most V_2. By the remark, $V_1 = V_2$, and therefore these are optimal.

(iii) implies (i): Let $V = \mathbf{x}^T A \mathbf{y}$ be the value of the game. Since playing $\mathbf{x}$ guarantees I a gain of a least V, player II has no incentive to deviate from $\mathbf{y}$. Similarly for player I. □

2.5.1. A first glimpse of incomplete information.

EXAMPLE 2.5.5 (**A random game**). Consider the zero-sum two-player game in which the game to be played is randomized by a fair coin toss. If the toss comes up heads, the payoff matrix is given by A^H, and if tails, it is given by A^T.

$$
A^H = \begin{array}{c} \\ \text{player I} \end{array}
\begin{array}{c} \text{player II} \\ \begin{array}{c|cc} & L & R \\ \hline U & 8 & 2 \\ D & 6 & 0 \end{array} \end{array}
\qquad
A^T = \begin{array}{c} \\ \text{player I} \end{array}
\begin{array}{c} \text{player II} \\ \begin{array}{c|cc} & L & R \\ \hline U & 2 & 6 \\ D & 4 & 10 \end{array} \end{array}
$$

If the players don't know the outcome of the coin flip before playing, they are merely playing the game given by the average matrix

$$\frac{1}{2}A^H + \frac{1}{2}A^T = \begin{pmatrix} 5 & 4 \\ 5 & 5 \end{pmatrix},$$

which has a value of 5. (For this particular matrix, the value does not change if I is required to reveal her move first.)

Now suppose that I (but not II) is told the result of the coin toss and is required to reveal her move first. If I adopts the simple strategy of picking the best row in whichever game is being played and II realizes this and counters, then I has an expected payoff of only 3, less than the expected payoff if she ignores the extra information! See Exercise 6.3 and §6.3.3 for a detailed analysis of this and related games.

This example demonstrates that sometimes the best strategy is to ignore extra information and play as if it were unknown. A related example arose during World War II. Polish and British cryptanalysts had broken the secret code the Germans were using (the Enigma machine) and could therefore decode the Germans' communications. This created a challenging dilemma for the Allies: Acting on the decoded information could reveal to the Germans that their code had been broken, which could lead them to switch to more secure encryption.

EXERCISE 2.d. What is the value of the game if both players know the outcome of the coin flip?

2.6. Proof of von Neumann's Minimax Theorem*

We now prove the von Neumann Minimax Theorem. A different, constructive, proof is given in §18.4.3. The proof will rely on a basic theorem from convex geometry.

Recall first that the **(Euclidean) norm of a vector** $\mathbf{v}$ is the (Euclidean) distance between $\mathbf{0}$ and $\mathbf{v}$ and is denoted by $\|\mathbf{v}\|$. Thus $\|\mathbf{v}\| = \sqrt{\mathbf{v}^T\mathbf{v}}$. A subset of a metric space is **closed** if it contains all its limit points, and **bounded** if it is contained inside a ball of some finite radius R.

DEFINITION 2.6.1. A set $K \subseteq \mathbb{R}^d$ is **convex** if, for any two points $\mathbf{a}, \mathbf{b} \in K$, the line segment connecting them also lies in K. In other words, for every $\mathbf{a}, \mathbf{b} \in K$ and $p \in [0, 1]$

$$p\,\mathbf{a} + (1 - p)\mathbf{b} \in K.$$

THEOREM 2.6.2 (**The Separating Hyperplane Theorem**). *Suppose that* $K \subseteq \mathbb{R}^d$ *is closed and convex. If* $\mathbf{0} \notin K$, *then there exist* $\mathbf{z} \in \mathbb{R}^d$ *and* $c \in \mathbb{R}$ *such that*

$$0 < c < \mathbf{z}^T\mathbf{v}$$

for all $\mathbf{v} \in K$.

Here $\mathbf{0}$ denotes the vector of all 0's. The theorem says that there is a **hyperplane** (a line in two dimensions, a plane in three dimensions, or, more generally, an affine $\mathbb{R}^{d-1}$-subspace in $\mathbb{R}^d$) that separates $\mathbf{0}$ from K. In particular, on any continuous path from $\mathbf{0}$ to K, there is some point that lies on this hyperplane. The separating hyperplane is given by $\{\mathbf{x} \in \mathbb{R}^d : \mathbf{z}^T\mathbf{x} = c\}$. The point $\mathbf{0}$ lies in the half-space $\{\mathbf{x} \in \mathbb{R}^d : \mathbf{z}^T\mathbf{x} < c\}$, while the convex body K lies in the complementary half-space $\{\mathbf{x} \in \mathbb{R}^d : \mathbf{z}^T\mathbf{x} > c\}$.

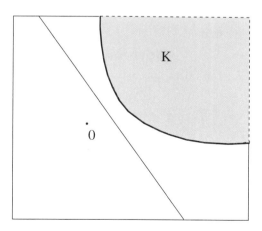

FIGURE 2.4. Hyperplane separating the closed convex body K from $\mathbf{0}$.

In what follows, the metric is the Euclidean metric.

PROOF OF THEOREM 2.6.2. Choose r so that the ball $B_r = \{\mathbf{x} \in \mathbb{R}^d : \|\mathbf{x}\| \leq r\}$ intersects K. Then the function $\mathbf{w} \mapsto \|\mathbf{w}\|$, considered as a map from $K \cap B_r$

to $[0, \infty)$, is continuous, with a domain that is nonempty, closed, and bounded (see Figure 2.5). Thus the map attains its infimum at some point $\mathbf{z}$ in K. For this $\mathbf{z} \in K$, we have

$$\|\mathbf{z}\| = \inf_{\mathbf{w} \in K} \|\mathbf{w}\|.$$

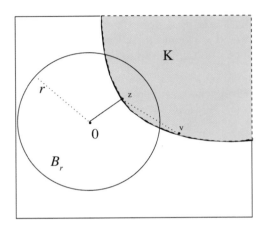

FIGURE 2.5. Intersecting K with a ball to get a nonempty closed bounded domain.

Let $\mathbf{v} \in K$. Because K is convex, for any $\varepsilon \in (0,1)$, we have that $\varepsilon \mathbf{v} + (1-\varepsilon)\mathbf{z} = \mathbf{z} - \varepsilon(\mathbf{z} - \mathbf{v}) \in K$. Since $\mathbf{z}$ has the minimum norm of any point in K,

$$\|\mathbf{z}\|^2 \le \|\mathbf{z} - \varepsilon(\mathbf{z} - \mathbf{v})\|^2 = \|\mathbf{z}\|^2 - 2\varepsilon \mathbf{z}^T(\mathbf{z} - \mathbf{v}) + \varepsilon^2 \|\mathbf{z} - \mathbf{v}\|^2.$$

Rearranging terms, we get

$$2\varepsilon \mathbf{z}^T(\mathbf{z} - \mathbf{v}) \le \varepsilon^2 \|\mathbf{z} - \mathbf{v}\|^2, \quad \text{that is,} \quad \mathbf{z}^T(\mathbf{z} - \mathbf{v}) \le \frac{\varepsilon}{2}\|\mathbf{z} - \mathbf{v}\|^2.$$

Letting ε approach 0, we find

$$\mathbf{z}^T(\mathbf{z} - \mathbf{v}) \le 0, \quad \text{which means that} \quad \|\mathbf{z}\|^2 \le \mathbf{z}^T \mathbf{v}.$$

Since $z \in K$ and $\mathbf{0} \notin K$, the norm $\|\mathbf{z}\| > 0$. Choosing $c = \frac{1}{2}\|\mathbf{z}\|^2$, we get $0 < c < \mathbf{z}^T \mathbf{v}$ for all $\mathbf{v} \in K$. $\qquad\square$

We will also need the following simple lemma:

LEMMA 2.6.3. *Let X and Y be closed and bounded sets in $\mathbb{R}^d$. Let $f : X \times Y \to \mathbb{R}$ be continuous. Then*

$$\max_{\mathbf{x} \in X} \min_{\mathbf{y} \in Y} f(\mathbf{x}, \mathbf{y}) \le \min_{\mathbf{y} \in Y} \max_{\mathbf{x} \in X} f(\mathbf{x}, \mathbf{y}). \tag{2.13}$$

PROOF. We first prove the lemma for the case where X and Y are finite sets (with no assumptions on f). Let $(\tilde{\mathbf{x}}, \tilde{\mathbf{y}}) \in X \times Y$. Clearly

$$\min_{\mathbf{y} \in Y} f(\tilde{\mathbf{x}}, \mathbf{y}) \le f(\tilde{\mathbf{x}}, \tilde{\mathbf{y}}) \le \max_{\mathbf{x} \in X} f(\mathbf{x}, \tilde{\mathbf{y}}).$$

Because the inequality holds for any $\tilde{\mathbf{x}} \in X$,

$$\max_{\tilde{\mathbf{x}} \in X} \min_{\mathbf{y} \in Y} f(\tilde{\mathbf{x}}, \mathbf{y}) \le \max_{\mathbf{x} \in X} f(\mathbf{x}, \tilde{\mathbf{y}}).$$

Minimizing over $\tilde{\mathbf{y}} \in Y$, we obtain (2.13).

To prove the lemma in the general case, we just need to verify the existence of the relevant maxima and minima. Since continuous functions achieve their minimum on compact sets, $g(\mathbf{x}) = \min_{\mathbf{y} \in Y} f(\mathbf{x}, \mathbf{y})$ is well-defined. The continuity of f and compactness of $X \times Y$ imply that f is uniformly continuous on $X \times Y$. In particular,

$$\forall \epsilon \; \exists \delta \; : \; |\mathbf{x}_1 - \mathbf{x}_2| < \delta \Longrightarrow |f(\mathbf{x}_1, \mathbf{y}) - f(\mathbf{x}_2, \mathbf{y})| \leq \epsilon$$

and hence $|g(\mathbf{x}_1) - g(\mathbf{x}_2)| \leq \epsilon$. Thus, $g : X \to \mathbb{R}$ is a continuous function and $\max_{\mathbf{x} \in X} g(\mathbf{x})$ exists. $\qquad \square$

We can now prove

THEOREM 2.3.1 (**Von Neumann's Minimax Theorem**). *Let A be an $m \times n$ payoff matrix, and let $\Delta_m = \{\mathbf{x} \in \mathbb{R}^m : \mathbf{x} \geq 0, \sum_i x_i = 1\}$ and $\Delta_n = \{\mathbf{y} \in \mathbb{R}^n : \mathbf{y} \geq 0, \sum_j y_j = 1\}$. Then*

$$\max_{\mathbf{x} \in \Delta_m} \min_{\mathbf{y} \in \Delta_n} \mathbf{x}^T A \mathbf{y} = \min_{\mathbf{y} \in \Delta_n} \max_{\mathbf{x} \in \Delta_m} \mathbf{x}^T A \mathbf{y}. \tag{2.14}$$

*As we discussed earlier, this quantity is called the **value** of the two-person zero-sum game with payoff matrix A.*

PROOF. The inequality

$$\max_{\mathbf{x} \in \Delta_m} \min_{\mathbf{y} \in \Delta_n} \mathbf{x}^T A \mathbf{y} \leq \min_{\mathbf{y} \in \Delta_n} \max_{\mathbf{x} \in \Delta_m} \mathbf{x}^T A \mathbf{y}$$

follows immediately from Lemma 2.6.3 because $f(\mathbf{x}, \mathbf{y}) = \mathbf{x}^T A \mathbf{y}$ is a continuous function in both variables and $\Delta_m \subset \mathbb{R}^m$, $\Delta_n \subset \mathbb{R}^n$ are closed and bounded.

To prove that the left-hand side of (2.14) is at least the right-hand side, suppose that

$$\lambda < \min_{\mathbf{y} \in \Delta_n} \max_{\mathbf{x} \in \Delta_m} \mathbf{x}^T A \mathbf{y}. \tag{2.15}$$

Define a new game with payoff matrix $\hat{A}$ given by $\hat{a}_{i,j} = a_{ij} - \lambda$. For this new game

$$0 < \min_{\mathbf{y} \in \Delta_n} \max_{\mathbf{x} \in \Delta_m} \mathbf{x}^T \hat{A} \mathbf{y}. \tag{2.16}$$

Each mixed strategy $\mathbf{y} \in \Delta_n$ for player II yields a gain vector $\hat{A} \mathbf{y} \in \mathbb{R}^m$. Let K denote the set of all vectors which dominate some gain vector $\hat{A} \mathbf{y}$; that is,

$$K = \left\{ \hat{A} \mathbf{y} + \mathbf{v} : \mathbf{y} \in \Delta_n, \, \mathbf{v} \in \mathbb{R}^m, \mathbf{v} \geq \mathbf{0} \right\}.$$

The set K is convex and closed since Δ_n is closed, bounded, and convex, and the set $\{\mathbf{v} \in \mathbb{R}^m, \mathbf{v} \geq \mathbf{0}\}$ is closed and convex. (See Exercise 2.19.) Also, K cannot contain the $\mathbf{0}$ vector, because if $\mathbf{0}$ was in K, there would be some mixed strategy $\mathbf{y} \in \Delta_n$ such that $\hat{A} \mathbf{y} \leq \mathbf{0}$. But this would imply that $\max_{\mathbf{x} \in \Delta_m} \mathbf{x}^T \hat{A} \mathbf{y} \leq 0$, contradicting (2.16).

Thus K satisfies the conditions of the Separating Hyperplane Theorem (Theorem 2.6.2), which gives us $\mathbf{z} \in \mathbb{R}^m$ and $c > 0$ such that $\mathbf{z}^T \mathbf{w} > c > 0$ for all $\mathbf{w} \in K$. That is,

$$\mathbf{z}^T (\hat{A} \mathbf{y} + \mathbf{v}) > c > 0 \text{ for all } \mathbf{y} \in \Delta_n \text{ and } \mathbf{v} \geq \mathbf{0}. \tag{2.17}$$

We claim that $\mathbf{z} \geq \mathbf{0}$. If not, say $z_j < 0$ for some j, then for $\mathbf{v} \in \mathbb{R}^m$ with v_j sufficiently large and $v_i = 0$ for all $i \neq j$, we would have $\mathbf{z}^T (\hat{A} \mathbf{y} + \mathbf{v}) = \mathbf{z}^T \hat{A} \mathbf{y} + z_j v_j < 0$ for some $\mathbf{y} \in \Delta_n$, which would contradict (2.17).

It also follows from (2.17) that $\mathbf{z} \neq \mathbf{0}$. Thus $s = \sum_{i=1}^{m} z_i$ is strictly positive, so $\tilde{\mathbf{x}} = \frac{1}{s}(z_1, \ldots, z_m)^T = \mathbf{z}/s \in \Delta_m$ satisfies $\tilde{\mathbf{x}}^T \hat{A} \mathbf{y} > c/s > 0$ for all $\mathbf{y} \in \Delta_n$. Therefore, $\min_{\mathbf{y} \in \Delta_n} \tilde{\mathbf{x}}^T A \mathbf{y} > \lambda$, whence

$$\max_{\mathbf{x} \in \Delta_m} \min_{\mathbf{y} \in \Delta_n} \mathbf{x}^T A \mathbf{y} > \lambda.$$

Since this holds for every λ satisfying (2.15), the theorem follows. $\square$

2.7. Zero-sum games with infinite action spaces*

THEOREM 2.7.1. *Consider a zero-sum game in which the players' action spaces are $[0, 1]$ and the gain is $A(x, y)$ when player I chooses action x and player II chooses action y. Suppose that $A(x, y)$ is continuous on $[0, 1]^2$. Let $\Delta = \Delta_{[0,1]}$ be the space of probability distributions on $[0, 1]$. Then*

$$\max_{F \in \Delta} \min_{G \in \Delta} \int \int A(x, y) dF(x) dG(y) = \min_{G \in \Delta} \max_{F \in \Delta} \int \int A(x, y) dF(x) dG(y). \quad (2.18)$$

PROOF. If there is a matrix (a_{ij}) for which

$$A(x, y) = a_{\lceil nx \rceil, \lceil ny \rceil}, \quad (2.19)$$

then (2.18) reduces to the finite case. If A is continuous, then there are functions A_0 and A_1 of the form (2.19) so that $A_0 \leq A \leq A_1$ and $|A_1 - A_0| \leq \epsilon$. This implies (2.18) with infs and sups in place of min and max. The existence of the maxima and minima follows from compactness of $\Delta_{[0,1]}$ as in the proof of Lemma 2.6.3. $\square$

REMARK 2.7.2. The previous theorem applies in any setting where the action spaces are compact metric spaces and the payoff function is continuous.

S   EXERCISE 2.e. Two players each choose a number in $[0, 1]$. If they choose the same number, the payoff is 0. Otherwise, the player that chose the lower number pays \$1 to the player who chose the higher number, unless the higher number is 1, in which case the payment is reversed. Show that this game has no mixed Nash equilibrium. Show that the safety values for players I and II are -1 and 1, respectively.

REMARK 2.7.3. The game from the previous exercise shows that the continuity assumption on the payoff function $A(x, y)$ cannot be removed. See also Exercise 2.23.

Notes

The theory of two-person zero-sum games was first laid out in a 1928 paper by John von Neumann [vN28], where he proved the Minimax Theorem (Theorem 2.6). The foundations were further developed in the book by von Neumann and Morgenstern [vNM53], first published in 1944.

The original proof of the Minimax Theorem used a fixed point theorem. A proof based on the Separating Hyperplane Theorem (Theorem 2.6.2) was given by Weyl [Wey50], and an inductive proof was given by Owen [Owe67]. Subsequently, many other minimax theorems were proved, such as Theorem 2.7.1, due to Glicksberg [Gli52], and Sion's minimax theorem [Sio58]. An influential example of a zero-sum game on the unit square with discontinuous payoff functions and without a value is in [SW57]. Games of timing also have discontinuous payoff functions, but do have a value. See, e.g., [Gar00].

John von Neumann

Oskar Morgenstern

Given an $m \times n$ payoff matrix A, the optimal strategy for player I might be supported on all m rows. However, Lipton and Young [LY94] showed that (assuming $0 \leq a_{ij} \leq 1$ for all i, j), player I has an ϵ-optimal strategy supported on $k = \lceil \frac{\ln n}{2\epsilon^2} \rceil$ rows. This follows by sampling k rows at random from the optimal mixed strategy and applying the Hoeffding-Azuma Inequality (Theorem B.2.2).

Exercise 2.2 is from [Kar59]. Exercise 2.17 comes from [HS89]. Exercise 2.18 is an example of a class of recursive games studied in [Eve57].

More detailed accounts of the material in this chapter can be found in Ferguson [Fer08], Karlin [Kar59], and Owen [Owe95], among others.

In §2.4, we present techniques for simplifying and solving zero-sum games by hand. However, for large games, there are efficient algorithms for finding optimal strategies and the value of the game based on linear programming. A brief introduction to linear programming can be found in Appendix A. There are also many books on the topic including, for example, [MG07].

The Minimax Theorem played a key role in the development of linear programming. George Dantzig, one of the pioneers of linear programming, relays the following story about his first meeting with John von Neumann [Dan82].

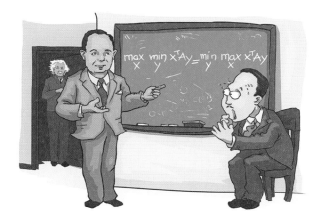

FIGURE 2.6. Von Neumann explaining duality to Dantzig.

On October 3, 1947, I visited him (von Neumann) for the first time at the Institute for Advanced Study at Princeton. I remember trying to describe to von Neumann, as I would to an ordinary mortal, the Air Force problem. I began with the formulation of the linear programming model in terms of activities and items, etc. Von Neumann did something which I believe was uncharacteristic of him. "Get to the point," he said impatiently. Having at times a somewhat low kindling-point, I said to myself "O.K., if he wants it quicky, then that's what he will get." In under one minute I slapped the geometric and algebraic version of the problem on the blackboard. Von Neumann stood up and said "Oh that!" Then for the next hour and a half, he proceeded to give me a lecture on the mathematical theory of linear programs.

At one point seeing me sitting there with my eyes popping and my mouth open (after I had searched the literature and found nothing), von Neumann said: "I don't want you to think I am pulling all this out of my sleeve at the spur of the moment like a magician. I have just recently completed a book with Oskar Morgenstern on the theory of games. What I am doing is conjecturing that the two problems are equivalent. The theory that I am outlining for your problem is an analogue to the one we have developed for games." Thus I learned about Farkas' Lemma, and about duality for the first time.

Exercises

2.1.   Show that all saddle points in a zero-sum game (assuming there is at least one) result in the same payoff to player I.

2.2.   Show that if a zero-sum game has a saddle point in every 2×2 submatrix, then it has a saddle point.

2.3.   Find the value of the following zero-sum game and determine some optimal strategies for each of the players:

$$\begin{pmatrix} 8 & 3 & 4 & 1 \\ 4 & 7 & 1 & 6 \\ 0 & 3 & 8 & 5 \end{pmatrix}.$$

2.4.   Find the value of the zero-sum game given by the following payoff matrix, and determine some optimal strategies for each of the players:

$$\begin{pmatrix} 0 & 9 & 1 & 1 \\ 5 & 0 & 6 & 7 \\ 2 & 4 & 3 & 3 \end{pmatrix}.$$

2.5.   Find the value of the zero-sum game given by the following payoff matrix and determine all optimal strategies for both players:

$$\begin{pmatrix} 3 & 0 \\ 0 & 3 \\ 2 & 2 \end{pmatrix}.$$

S 2.6.   Given a 5×5 zero-sum game, such as the following, how would you quickly
           determine by hand if it has a saddle point:

$$\begin{pmatrix} 20 & 1 & 4 & 3 & 1 \\ 2 & 3 & 8 & 4 & 4 \\ 10 & 8 & 7 & 6 & 9 \\ 5 & 6 & 1 & 2 & 2 \\ 3 & 7 & 9 & 1 & 5 \end{pmatrix}?$$

2.7.   Give an example of a two-person zero-sum game where there are no pure
          Nash equilibria. Can you give an example where all the entries of the payoff
          matrix are different?

2.8.   Define a zero-sum game in which one player's unique optimal strategy is
          pure and all of the other player's optimal strategies are mixed.

2.9.   Player II is moving an important item in one of three cars, labeled 1, 2,
          and 3. Player I will drop a bomb on one of the cars of her choosing. She
          has no chance of destroying the item if she bombs the wrong car. If she
          chooses the right car, then her probability of destroying the item depends
          on that car. The probabilities for cars 1, 2, and 3 are equal to 3/4, 1/4,
          and 1/2.
              Write the 3×3 payoff matrix for the game, and find an optimal strat-
          egy for each player.

2.10.   Using the result of Proposition 2.5.3, give an exponential time algorithm to
          solve an $n \times m$ two-person zero-sum game. Hint: Consider each possibility
          for which subset S of player I strategies have $x_i > 0$ and which subset T of
          player II strategies have $y_j > 0$.

2.11.   Consider the following two-person zero-sum game. Both players simulta-
          neously call out one of the numbers $\{2, 3\}$. Player I wins if the sum of the
          numbers called is odd and player II wins if their sum is even. The loser
          pays the winner the product of the two numbers called (in dollars). Find
          the payoff matrix, the value of the game, and an optimal strategy for each
          player.

2.12.   Consider the four-mile stretch of road shown in Figure 2.7. There are three
          locations at which restaurants can be opened: Left, Central, and Right.
          Company I opens a restaurant at one of these locations and company II
          opens two restaurants (both restaurants can be at the same location). A
          customer is located at a uniformly random location along the four-mile

stretch. He walks to the closest location at which there is a restaurant and then into one of the restaurants there, chosen uniformly at random. The payoff to company I is the probability that the customer visits a company I restaurant. Determine the value of the game, and find some optimal mixed strategies for the companies.

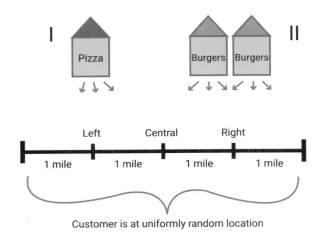

FIGURE 2.7. Restaurant location game.

2.13.   Bob has a concession at Yankee Stadium. He can sell 500 umbrellas at $10 each if it rains. (The umbrellas cost him $5 each.) If it shines, he can sell only 100 umbrellas at $10 each and 1000 sunglasses at $5 each. (The sunglasses cost him $2 each.) He has $2500 to invest in one day, but everything that isn't sold is trampled by the fans and is a total loss.

This is a game against nature. Nature has two strategies: rain and shine. Bob also has two strategies: buy for rain or buy for shine.

Find the optimal strategy for Bob assuming that the probability for rain is 50%.

2.14.   The Number Picking Game: Two players I and II pick a positive integer each. If the two numbers are the same, no money changes hands. If the players' choices differ by 1, the player with the lower number pays $1 to the opponent. If the difference is at least 2, the player with the higher number pays $2 to the opponent. Find the value of this zero-sum game and determine optimal strategies for both players. (Hint: Use domination.)

2.15.   Show that in Submarine Salvo the submarine has an optimal strategy where all choices containing a corner and a clockwise adjacent site are excluded.

2.16.   A zebra has four possible locations to cross the Zambezi River; call them a, b, c, and d, arranged from north to south. A crocodile can wait (undetected) at one of these locations. If the zebra and the crocodile choose the same location, the payoff to the crocodile (that is, the chance it will catch the zebra) is 1. The payoff to the crocodile is $1/2$ if they choose adjacent locations, and 0 in the remaining cases, when the locations chosen are distinct and nonadjacent.

(a) Write the payoff matrix for this game.

(b) Can you reduce this game to a 2×2 game?

(c) Find the value of the game (to the crocodile) and optimal strategies for both.

S 2.17.   Generalized Matching Pennies: Consider a directed graph $G = (V, E)$ with nonnegative weights w_{ij} on each edge (i, j). Let $W_i = \sum_j w_{ij}$. Each player chooses a vertex, say i for player I and j for player II. Player I receives a payoff of w_{ij} if $i \neq j$ and loses $W_i - w_{ii}$ if $i = j$. Thus, the payoff matrix A has entries $a_{ij} = w_{ij} - 1_{\{i=j\}} W_i$. If $n = 2$ and the w_{ij}'s are all 1, this game is called Matching Pennies.

- Show that the game has value 0.
- Deduce that for some $x \in \Delta_n$, $\mathbf{x}^T A = 0$.

2.18.   A recursive zero-sum game: Trumm Seafood has wild salmon on the menu. Each day, the owner, Mr. Trumm, decides whether to cheat and serve the cheaper farmed salmon instead. An inspector selects a day in $1, \ldots, n$ and inspects the restaurant on that day. The payoff to the inspector is 1 if he inspects while Trumm is cheating. The payoff is -1 if the Trumm cheats and is not caught. The payoff is also -1 if the inspector inspects but Trumm did not cheat and there is at least one day left. This leads to the following matrices Γ_n for the game with n days: The matrix Γ_1 is shown on the left, and the matrix Γ_n is shown on the right.

		Trumm	
inspector		cheat	honest
	inspect	1	0
	wait	−1	0

		Trumm	
inspector		cheat	honest
	inspect	1	−1
	wait	−1	Γ_{n-1}

Find the optimal strategies and the value of Γ_n.

S 2.19.   Prove that if set $G \subseteq \mathbb{R}^d$ is compact and $H \subseteq \mathbb{R}^d$ is closed, then $G + H$ is closed. (This fact is used in the proof of the Minimax Theorem to show that the set K is closed.)

S 2.20.   Find two closed sets $F_1, F_2 \subset \mathbb{R}^2$ such that $F_1 - F_2$ is not closed.

S * 2.21.   Consider a zero-sum game A and suppose that π and σ are permutations of I's strategies $\{1, \ldots, m\}$ and player II's strategies $\{1, \ldots, n\}$, respectively, such that

$$a_{\pi(i)\sigma(j)} = a_{ij} \tag{2.20}$$

for all i and j. Show that there exist optimal strategies $\mathbf{x}^*$ and $\mathbf{y}^*$ such that $x_i^* = x_{\pi(i)}^*$ for all i and $y_j^* = y_{\sigma(j)}^*$ for all j.

S 2.22. Player I chooses a positive integer $x > 0$ and player II chooses a positive integer $y > 0$. The player with the lower number pays a dollar to the player with the higher number unless the higher number is more than twice larger in which case the payments are reversed.

$$A(x,y) = \begin{cases} 1 & \text{if } y < x \le 2y \text{ or } x < y/2, \\ -1 & \text{if } x < y \le 2x \text{ or } y < x/2, \\ 0 & \text{if } x = y. \end{cases}$$

Find the unique optimal strategy in this game.

2.23. Two players each choose a positive integer. The player that chose the lower number pays \$1 to the player who chose the higher number (with no payment in case of a tie). Show that this game has no Nash equilibrium. Show that the safety values for players I and II are -1 and 1 respectively.

2.24. Two players each choose a number in $[0,1]$. Suppose that $A(x,y) = |x-y|$.
   - Show that the value of the game is $1/2$.
   - More generally, suppose that $A(x,y)$ is a convex function in each of x and y and that it is continuous. Show that player I has an optimal strategy supported on 2 points and player II has an optimal pure strategy.

2.25. Consider a zero-sum game in which the strategy spaces are $[-1,1]$ and the gain of player I when she plays x and player II plays y is

$$A(x,y) = \log \frac{1}{|x-y|}.$$

Show that I picking $X = \cos\Theta$, where Θ is uniform on $[-1,1]$, and II using the same strategy is a pair of optimal strategies.

CHAPTER 3

Zero-sum games on graphs

In this chapter, we consider a number of graph-theoretic zero-sum games.

3.1. Games in series and in parallel

EXAMPLE 3.1.1 (**Hannibal and the Romans**). Hannibal and his army (player I) and the Romans (player II) are on opposite sides of a mountain. There are two routes available for crossing the mountain. If they both choose the wide mountain pass, their confrontation is captured by a zero-sum game G_1. If they both choose the narrow mountain pass, their confrontation is captured by zero-sum game G_2, with different actions and payoffs (e.g., the elephants cannot cross the narrow pass). If they choose different passes, no confrontation occurs and the payoff is 0. We assume that the value of both G_1 and G_2 is positive. The resulting game is a *parallel-sum* of G_1 and G_2.

FIGURE 3.1. Hannibal approaching battle.

In the second scenario, Hannibal has two separate and consecutive battles with two Roman armies. Again, each battle is captured by a zero-sum game, the first G_1 and the second G_2. This is an example of a *series-sum game*.

DEFINITION 3.1.2. Given two zero-sum games G_1 and G_2, their **series-sum game** corresponds to playing G_1 and then G_2. In a **parallel-sum game**, each player chooses either G_1 or G_2 to play. If each picks the same game, then it is that game which is played. If they differ, then no game is played, and the payoff is zero.

EXERCISE 3.a. Show that if G_i has value v_i for $i = 1, 2$, then their series-sum game has value $v_1 + v_2$.

To solve for optimal strategies in the parallel-sum gum, we write a big payoff matrix, in which player I's strategies are the union of her strategies in G_1 and her strategies in G_2 as follows:

	player II	
	pure strategies of G_1	pure strategies of G_2
pure strategies of G_1	G_1	0
pure strategies of G_2	0	G_2

(player I)

In this payoff matrix, we have abused notation and written G_1 and G_2 inside the matrix to denote the payoff matrices of G_1 and G_2, respectively. If the two players play G_1 and G_2 optimally, the payoff matrix can be reduced to:

	player II	
	play in G_1	play in G_2
play in G_1	v_1	0
play in G_2	0	v_2

(player I)

Thus to find optimal strategies, the players just need to determine with what probability they should play G_1 and with what probability they should play G_2. If both payoffs v_1 and v_2 are positive, the optimal strategy for each player consists of playing G_1 with probability $v_2/(v_1 + v_2)$ and G_2 with probability $v_1/(v_1 + v_2)$. The value of the parallel-sum game is

$$\frac{v_1 v_2}{v_1 + v_2} = \frac{1}{1/v_1 + 1/v_2}.$$

Those familiar with electrical networks will note that the rules for computing the value of series or parallel games in terms of the values of the component games are precisely the same as the rules for computing the effective resistance of a pair of resistors in series or in parallel. In the next section, we explore a game that exploits this connection.

3.1.1. Resistor networks and troll games.

EXAMPLE 3.1.3 (**Troll and Traveler**). A troll (player I) and a traveler (player II) will each choose a route along which to travel from Syracuse (s) to Troy (t) and then they will disclose their routes. Each road has an associated toll. In each case where the troll and the traveler have chosen the same road, the traveler pays the toll to the troll.

In the special case where there are exactly two parallel roads from A to B (or two roads in series), this is the parallel-sum (respectively, series-sum) game we saw earlier. For graphs that are constructed by repeatedly combining graphs in series or in parallel, there is an elegant and general way to solve the Troll and Traveler game, by interpreting the road network as an electrical network and the tolls as resistances.

DEFINITION 3.1.4. An **electrical network** is a finite connected graph G with positive edge labels (representing edge resistances) and a specified source s and sink t.

FIGURE 3.2. The Troll and the Traveler.

We combine two networks G_1 and G_2, with sources s_i and sinks t_i, for $i = 1, 2$, either **in series**, by identifying t_1 with s_2, or **in parallel**, by identifying s_1 with s_2, and t_1 with t_2.

A network G is a **series-parallel network** if it is either a single directed edge from the source to the sink or it is obtained by combining two series-parallel networks G_1 and G_2 in series or in parallel.

See the upper left of Figure 3.4 for a graph constructed by combining two edges in series and in parallel, and see Figure 3.16 for a more complex series-parallel graph.

Recall that if two nodes are connected by a resistor with **resistance R** and there is a voltage drop of V across the two nodes, then the current that flows through the resistor is V/R. The **conductance** is the reciprocal of the resistance. When the pair of nodes are connected by a pair of resistors with resistances R_1 and R_2 arranged in series (see the top of Figure 3.3), the **effective resistance** between the nodes is $R_1 + R_2$, because the current that flows through the resistors is $V/(R_1 + R_2)$. When the resistors are arranged in parallel (see the bottom of Figure 3.3), it is the conductances that add; i.e., the **effective conductance** between the nodes is $1/R_1 + 1/R_2$ and the effective resistance is

$$\frac{1}{1/R_1 + 1/R_2} = \frac{R_1 R_2}{R_1 + R_2}.$$

These series and parallel rules for computing the effective resistance can be used repeatedly to compute the effective resistance of any series-parallel network, as illustrated in Figure 3.4. Applying this argument inductively yields the following claim.

CLAIM 3.1.5. *The value of the Troll and Traveler game played on a series-parallel network G with source s and sink t is the effective resistance between s and t. Optimal strategies in the Troll and Traveler game are defined as follows: If G is obtained by combining G_1 and G_2 in series, then each player plays his or her optimal strategy in G_1 followed by his optimal strategy in G_2. If G is obtained by*

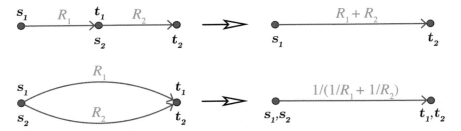

FIGURE 3.3. In a network consisting of two resistors with resistances R_1 and R_2 in series (shown on top), the effective resistance is $R_1 + R_2$. When the resistors are in parallel, the effective conductance is $1/R_1 + 1/R_2$, so the effective resistance is $1/(1/R_1 + 1/R_2) = R_1R_2/(R_1 + R_2)$. If these figures represent the roads leading from s to t in the Troll and Traveler game and the toll on each road corresponds to the resistance on that edge, then the effective resistance is the value of the game, and the optimal strategy for each player is to move along an edge with probability proportional to the conductance on that edge.

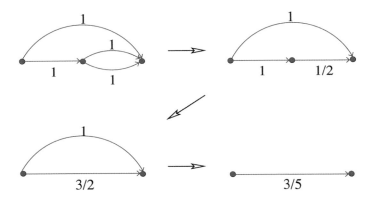

FIGURE 3.4. A resistor network, with resistances all equaling to 1, has an effective resistance of $3/5$. Here the parallel rule was used first, then the series rule, and then the parallel rule again.

combining G_1 and G_2 (with sources s_1, s_2 and sinks t_1, t_2) in parallel, then each player plays his optimal strategy in G_i with probability $C_i/(C_1 + C_2)$, where C_i is the effective conductance between s_i and t_i in G_i.

3.2. Hide and Seek games

EXAMPLE 3.2.1 (**Hide and Seek**). A robber, player II, hides in one of a set of safehouses located at certain street/avenue intersections in Manhattan. A cop, player I, chooses one of the avenues or streets to travel along. The cop wins a unit payoff if she travels on a road that intersects the robber's location and nothing otherwise.

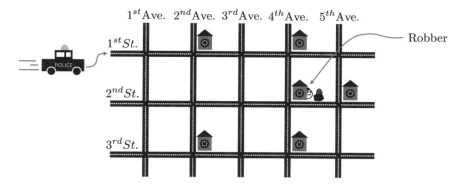

FIGURE 3.5. The figure shows an example scenario for the Hide and Seek game. In this example, the robber chooses to hide at the safehouse at the intersection of 2nd St. and 4th Ave., and the cop chooses to travel along 1st St. Thus, the payoff to the cop is 0.

We represent this situation with a 0/1 matrix H where $h_{ij} = 1$ if there is a safehouse at the intersection of street i and avenue j, and $h_{ij} = 0$ otherwise. The following matrix corresponds to Figure 3.5:

$$\begin{pmatrix} 0 & 1 & 0 & 1 & 0 \\ 0 & 0 & 0 & 1 & 1 \\ 0 & 1 & 0 & 1 & 0 \end{pmatrix}.$$

The cop's actions correspond to choosing a row or column of this matrix and the robber's actions correspond to picking a 1 in the matrix.

Clearly, it is useless for the cop to choose a road that doesn't contain a safehouse; a natural strategy for her is to find a smallest set of roads that contain all safehouses and choose one of these at random. Formally, a **line-cover** of the matrix H is a set of lines (rows and columns) that cover all nonzero entries of H. The proposed cop strategy is to fix a minimum-sized line-cover $\mathcal{C}$ and choose one of the lines in $\mathcal{C}$ uniformly at random. This guarantees the cop an expected gain of at least $1/|\mathcal{C}|$ against any robber strategy.

Next we consider robber strategies. A vulnerable strategy would be to choose from among a set of safehouses that all lie on the same road. The "opposite" of that is to find a maximum-sized set $\mathcal{M}$ of safehouses, where no two lie on the same road, and choose one of these uniformly at random. This guarantees that the cop's expected gain is at most $1/|\mathcal{M}|$.

It is not obvious that the proposed strategies are optimal. However, in the next section, we prove that

$$|\mathcal{C}| = |\mathcal{M}|. \tag{3.1}$$

This implies that the proposed pair of strategies is jointly optimal for Hide and Seek.

3.2.1. Maximum matching and minimum covers. Given a set of boys B and a set of girls G, draw an edge between a boy and a girl if they know each other. The resulting graph is called a **bipartite graph** since there are two disjoint sets of nodes and all edges go between them. Bipartite graphs are ubiquitous. For instance, there is a natural bipartite graph where one set of nodes represents workers, the other set represents jobs, and an edge from worker w to job j means that worker w can perform job j. Other examples involve customers and suppliers, or students and colleges.

A **matching** in a bipartite graph is a collection of disjoint edges, e.g., a set of boy-girl pairs that know each other, where every individual occurs in at most one pair. (See Figure 3.6.)

Suppose $|B| \leq |G|$. Clearly there cannot be a matching that includes more than $|B|$ edges. Under what condition is there a matching of size $|B|$, i.e. a matching in which every boy is matched to a girl he knows?

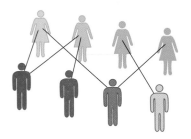 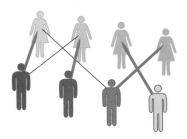

FIGURE 3.6. On the left is a bipartite graph where an edge between a boy and a girl means that they know each other. The edges in a matching are shown by bold lines in the figure on the right.

An obvious necessary condition, known as **Hall's condition**, is that each subset B' of the boys collectively knows enough girls, at least $|B'|$ of them. Hall's Marriage Theorem asserts that this condition is also sufficient.

THEOREM 3.2.2 (**Hall's Marriage Theorem**). *Suppose that B is a finite set of boys and G is a finite set of girls. For any particular boy $b \in B$, let $f(b)$ denote the set of girls that b knows. For a subset $B' \subseteq B$ of the boys, let $f(B')$ denote the set of girls that boys in B' collectively know; i.e., $f(B') = \bigcup_{b \in B'} f(b)$. There is a matching of size $|B|$ if and only if Hall's condition holds: Every subset $B' \subseteq B$ satisfies $|f(B')| \geq |B'|$.*

PROOF. We need only prove that Hall's condition is sufficient, which we do by induction on the number of boys. The case $|B| = 1$ is straightforward. For the induction step, we consider two cases:

Case 1: $|f(B')| > |B'|$ for each nonempty $B' \subsetneq B$. Then we can just match an arbitrary boy b to any girl he knows. The set of remaining boys and girls still satisfies Hall's condition, so by the inductive hypothesis, we can match them up.

Case 2: There is a nonempty $B' \subsetneq B$ for which $|f(B')| = |B'|$. By the inductive hypothesis, there is a matching of size $|B'|$ between B' and $f(B')$. Once we show that Hall's condition holds for the bipartite graph between $B \setminus B'$ and $G \setminus f(B')$, another application of the inductive hypothesis will yield the theorem.

Suppose Hall's condition fails; i.e., there is a set A of boys disjoint from B' such that the set $S = f(A) \setminus f(B')$ of girls they know outside $f(B')$ has $|S| < |A|$. (See Figure 3.7.) Then

$$|f(A \cup B')| = |S \cup f(B')| < |A| + |B'|,$$

violating Hall's condition for the full graph, a contradiction.   □

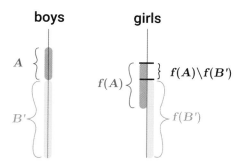

FIGURE 3.7. Hall's Marriage Theorem: Case 2 of the inductive argument. By hypothesis there is a matching of size $|B'|$ between B' and $f(B')$. If $|S| = |f(A) \setminus f(B')| < |A|$ for $A \subset B \setminus B'$, then the set $A \cup B'$ violates Hall's condition.

As we saw earlier, a useful way to represent a bipartite graph whose edges go between vertex sets I and J is via its **adjacency matrix** H. This is a $0/1$ matrix where the rows correspond to vertices in I, the columns to vertices in J, and $h_{ij} = 1$ if and only if there is an edge between i and j. Conversely, any $0/1$ matrix is the adjacency matrix of a bipartite graph. A set of pairs $\mathcal{S} \subset I \times J$ is a **matching** for the adjacency matrix H if $h_{ij} = 1$ for all $(i, j) \in \mathcal{S}$ and no two elements of $\mathcal{S}$ are in the same row or column. This corresponds to a matching between I and J in the graph represented by H.

For example, the following matrix is the adjacency matrix for the bipartite graph shown in Figure 3.6, with the edges corresponding to the matching in bold red in the matrix. (Rows represent boys from top to bottom and columns represent girls from left to right.)

$$\begin{pmatrix} 1 & 1 & 0 & 0 \\ 0 & 1 & 0 & 0 \\ 1 & 0 & 0 & 1 \\ 0 & 0 & 1 & 0 \end{pmatrix}$$

We restate Hall's Marriage Theorem in matrix language and in graph language.

THEOREM 3.2.3 (**Hall's Marriage Theorem – matrix version**). *Let H be an $m \times n$ nonnegative $0/1$ matrix with $m \leq n$. Given a set S of rows, say column j intersects S positively if $h_{ij} = 1$ for some $i \in S$. Suppose that for any set S of rows in H, there are at least $|S|$ columns in H that intersect S positively. Then there is a matching of size m in H.*

THEOREM 3.2.4 (**Hall's Marriage Theorem – graph version**). *Let $G = (U, V, E)$ be a bipartite graph, with $|U| = m$, $|V| = n$, with $m \leq n$. Suppose that the neighborhood [1] of each subset of vertices $S \subseteq U$ has size at least $|S|$. Then there is a matching of size m in G.*

---

[1]The neighborhood of a set S of vertices in a graph is $\{v \mid \exists u \in S \text{ such that } (u, v) \in E\}$.

LEMMA 3.2.5 (**König's lemma**). *Given an $m \times n$ 0/1 matrix H, the size of the maximum matching is equal to the size of the minimum line-cover*[2].

PROOF. Suppose the maximum matching has size k and the minimum line-cover $\mathcal{C}$ has size ℓ. At least one member of each pair in the matching has to be in $\mathcal{C}$ and therefore $k \leq \ell$.

For the other direction, we use Hall's Marriage Theorem. Suppose that there are r rows and c columns in the minimum line-cover $\mathcal{C}$, so $r + c = \ell$. Let S be a subset of rows in $\mathcal{C}$, and let T be the set of columns outside $\mathcal{C}$ that have a positive intersection with some row of S. Then $(\mathcal{C} \setminus S) \cup T$ is also a line-cover, so by the minimality of $\mathcal{C}$, we have $|T| \geq |S|$. Thus, Hall's condition is satisfied for the rows in $\mathcal{C}$ and the columns outside $\mathcal{C}$, and hence there is a matching M of size r in this submatrix. Similarly, there is a matching M' of size c in the submatrix defined by the rows outside $\mathcal{C}$ and the columns in $\mathcal{C}$. Therefore, $M \cup M'$ is a matching of size at least ℓ, and hence $\ell \leq k$, completing the proof. See Figure 3.8. $\square$

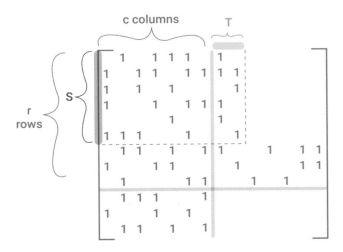

FIGURE 3.8. An illustration of the last part of the proof of Lemma 3.2.5. The first r rows and c columns in the matrix are in the cover $\mathcal{C}$. If T, as defined in the proof, was smaller than S, this would contradict the minimality of $\mathcal{C}$.

COROLLARY 3.2.6. *For the Hide and Seek game, an optimal strategy for the cop is to choose uniformly at random a line in a minimum line-cover. An optimal strategy for the robber is to hide at a uniformly random safehouse in a maximum matching.*

3.3. A pursuit-evasion game: Hunter and Rabbit[*]

Consider the following game[3]: A hunter (player I) is chasing a rabbit (player II). At every time step, each player occupies a vertex of the cycle $\mathbb{Z}_n$. At time 0, the

---

[2]This is also called a *cover* or a *vertex cover*.
[3]This game was first analyzed in [ARS+03]; the exposition here follows, almost verbatim, the paper [BPP+14].

hunter and rabbit choose arbitrary initial positions. At each subsequent step, the hunter may move to an adjacent vertex or stay where she is; simultaneously, the rabbit may stay where he is or jump to any vertex on the cycle. The hunter captures the rabbit at time t if both players occupy the same vertex at that time. Neither player can see the other's position unless they occupy the same vertex. The payoff to the hunter is 1 if she captures the rabbit in the first n steps, and 0 otherwise.

FIGURE 3.9. The hunter and the rabbit.

Clearly, the value of the game is the probability p_n of capture under optimal play.

Here are some possible rabbit strategies:

- If the rabbit chooses a random node and stays there, the hunter can sweep the cycle and capture the rabbit with probability 1.
- If the rabbit jumps to a uniformly random node at every step, he will be caught with probability $1/n$ at each step. Thus, the probability of capture in n steps is $1 - (1 - 1/n)^n \to 1 - 1/e$ as $n \to \infty$.

The Sweep strategy for the hunter consists of choosing a uniformly random starting point and a random direction and then walking in that direction. A rabbit counterstrategy is the following: From a random starting node, walk $\sqrt{n}$ steps to the right, then jump $2\sqrt{n}$ steps to the left; repeat. Figure 3.10 shows a representation of the rabbit counterstrategy to Sweep. Consider the space-time integer lattice, where the vertical coordinate represents time t and the horizontal coordinate represents the position x on the circle. Sincem during the n steps, the rabbit's space-time path will intersect $\Theta(\sqrt{n})$ diagonal lines (lines of the form $x = t + i \mod n$) and the hunter traverses exactly one random diagonal line in space-time, the probability of capture is $\Theta(1/\sqrt{n})$. In fact, the Sweep strategy guarantees the hunter a probability of capture $\Omega(1/\sqrt{n})$ against any rabbit strategy. (See Exercise 3.6.)

It turns out that, to within a constant factor, the best the hunter can do is to start at a random point and move at a random speed. We analyze this strategy in the next section and show that it increases the probability of capture to $\Theta(1/\log(n))$.

3.3.1. Towards optimal strategies. Let H_t be the position of the hunter at time t. Then the set of pure strategies $\mathcal{H}$ available to her are

$$\mathcal{H} = \{(H_t)_{t=0}^{n-1} : H_t \in \mathbb{Z}_n, |H_{t+1} - H_t| \leq 1\}.$$

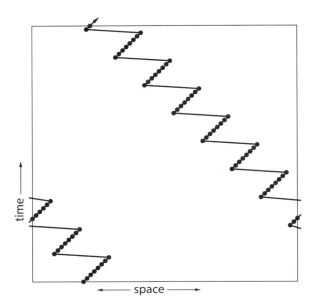

FIGURE 3.10. The figure shows a typical path in space-time for a rabbit employing the counterstrategy to Sweep.

Similarly, if R_t is the position of the rabbit at time t, then $\mathcal{R}$ is the set of pure strategies available to him:

$$\mathcal{R} = \{(R_t)_{t=0}^{n-1} : R_t \in \mathbb{Z}_n\}.$$

If V is the value of the game, then there exists a randomized strategy for the hunter so that against every strategy of the rabbit the probability that they collide in the first n steps is at least V; and there exists a randomized strategy for the rabbit, so that against every strategy of the hunter, the probability that they collide is at most V.

THEOREM 3.3.1. *There are positive constants c and c' such that*

$$\frac{c}{\log n} \leq V \leq \frac{c'}{\log n}. \tag{3.2}$$

3.3.2. The hunter's strategy. Consider the following **Random Speed** strategy for the hunter: Let a, b be independent random variables uniformly distributed on $[0, 1]$ and define $H_t = \lfloor an + bt \rfloor \bmod n$ for $0 \leq t < n$.

PROPOSITION 3.3.2. *If the hunter employs the Random Speed strategy, then against any rabbit strategy,*

$$\mathbb{P}(\text{capture}) \geq \frac{c}{\log n},$$

where c is a universal positive constant.

Proof. Let R_t be the location of the rabbit on the cycle at time t; i.e., $R_t \in \{0, \ldots, n-1\}$. Denote by K_n the number of collisions before time n; i.e., $K_n = \sum_{t=0}^{n-1} I_t$, where

$$I_t = \mathbb{1}_{\{H_t = R_t\}} = \{an + bt \in [R_t, R_t + 1) \cup [R_t + n, R_t + n + 1)\}. \tag{3.3}$$

By Lemma B.1.1, we have

$$\mathbb{P}(K_n > 0) \geq \frac{(\mathbb{E}[K_n])^2}{\mathbb{E}[K_n^2]}. \tag{3.4}$$

For any fixed b and t, the random variable $(an + bt) \bmod n$ is uniformly distributed in $[0, n)$. Therefore $\lfloor an + bt \rfloor \bmod n$ is uniform on $\{0, \ldots, n-1\}$, so $\mathbb{P}(I_t) = 1/n$. Thus,

$$\mathbb{E}[K_n] = \sum_{t=0}^{n-1} \mathbb{P}(I_t) = 1. \tag{3.5}$$

Next, we estimate the second moment:

$$\mathbb{E}[K_n^2] = \mathbb{E}\left[\left(\sum_{t=0}^{n-1} I_t\right)^2\right] = \mathbb{E}[K_n] + \sum_{t \neq m} \mathbb{E}[I_t \cap I_m]$$

$$= 1 + 2\sum_{t=0}^{n-1}\sum_{j=1}^{n-t-1} \mathbb{P}(I_t \cap I_{t+j}). \tag{3.6}$$

To bound $\mathbb{P}(I_t \cap I_{t+j})$, observe that for any r, s, the relations $an + bt \in [r, r+1)$ and $an + b(t+j) \in [s, s+1)$ together imply that $bj \in [s-r-1, s-r+1]$, so

$$\mathbb{P}(an + bt \in [r, r+1) \text{ and } an + b(t+j) \in [s, s+1))$$

$$\leq \mathbb{P}(bj \in [s-r-1, s-r+1]) \cdot \max_b \mathbb{P}(an + bt \in [r, r+1) \mid b) \leq \frac{2}{j} \cdot \frac{1}{n}.$$

Summing over $r \in \{R_t, n + R_t\}$ and $s \in \{R_{t+j}, n + R_{t+j}\}$, we obtain

$$\mathbb{P}(I_t \cap I_{t+j}) \leq \frac{8}{jn}.$$

Plugging back into (3.6), we obtain

$$\mathbb{E}[K_n^2] \leq 1 + 2\sum_{t=0}^{n-1}\sum_{j=1}^{n} \frac{8}{jn} \leq C \log n, \tag{3.7}$$

for a positive constant C. Combining (3.5) and (3.7) using (3.4), we obtain

$$\mathbb{P}(\text{capture}) = \mathbb{P}(K_n > 0) \geq \frac{1}{C \log n},$$

completing the proof of the proposition. $\qquad \square$

3.3.3. The rabbit's strategy. In this section we prove the upper bound in Theorem 3.3.1 by constructing a randomized strategy for the rabbit. It is natural for the rabbit to try to maximize the uncertainty the hunter has about his location. Thus, he will choose a strategy in which, at each point in time, the probability that he is at any particular location on the cycle is $1/n$. With such a strategy, the expected number of collisions of the hunter and the rabbit over the n steps is 1. However, the rabbit's goal is to ensure that the *probability* of collision is low; this will be obtained by concentrating the collisions on a small number of paths.

As before, let K_t be the number of collisions during the first t time steps. As a computational device, we will extend both players' strategies to $2n$ steps. Since

$$\mathbb{E}[K_{2n}] \geq \mathbb{E}[K_{2n} \mid K_n > 0] \cdot \mathbb{P}(K_n > 0)$$

and $\mathbb{E}[K_{2n}] = 2$, we have

$$\mathbb{P}(K_n > 0) \leq \frac{2}{\mathbb{E}[K_{2n}|K_n > 0]}. \tag{3.8}$$

To keep $\mathbb{P}(K_n > 0)$ small, we will construct the rabbit's strategy so that $\mathbb{E}[K_{2n}|K_n > 0]$ is large; that is, given that the rabbit and the hunter meet in the first n steps, the expected number of meetings in $2n$ steps is large.

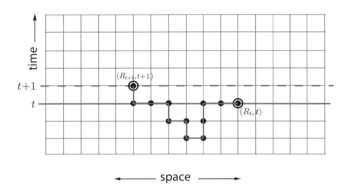

FIGURE 3.11. The figure shows how the random walk determines R_{t+1} from R_t.

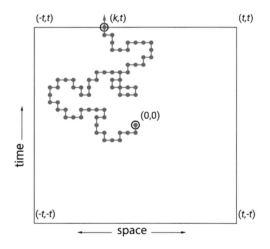

FIGURE 3.12. The figure shows a random walk escaping the $2t \times 2t$ square. Since the walk first hits the line $y = t$ at position (k, t), at time t, the rabbit is in position k.

The strategy is easiest to describe via the following thought experiment: Again, draw the space-time integer lattice, where the y-coordinate represents time and the x-position represents the position on the circle. We identify all the points $(x+in, y)$ for integer i. Suppose that at time t, the rabbit is at position R_t. Execute a simple random walk on the 2D lattice starting at $(x, y) = (R_t, t)$. At the next step, the rabbit jumps to R_{t+1}, where $(R_{t+1}, t+1)$ is the first point at which the random walk hits the line $y = t + 1$. See Figure 3.12.

LEMMA 3.3.3. *There is a constant $c > 0$ such that for all $\ell \in [-t, t]$*

$$\mathbb{P}(R_t = (R_0 + \ell) \bmod n) \geq \frac{c}{t}. \tag{3.9}$$

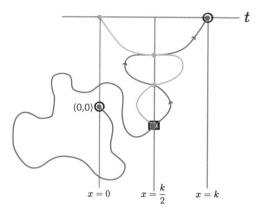

FIGURE 3.13. The figure illustrates the reflection principle. The purple (darker) path represents the random walk. The pink (lighter) line is the reflection of that path across the line $x = k/2$ starting from the first time the random walk hits the line.

PROOF. Let $(\tilde{R}_t, t)$ be the first point that a simple random walk starting at $(0,0)$ hits on the line $y = t$. Let S_t be the square $[-t, t]^2$. First, since a random walk starting at $(0,0)$ is equally likely to exit this square on each of the four sides, we have

$$\mathbb{P}(\text{random walk exits } S_t \text{ on top}) = \frac{1}{4}. \tag{3.10}$$

(See Figure 3.12.) Therefore

$$\sum_{k=-t}^{t} \mathbb{P}\left(\tilde{R}_t = k\right) \geq \frac{1}{4}. \tag{3.11}$$

Next, we show that

$$\mathbb{P}\left(\tilde{R}_t = 0\right) \geq \mathbb{P}\left(\tilde{R}_t = k\right) \tag{3.12}$$

for all $k \in [-t, t]$. First consider the case where k is even. An application of the reflection principle (see Figure 3.13) shows that for each path from $[0, 0]$ to $[k, t]$ there is another equally likely path from $[0, 0]$ to $[0, t]$. To handle the case of k odd, we extend our lattice by adding a mid-point along every edge. This slows down the random walk, but does not change the distribution of $\tilde{R}_k$ and allows the reflection argument to go through.

Inequalities (3.11) and (3.12) together imply that

$$\mathbb{P}\left(\tilde{R}_t = 0\right) \geq \frac{1}{4(2t+1)}. \tag{3.13}$$

To prove that there is a constant c such that

$$\mathbb{P}\left(\tilde{R}_t = k\right) \geq \frac{c}{t},$$

consider the smaller square from $-k$ to k. With probability $1/4$, the random walk first hits the boundary of the square on the right side. If so, run the argument used to show (3.13) starting from this hitting point. See Figure 3.14. □

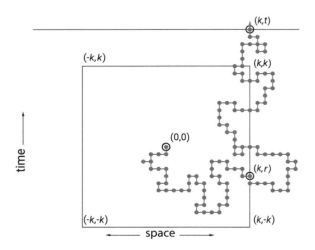

FIGURE 3.14. The figure shows a random walk escaping the $2k \times 2k$ square.

COROLLARY 3.3.4. *There is a constant c such that*

$$\mathbb{E}[K_{2n}|K_n > 0] \geq c \log(n).$$

PROOF. Suppose that the hunter and the rabbit both start out at position $(0,0)$ together. Then the position of the hunter on the cycle at time t must be in $\{-t, t\}$. The lemma then implies that the probability of a collision at time t is at least c/t. Thus, in T steps, the expected number of collisions is at least $c \log(T)$. Let F_t denote the event that the hunter and the rabbit *first* collide at time t. Observing that $\sum_{0 \leq t < n} \mathbb{P}(F_t|K_n > 0) = 1$, we have

$$\mathbb{E}[K_{2n}|K_n > 0] = \sum_{0 \leq t < n} \mathbb{E}[K_{2n}|F_t]\,\mathbb{P}(F_t|K_n > 0)$$

$$\geq \sum_{0 \leq t < n} c \log(2n - t)\mathbb{P}(F_t|K_n > 0)$$

$$\geq c \log n. \qquad \square$$

Substituting the result of this corollary into (3.8) completes the proof of the upper bound in Theorem 3.3.1.

3.4. The Bomber and Battleship game

In this family of games, a battleship is initially located at the origin in $\mathbb{Z}$. At each time step in $\{0, 1, \ldots\}$, the ship moves either left or right to a new site where it remains until the next time step. The bomber (player I), who can see the current location of the battleship (player II), drops one bomb at some time j over some site in $\mathbb{Z}$. The bomb arrives at time $j + 2$ and destroys the battleship if it hits it. (The battleship cannot see the bomber or the bomb in time to change course.) For the game G_n, the bomber has enough fuel to drop its bomb at any time $j \in \{0, 1, \ldots, n\}$. What is the value of the game?

EXERCISE 3.b. (i) Show that the value of G_0 is $1/3$. (ii) Show that the value of G_1 is also $1/3$. (iii) Show that the value of G_2 is greater than $1/3$.

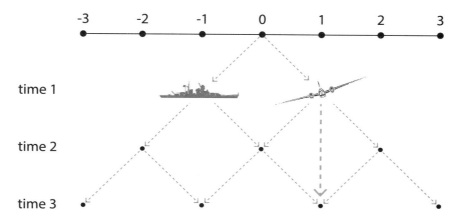

FIGURE 3.15. The bomber drops her bomb where she hopes the battleship will be two time units later. In the picture the bomber is at position 1 at time 1 and drops a bomb. The battleship does not see the bomb coming and randomizes his path to avoid the bomb, but if he is at position 1 at time 3, he will be hit.

Consider the following *p-reversal* strategy for G_n. On the first move, go left with probability p and right with probability $1 - p$. From then on, at each step reverse direction with probability $1 - p$ and keep going with probability p.

The battleship chooses p to maximize the probability of survival. Its probabilities of arrival at sites -2, 0, or 2 at time 2 are p^2, $1 - p$, and $p(1 - p)$. Thus, p will be chosen so that $\max\{p^2, 1 - p\}$ is minimized. This is attained when $p^2 = 1 - p$, whose solution in $(0, 1)$ is given by $p = 2/(1 + \sqrt{5})$. For any time j that the bomber chooses to drop a bomb, the battleship's relative position two time steps later has the same distribution. Therefore, the payoff for the bomber against this strategy is at most $1 - p$, so $v(G_n) \leq 1 - p$ for every n. While there is no value of n for which the p-reversal strategy is optimal in G_n, it is asymptotically optimal, i.e. $\lim_{n \to \infty} v(G_n) = 1 - p = (\sqrt{5} - 1)/(\sqrt{5} + 1)$. See the notes.

Notes

A generalization of the Troll and Traveler game from §3.1.1 can be played on an arbitrary (not necessarily series-parallel) undirected graph with two distinguished vertices

s and t: If the troll and the traveler traverse an edge in the same direction, the traveler pays the cost of the road to the troll, whereas if they traverse a road in opposite directions, then the troll pays the cost of the road to the traveler. If we interpret the cost of each road e as an edge resistance R_e, then the value of the game turns out to be the effective resistance between s and t: There is a unique unit flow F from s to t (called the unit current flow) that satisfies the cycle law $\sum_e R_e F_e = 0$ along any directed cycle. This flow can be decomposed into a convex combination of path flows from s to t. Let p_γ be the weight for path γ in this convex combination. Then an optimal strategy for each player is to choose γ with probability p_γ. For more details on effective resistance and current flows, see, e.g., [DS84, LPW09].

The Hide and Seek game in §3.2 comes from [vN53]. The theory of maximum matching and minimum covers in §3.2.1 was first developed by Frobenius [Fro17] and König [Kön31], and rediscovered by Hall [Hal35]. For a detailed history, see Section 16.7h in Schrijver [Sch03], and for a detailed exposition of these topics, see, e.g., [LP09, vLW01, Sch03].

As noted in §3.3, the Hunter and Rabbit game discussed there was first analyzed in [ARS+03]; our exposition follows verbatim the paper [BPP+14].

An interesting open problem is whether there is a hunter strategy that captures a weak rabbit (that can only jump to adjacent nodes) in n steps with constant probability on any n vertex graph.

The Hunter and Rabbit game is an example of a search game. These games are the subject of the book [AG03]. See also the classic [Isa65].

The Bomber and Battleship game of §3.4 was proposed by R. Isaacs [Isa55], who devised the battleship strategy discussed in the text. The value of the game was determined by Dubins [Dub57] and Karlin [Kar57].

Exercise 3.5 is from Kleinberg and Tardos [KT06]. Exercise 3.4 is from [vN53].

Exercises

3.1.   Solve Troll and Traveler on the graph in Figure 3.16 assuming that the toll on each edge is 1.

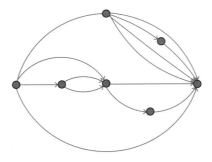

FIGURE 3.16. A series-parallel graph.

3.2.   Prove that every k-regular bipartite graph has a perfect matching. (A bipartite graph is k-regular if it is $n \times n$, and each vertex has exactly k edges incident to it.)

3.3. Let M be the 0-1 matrix corresponding to a bipartite graph G. Show that if there is no perfect matching in M, then there is a set I of rows of M and a set J of columns of M such that $|I| + |J| > n$ and there is no edge from I to J.

3.4. **Birkhoff-von Neumann Theorem**: Prove that every doubly stochastic $n \times n$ matrix is a convex combination of permutation matrices.
Hint: Use Hall's Marriage Theorem and induction on the number of nonzero entries.

3.5. Let G be an $n \times n$ bipartite graph, where vertices on one side correspond to actors, vertices on the other side correspond to actresses, and there is an edge between actor i and actress j if they have starred in a movie together. Consider a game where the players alternate naming names, with player I naming an actor i from the left side of the graph, then player II naming an actress j that has starred with i from the right side, then player I naming an actor i' from the left side that has starred with j, and so on. No repetition of names is allowed. The player who cannot respond with a new name loses the game. For an example, see Figure 3.17. Show that II has a winning strategy if the graph G has a perfect matching, and I has a winning strategy otherwise. *Hint:* Player I's winning strategy in the latter case is to find a maximum matching in the graph, and then begin by naming an actor that is not in that maximum matching.

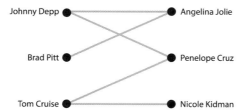

FIGURE 3.17. In this game, if player I names Johnny Depp, then player II must name either Angelina Jolie or Penelope Cruz. In the latter case, player I must then name Tom Cruise, and player II must name Nicole Kidman. At this point, player I cannot name a new actor and loses the game.

3.6. Show that the Sweep strategy described in §3.3 guarantees the hunter a probability of capture $\Omega(1/\sqrt{n})$ against any rabbit strategy. *Hint:* Project the space-time path of the rabbit on both diagonals. One of these projections must have size $\Omega(\sqrt{n})$.

3.7.   Prove that the following hunter strategy also guarantees probability of cap-
        ture $\Omega(1/\log(n))$. Pick a uniform $u \in [0, 1]$ and a random starting point.
        At each step, walk to the right with probability u, and stay in place other-
        wise.

3.8.   Suppose that the Hunter and Rabbit game runs indefinitely. Consider the
        zero-sum game in which the gain to the rabbit is the time until he is cap-
        tured. Show that the hunter has a strategy guaranteeing that the expected
        capture time is $O(n \log n)$ and that the rabbit has a strategy guaranteeing
        that the expected capture time is is $\Omega(n \log n)$.

3.9.   Consider the Hunter and Rabbit game on an arbitrary undirected n-vertex
        graph G. Show that there is a hunter strategy guaranteeing that
        $$\mathbb{P}(\text{capture in } n \text{ steps}) \geq \frac{c}{\log n},$$
        where $c > 0$ is an absolute constant. *Hint:* Construct a spanning tree of
        the graph and then reduce to the cycle case by traversing the spanning tree
        in a depth-first order.

*3.10.   Show that for the Hunter and Rabbit game on a $\sqrt{n} \times \sqrt{n}$ grid,
        $$\mathbb{P}(\text{capture in } n \text{ steps}) \geq c > 0$$
        for some hunter strategy. *Hint:* Random direction, random speed.

3.11.   Show that a weak rabbit that can only jump to adjacent nodes will be
        caught in n steps on the n-cycle with probability at least $c > 0$ by a sweep-
        ing hunter.

3.12.   A pirate hides a treasure somewhere on a circular desert island of radius r.
        The police fly across the island k times in a straight path. They will locate
        the treasure (and win the game) if their path comes within distance w of it.
        See Figure 3.18. Find the value of the game and some optimal strategies.
        Hint: Archimedes showed that the surface area on a sphere in $\mathbb{R}^3$ that lies
        between two parallel planes that intersect the sphere is proportional to the
        distance between the planes. Verify this (if you are so inclined) and use it
        to construct an optimal strategy for the thief.

3.13.   Set up the payoff matrix for the Bomber and Battleship game from §3.4
        and find the value of the game G_2.

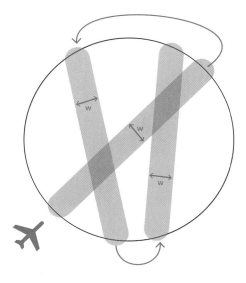

FIGURE 3.18. Plane criss-crossing a desert island.

CHAPTER 4

General-sum games

We now turn to the theory of **general-sum games**. Such a game is specified by two matrices $A = (a_{ij})$ and $B = (b_{ij})$. If player I chooses action i and player II chooses action j, their payoffs are a_{ij} and b_{ij}, respectively. In contrast to zero-sum games, there is no reasonable definition of "optimal strategies". Safety strategies still exist, but they no longer correspond to equilibria. The most important notion is that of a **Nash equilibrium**, i.e., a pair of strategies, one per player, such that each is a best response to the other. General-sum games and the notion of Nash equilibrium extend naturally to more than two players.

4.1. Some examples

EXAMPLE 4.1.1 (**Prisoner's Dilemma**). Two suspects are imprisoned by the police who ask each of them to confess. The charge is serious, but there is not enough evidence to convict. Separately, each prisoner is offered the following plea deal. If he confesses and the other prisoner remains silent, the confessor goes free, and his confession is used to sentence the other prisoner to ten years in jail. If both confess, they will both spend eight years in jail. If both remain silent, the sentence is one year to each for the minor crime that can be proved without additional evidence.

FIGURE 4.1. Two prisoners considering whether to confess or remain silent.

The following matrix summarizes the payoffs, where negative numbers represent years in jail, and an entry $(-10, 0)$ means payoff -10 to prisoner I and 0 to prisoner II.

		prisoner II	
		silent	confess
prisoner I	silent	$(-1, -1)$	$(-10, 0)$
	confess	$(0, -10)$	$(-8, -8)$

In this game, the prisoners are better off if both of them remain silent than they are if both of them confess. However, the two prisoners select their actions separately, and for each possible action of one prisoner, the other is better off confessing; i.e., confessing is a **dominant strategy**.

The same phenomenon occurs even if the players play this game a fixed number of times. This can be shown by a backwards induction argument. (See Exercise 4.4.) However, as we shall see in §6.4, if the game is played repeatedly, but play ends at a random time, then the mutually preferable solution may become an equilibrium.

EXAMPLE 4.1.2 (**Stag Hunt**). Two hunters are following a stag when a hare runs by. Each hunter has to make a split-second decision: to chase the hare or to continue tracking the stag. The hunters must cooperate to catch the stag, but each hunter can catch the hare on his own. (If they both go for the hare, they share it.) A stag is worth four times as much as a hare. This leads to the following payoff matrix

		Hunter II	
		Stag (S)	Hare (H)
Hunter I	Stag (S)	(4, 4)	(0, 2)
	Hare (H)	(2, 0)	(1, 1)

What are good strategies for the hunters? We begin by considering safety strategies.[1] For each player, H is the unique safety strategy and yields a payoff of 1. The strategy pair (H, H) is also a pure Nash equilibrium, since given the choice by the other hunter to pursue a hare, a hunter has no incentive to continue tracking the stag. There is another pure Nash equilibrium, (S, S), which yields both players a payoff of 4. Finally, there is a mixed Nash equilibrium, in which each player selects S with probability $1/3$. This results in an expected payoff of $4/3$ to each player.

This example illustrates a phenomenon that doesn't arise in zero-sum games: *a multiplicity of equilibria with different expected payoffs to the players.*

EXAMPLE 4.1.3 (**War and Peace**). Two countries in conflict have to decide between diplomacy and military action. One possible payoff matrix is:

		Firm II	
		diplomacy	attack
Firm I	diplomacy	(2, 2)	(-2, 0)
	attack	(0, -2)	(-1, -1)

---

[1] A safety strategy for player I is defined as in Definition 2.2.1. For player II the same definition applies with the payoff matrix B replacing A.

FIGURE 4.2. Stag Hunt.

Like Stag Hunt, this game has two pure Nash equilibria, where one arises from safety strategies, and the other yields higher payoffs. In fact, this payoff matrix is the Stag Hunt matrix, with all payoffs reduced by 2.

EXAMPLE 4.1.4 (**Driver and Parking Inspector**). Player I is choosing between parking in a convenient but illegal parking spot (payoff 10 if she's not caught) and parking in a legal but inconvenient spot (payoff 0). If she parks illegally and is caught, she will pay a hefty fine (payoff -90). Player II, the inspector representing the city, needs to decide whether to check for illegal parking. There is a small cost (payoff -1) to inspecting. However, there is a greater cost to the city if player I has parked illegally since that can disrupt traffic (payoff -10). This cost is partially mitigated if the inspector catches the offender (payoff -6).

The resulting payoff matrix is the following:

		Inspector	
		Don't Inspect	Inspect
Driver	Legal	$(0,0)$	$(0,-1)$
	Illegal	$(10,-10)$	$(-90,-6)$

In this game, the safety strategy for the driver is to park legally (guaranteeing her a payoff of 0), and the safety strategy for the inspector is to inspect (guaranteeing him/the city a payoff of -6). However, the strategy pair (legal, inspect) is *not* a Nash equilibrium. Indeed, knowing the driver is parking legally, the inspector's best response is not to inspect. It is easy to check that this game has no Nash equilibrium in which either player uses a pure strategy.

There is, however, a mixed Nash equilibrium. Suppose that the strategy pair $(x, 1-x)$ for the driver and $(y, 1-y)$ for the inspector is a Nash equilibrium. If $0 < y < 1$, then both possible actions of the inspector must yield him the same payoff. If, for instance, inspecting yielded a higher payoff, then $(0, 1)$ would be a better strategy than $(y, 1 - y)$. Thus, $-10(1 - x) = -x - 6(1 - x)$. Similarly, $0 < x < 1$ implies that $0 = 10y - 90(1 - y)$. These equations yield $x = 0.8$ (the driver parks legally with probability 0.8 and obtains an expected payoff of 0) and $y = 0.9$ (the inspector inspects with probability 0.1 and obtains an expected payoff of -2).

4.2. Nash equilibria

A two-person general-sum game can be represented by a pair[2] of $m \times n$ **payoff matrices** $A = (a_{ij})$ and $B = (b_{ij})$, whose rows are indexed by the m possible actions of player I and whose columns are indexed by the n possible actions of player II. Player I selects an action i and player II selects an action j, each unaware of the other's selection. Their selections are then revealed and player I receives a payoff of a_{ij} and player II receives a payoff of b_{ij}.

A **mixed strategy** for player I is determined by a vector $(x_1, \ldots, x_m)^T$ where x_i represents the probability that player I plays action i and a mixed strategy for player II is determined by a vector $(y_1, \ldots, y_n)^T$ where y_j is the probability that player II plays action j. A mixed strategy in which a particular action is played with probability 1 is called a **pure strategy**.

DEFINITION 4.2.1 (**Nash equilibrium**). A pair of mixed strategy vectors $(\mathbf{x}^*, \mathbf{y}^*)$ with $\mathbf{x}^* \in \Delta_m$ (where $\Delta_m = \{\mathbf{x} \in \mathbb{R}^m : x_i \geq 0, \sum_{i=1}^m x_i = 1\}$) and $\mathbf{y}^* \in \Delta_n$ is a **Nash equilibrium** if no player gains by unilaterally deviating from it. That is,

$$(\mathbf{x}^*)^T A \mathbf{y}^* \geq \mathbf{x}^T A \mathbf{y}^*$$

for all $\mathbf{x} \in \Delta_m$ and

$$(\mathbf{x}^*)^T B \mathbf{y}^* \geq (\mathbf{x}^*)^T B \mathbf{y}$$

for all $\mathbf{y} \in \Delta_n$.

The game is called **symmetric** if $m = n$ and $a_{i,j} = b_{j,i}$ for all $i, j \in \{1, 2, \ldots, n\}$. A pair $(\mathbf{x}, \mathbf{y})$ of strategies in Δ_n is called **symmetric** if $x_i = y_i$ for all $i = 1, \ldots, n$.

One reason that Nash equilibria are important is that any strategy profile that is *not* a Nash equilibrium is, by definition, *unstable*: There is always at least one player who prefers to switch strategies. We will see that there always exists a Nash equilibrium; however, there can be many of them, and they may yield different payoffs to the players. Thus, Nash equilibria do not have the predictive power in general-sum games that safety strategies have in zero-sum games. See the notes for a discussion of critiques of Nash equilibria.

EXAMPLE 4.2.2 (**Cheetahs and Antelopes**). Two cheetahs are chasing a pair of antelopes, one large and one small. Each cheetah has two possible strategies: Chase the large antelope (L) or chase the small antelope (S). The cheetahs will catch any antelope they choose, but if they choose the same one, they must share the spoils. Otherwise, the catch is unshared. The large antelope is worth ℓ and the small one is worth s. Here is the payoff matrix:

---

[2]In examples, we write one matrix whose entries are pairs (a_{ij}, b_{ij}).

	cheetah II	
	L	S
L	$(\ell/2, \ell/2)$	(ℓ, s)
S	(s, ℓ)	$(s/2, s/2)$

cheetah I

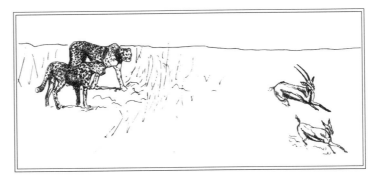

FIGURE 4.3. Cheetahs deciding whether to chase the large or the small antelope.

If the larger antelope is worth at least twice as much as the smaller ($\ell \geq 2s$), then strategy L dominates strategy S. Hence each cheetah should just chase the larger antelope. If $s < \ell < 2s$, then there are two pure Nash equilibria, (L, S) and (S, L). These pay off quite well for both cheetahs — but how would two healthy cheetahs agree which should chase the smaller antelope? Therefore it makes sense to look for symmetric mixed equilibria.

If the first cheetah chases the large antelope with probability x, then the expected payoff to the second cheetah from chasing the larger antelope is

$$L(x) = \frac{\ell}{2}x + (1 - x)\ell,$$

and the expected payoff from chasing the smaller antelope is

$$S(x) = xs + (1 - x)\frac{s}{2}.$$

These expected payoffs are equal when

$$x = x^* := \frac{2\ell - s}{\ell + s}. \tag{4.1}$$

For any other value of x, the second cheetah would prefer either the pure strategy L or the pure strategy S, and then the first cheetah would do better by simply playing pure strategy S or pure strategy L. But if both cheetahs chase the large antelope with probability x^* in (4.1), then neither has an incentive to deviate, so this is a (symmetric) Nash equilibrium.

There is a fascinating connection between symmetric mixed Nash equilibria in games such as this and equilibria in biological populations. Consider a population of cheetahs, and suppose a fraction x of them are greedy (i.e., play strategy L). Each time a cheetah plays this game, he plays it against a random cheetah in the population. Then a greedy cheetah obtains an expected payoff of $L(x)$, whereas a nongreedy cheetah obtains an expected payoff of $S(x)$. If $x > x^*$, then $S(x) > L(x)$ and nongreedy cheetahs have an advantage over greedy cheetahs. On the other

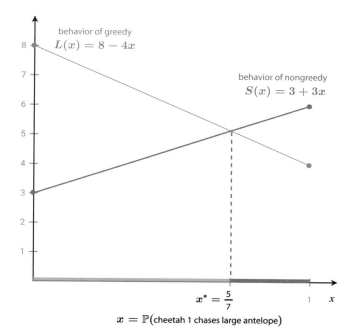

FIGURE 4.4. $L(x)$ (respectively, $S(x)$) is the payoff to cheetah II from chasing the large antelope worth $\ell = 8$ (respectively, the small antelope worth $s = 6$).

hand, if $x < x^*$, greedy cheetahs have an advantage. See Figure 4.4. Altogether, the population seems to be pushed by evolution towards the symmetric mixed Nash equilibrium $(x^*, 1 - x^*)$. Indeed, such phenomena have been observed in real biological systems. The related notion of an **evolutionarily stable strategy** is formalized in §7.1.

EXAMPLE 4.2.3 (**Chicken**). Two drivers speed head-on toward each other and a collision is bound to occur unless one of them chickens out and swerves at the last minute. If both swerve, everything is OK (in this case, they both get a payoff of 1). If one chickens out and swerves, but the other does not, then it is a great success for the player with iron nerves (yielding a payoff of 2) and a great disgrace for the chicken (a penalty of 1). If both players have iron nerves, disaster strikes (and both incur a large penalty M).

		player II	
		Swerve (S)	Drive (D)
player I	Swerve (S)	$(1, 1)$	$(-1, 2)$
	Drive (D)	$(2, -1)$	$(-M, -M)$

There are two pure Nash equilibria in this game, (S, D) and (D, S): if one player knows with certainty that the other will drive on (respectively, swerve), that player is better off swerving (respectively, driving on).

To determine the mixed equilibria, suppose that player I plays S with probability x and D with probability $1 - x$. This presents player II with expected payoffs of $x + (1 - x) \cdot (-1)$, i.e., $2x - 1$ if he plays S, and $2x + (1 - x) \cdot (-M) = (M + 2)x - M$

FIGURE 4.5. The game of Chicken.

if he plays D. We seek an equilibrium where player II has positive probability on each of S and D. Thus,

$$2x - 1 = (M + 2)x - M; \quad \text{i.e.,} \quad x = 1 - \frac{1}{M}.$$

The resulting payoff for player II is $2x - 1 = 1 - 2/M$.

REMARKS 4.2.4.

(1) Even though both payoff matrices decrease as M increases, the equilibrium payoffs increase. This contrasts with zero sum games where decreasing a player's payoff matrix can only lower her expected payoff in equilibrium.

(2) The payoff for a player is lower in the symmetric Nash equilibrium than it is in the pure equilibrium where that player plays D and the other plays S. One way for a player to ensure[3] that the higher payoff asymmetric Nash equilibrium is reached is to irrevocably commit to the strategy D, for example, by ripping out the steering wheel and throwing it out of the car. In this way, it becomes impossible for him to chicken out, and if the other player sees this and believes her eyes, then she has no other choice but to chicken out.

In a number of games, making this kind of **binding commitment** pushes the game into a pure Nash equilibrium, and the nature of that equilibrium strongly depends on who managed to commit first. Here, the payoff for the player who did not make the commitment is lower than the payoff in the unique mixed Nash equilibrium, while in some games it is higher (e.g., see the Battle of the Sexes in §7.2).

(3) An amusing real-life example of commitments in a different game[4] arises in a certain narrow two-way street in Jerusalem. Only one car at a time can pass. If two cars headed in opposite directions meet in the street,

---

[3]This assumes rationality of the other player – a dubious assumption for people who play Chicken.
[4]See War of Attrition in §14.4.3.

FIGURE 4.6. Ripping out the steering wheel is a binding commitment in Chicken.

the driver that can signal to the opponent that he will not yield will convincingly force the other to back out. Some drivers carry a newspaper with them, which they can strategically pull out to signal that they are not in a rush.

FIGURE 4.7. A driver signaling that he has all the time in the world.

4.3. General-sum games with more than two players

We now consider general-sum games with more than two players and generalize the notion of Nash equilibrium to this setting. Each player i has a set S_i of pure strategies. We are given payoff or **utility functions** $u_i : S_1 \times S_2 \times \cdots \times S_k \to \mathbb{R}$, for each player i, where $i \in \{1, \ldots, k\}$. If player j plays strategy $s_j \in S_j$ for each $j \in \{1, \ldots, k\}$, then player i has a payoff or utility of $u_i(s_1, \ldots, s_k)$.

EXAMPLE 4.3.1 (**One Hundred Gnus and a Lioness**). A lioness is chasing one hundred gnus. It seems clear that if the gnus cooperated, they could chase the lioness away, but typically they do not. Indeed, for all the gnus to run away is a

Nash equilibrium, since it would be suicidal for just one of them to confront the lioness. On the other hand, cooperating to attack the lioness is not an equilibrium, since an individual gnu would be better off letting the other ninety-nine attack.

FIGURE 4.8. One Hundred Gnus and a Lioness.

EXAMPLE 4.3.2 (**Pollution game**). Three firms will either pollute a lake in the following year or purify it. They pay 1 unit to purify, but it is free to pollute. If two or more pollute, then the water in the lake is useless, and each firm must pay 3 units to obtain the water that they need from elsewhere. If at most one firm pollutes, then the water is usable, and the firms incur no further costs.

If firm III purifies, the cost matrix (cost $= -$ payoff) is

		firm II	
		purify	pollute
firm I	purify	(1,1,1)	(1,0,1)
	pollute	(0,1,1)	(3,3,4)

If firm III pollutes, then it is

		firm II	
		purify	pollute
firm I	purify	(1,1,0)	(4,3,3)
	pollute	(3,4,3)	(3,3,3)

FIGURE 4.9. Three firms deciding whether or pollute a lake or not.

To discuss the game, we generalize the notion of Nash equilibrium to games with more players.

DEFINITION 4.3.3. For a vector $\mathbf{s} = (s_1, \ldots, s_n)$, we use $\mathbf{s}_{-i}$ to denote the vector obtained by excluding s_i; i.e.,

$$\mathbf{s}_{-i} = (s_1, \ldots, s_{i-1}, s_{i+1}, \ldots, s_n).$$

We interchangeably refer to the full vector $(s_1, \ldots, s_n)$ as either $\mathbf{s}$, or, slightly abusing notation, $(s_i, \mathbf{s}_{-i})$.

DEFINITION 4.3.4. A **pure Nash equilibrium** in a k-player game is a sequence of pure strategies $(s_1^*, \ldots, s_k^*) \in S_1 \times \cdots \times S_k$ such that for each player $j \in \{1, \ldots, k\}$ and each $s_j \in S_j$, we have

$$u_j(s_j^*, \mathbf{s}_{-j}^*) \geq u_j(s_j, \mathbf{s}_{-j}^*).$$

In other words, for each player j, his selected strategy s_j^* is a best response to the selected strategies $\mathbf{s}_{-j}^*$ of the other players.

DEFINITION 4.3.5. A (mixed) strategy profile in a k-player game is a sequence $(\mathbf{x}_1, \ldots, \mathbf{x}_k)$, where $\mathbf{x}_j \in \Delta_{|S_j|}$ is a mixed strategy for player j. A **mixed Nash equilibrium** is a strategy profile $(\mathbf{x}_1^*, \ldots, \mathbf{x}_k^*)$ such that for each player $j \in \{1, \ldots, k\}$ and each probability vector $\mathbf{x}_j \in \Delta_{|S_j|}$, we have

$$u_j(\mathbf{x}_j^*, \mathbf{x}_{-j}^*) \geq u_j(\mathbf{x}_j, \mathbf{x}_{-j}^*).$$

Here

$$u_j(\mathbf{x}_1, \mathbf{x}_2, \ldots, \mathbf{x}_k) := \sum_{s_1 \in S_1, \ldots, s_k \in S_k} \mathbf{x}_1(s_1) \cdots \mathbf{x}_k(s_k) u_j(s_1, \ldots, s_k),$$

where $\mathbf{x}_i(s)$ is the probability that player i assigns to pure strategy s in the mixed strategy $\mathbf{x}_i$.

We will prove the following result in §5.1.

THEOREM 4.3.6 (**Nash's Theorem**). *Every finite general-sum game has a Nash equilibrium.*

For determining Nash equilibria in (small) games, the following lemma (which we have already applied several times) is useful.

LEMMA 4.3.7. *Consider a k-player game where* $\mathbf{x}_i$ *is the mixed strategy of player i. For each i, let* $T_i = \{s \in S_i \mid \mathbf{x}_i(s) > 0\}$. *Then* $(\mathbf{x}_1, \ldots, \mathbf{x}_k)$ *is a Nash equilibrium if and only if for each i, there is a constant* c_i *such that*[5]

$$\forall s_i \in T_i \quad u_i(s_i, \mathbf{x}_{-i}) = c_i \quad and \quad \forall s_i \notin T_i \quad u_i(s_i, \mathbf{x}_{-i}) \leq c_i.$$

EXERCISE 4.a. Prove Lemma 4.3.7.

REMARK 4.3.8. Lemma 4.3.7 extends the equivalence (i) $\leftrightarrow$ (ii) of Proposition 2.5.3.

Returning to the Pollution game, it is easy to check that there are two pure equilibria. The first consists of all three firms polluting, resulting in a cost of 3 to each player, and the second consists of two firms purifying (at cost 1 each) and one firm polluting (at no cost). The symmetric polluting equilibrium is an example of the *Tragedy of the Commons*[6]: All three firms would prefer any of the asymmetric equilibria, but cannot unilaterally transition to these equilibria.

Next we consider mixed strategies. First, observe that if player III purifies, then it is a best response for each of player I and II to purify with probability 2/3 and pollute with probability 1/3. Conversely, it is a best response for player III to purify, since his cost is $1 \cdot 8/9 + 4 \cdot 1/9 = 12/9$ for purifying, but $0 \cdot 4/9 + 3 \cdot 5/9 = 15/9$ for polluting. Similarly, there are Nash equilibria where player I (resp. player II) purifies and the other two mix in the proportions $(2/3, 1/3)$.

Finally, we turn to fully mixed strategies. Suppose that player i's strategy is $\mathbf{x}_i = (p_i, 1 - p_i)$ (that is, i purifies with probability p_i). It follows from Lemma 4.3.7 that these strategies are a Nash equilibrium with $0 < p_i < 1$, if and only if

$$u_i(\text{purify}, \mathbf{x}_{-i}) = u_i(\text{pollute}, \mathbf{x}_{-i}).$$

Thus, if $0 < p_1 < 1$, then

$$p_2 p_3 + p_2(1 - p_3) + p_3(1 - p_2) + 4(1 - p_2)(1 - p_3)$$
$$= 3p_2(1 - p_3) + 3p_3(1 - p_2) + 3(1 - p_2)(1 - p_3),$$

or, equivalently,

$$1 = 3(p_2 + p_3 - 2p_2 p_3). \tag{4.2}$$

Similarly, if $0 < p_2 < 1$, then

$$1 = 3(p_1 + p_3 - 2p_1 p_3), \tag{4.3}$$

and if $0 < p_3 < 1$, then

$$1 = 3(p_1 + p_2 - 2p_1 p_2). \tag{4.4}$$

Subtracting (4.3) from (4.4), we get $0 = 3(p_2 - p_3)(1 - 2p_1)$. This means that if all three firms use mixed strategies, then either $p_2 = p_3$ or $p_1 = 1/2$. In the first case ($p_2 = p_3$), equation (4.2) becomes quadratic in p_2, with two solutions

---

[5]The notation $(s_i, \mathbf{x}_{-i})$ is an abbreviation where we identify the pure strategy s_i with the probability vector 1_{s_i} that assigns s_i probability 1.

[6]In games of this type, individuals acting in their own self-interest deplete a shared resource, (in this case, clean water) thereby making everybody worse off. See also Example 4.5.1.

$p_2 = p_3 = (3 \pm \sqrt{3})/6$, both in $(0,1)$. Substituting these solutions into the other equations yields $p_1 = p_2 = p_3$, resulting in two symmetric mixed equilibria. If, instead of $p_2 = p_3$, we let $p_1 = 1/2$, then (4.3) becomes $1 = 3/2$, which is nonsense. This means that there is no asymmetric equilibrium with three mixed (non-pure) strategies. One can also check that there is no equilibrium with two pure and one non-pure strategy. Thus the set of Nash equilibria consists of one symmetric and three asymmetric pure equilibria, three equilibria where one player has a pure strategy and the other two play the same mixed strategy, and two symmetric mixed equilibria.

The most reasonable interpretation of the symmetric mixed strategies involves population averages, as in Example 4.2.2. If p denotes the fraction of firms that purify, then the only stable values of p are $(3 \pm \sqrt{3})/6$, as shown in Exercise 4.15.

4.3.1. Symmetric games. With the exception of Example 4.1.4, all of the games in this chapter so far are *symmetric*. Indeed, they are unchanged by any relabeling of the players. Here is a game with more restricted symmetry.

EXAMPLE 4.3.9 (**Location-sensitive Pollution**). Four firms are located around a lake. Each one chooses to pollute or purify. It costs 1 unit to purify, but it is free to pollute. But if a firm i and its two neighbors $i \pm 1 (\mod 4)$ pollute, then the water is unusable to i and $u_i = -3$.

This game is symmetric under rotation. In particular, $u_1(s_1, s_2, s_3, s_4) = u_2(s_4, s_1, s_2, s_3)$. Consequently, $u_1(s, \mathbf{x}, \mathbf{x}, \mathbf{x}) = u_2(\mathbf{x}, s, \mathbf{x}, \mathbf{x})$ for all pure strategies s and mixed strategies $\mathbf{x}$.

This motivates the following general definition.

DEFINITION 4.3.10. Suppose that all players in a k-player game have the same set of pure strategies S. Denote by $u_j(s; \mathbf{x})$ the utility of player j when he plays pure strategy $s \in S$ and all other players play the mixed strategy $\mathbf{x}$. We say the game is **symmetric** if

$$u_i(s; \mathbf{x}) = u_j(s; \mathbf{x})$$

for every pair of players i, j, pure strategy s, and mixed strategy $\mathbf{x}$.

We will prove the following proposition in §5.1.

PROPOSITION 4.3.11. *In a symmetric game, there is a symmetric Nash equilibrium (where all players use the same strategy).*

4.4. Potential games

In this section, we consider a class of games that *always* have a pure Nash equilibrium. Moreover, we shall see that a Nash equilibrium in such a game can be reached by a series of best-response moves. We begin with an example.

EXAMPLE 4.4.1 (**Congestion Game**). There is a road network with R roads and k drivers, where the jth driver wishes to drive from point s_j to point t_j. Each driver, say the jth, chooses a path P_j from s_j to t_j and incurs a cost or latency due to the congestion on the path selected.

This cost is determined as follows. Suppose that the paths selected by the k drivers are $\mathbf{P} = (P_1, P_2, \ldots, P_k)$. For each road r, let $n_r(\mathbf{P})$ be the number of drivers j that use r; i.e., r is on P_j. Denote by $c_r(n)$ the cost incurred by a driver using road r when n drivers use that road. The total cost incurred by driver i taking path P_i is the sum of the costs on the roads he uses; i.e.,

$$\text{cost}_i(\mathbf{P}) = \sum_{r \in P_i} c_r(n_r(\mathbf{P})).$$

Imagine adding the players one at a time and looking at the cost each player incurs at the moment he is added. We claim that the sum of these quantities[7] is

$$\phi(\mathbf{P}) := \sum_{r=1}^{R} \sum_{\ell=1}^{n_r(\mathbf{P})} c_r(\ell). \tag{4.5}$$

Indeed, if player i is the last one to be added, the claim follows by induction from the identity

$$\phi(\mathbf{P}) = \phi(\mathbf{P}_{-i}) + \text{cost}_i(\mathbf{P}) \tag{4.6}$$

Moreover, $\phi(\mathbf{P})$ does not depend on the order in which the players are added, and (4.6) holds for all i. Thus, any player can be viewed as the last player.

COROLLARY 4.4.2. *Let ϕ be defined by (4.5). Fix a strategy profile $\mathbf{P} = (P_1, \ldots, P_k)$. If player i switches from path P_i to an alternative path P_i', then the change in the value of ϕ equals the change in the cost he incurs:*

$$\phi(P_i', \mathbf{P}_{-i}) - \phi(\mathbf{P}) = \text{cost}_i(P_i', \mathbf{P}_{-i}) - \text{cost}_i(\mathbf{P}). \tag{4.7}$$

(See Figure 4.10.)

We call ϕ a *potential function* for this congestion game. Corollary 4.4.2 implies that the game has a pure Nash equilibrium: If $\mathbf{P}$ minimizes ϕ, then it is a Nash equilibrium, since $\text{cost}_i(P_i', \mathbf{P}_{-i}) - \text{cost}_i(\mathbf{P}) \geq 0$.

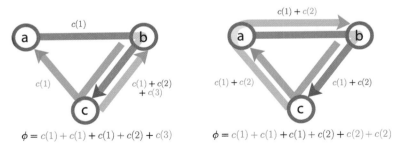

$$\phi = c(1) + c(1) + c(1) + c(2) + c(3) \qquad \phi = c(1) + c(1) + c(1) + c(2) + c(2) + c(2)$$

FIGURE 4.10. In this example, the cost is the same on all three roads, and is $c(n)$ if the number of drivers is n. The figure shows the potential ϕ before and after the player going from c to b switches from the direct path to the indirect path through a. The change in potential is the change in cost experienced by this player.

---

[7]Note that $\phi(\mathbf{P})$ is *not* the total cost incurred by the players.

4.4.1. The general notion. The congestion game we just saw is an example of a **potential game.** More generally, consider a k-player game, in which player j's strategy space is the finite set S_j. Let $u_i(s_1, s_2, \ldots, s_k)$ denote the payoff to player i when player j plays strategy s_j for each $j \in \{1, \ldots, k\}$. In a potential game, there is a function $\psi : S_1 \times \cdots \times S_k \to \mathbb{R}$ such that for each i and $\mathbf{s}_{-i} \in S_{-i}$

$$\psi(s_i, \mathbf{s}_{-i}) - u_i(s_i, \mathbf{s}_{-i}) \text{ is independent of } s_i. \tag{4.8}$$

Equivalently, for each i, $s_i, \tilde{s}_i \in S_i$ and $\mathbf{s}_{-i} \in S_{-i}$

$$\psi(\tilde{s}_i, \mathbf{s}_{-i}) - \psi(s_i, \mathbf{s}_{-i}) = u_i(\tilde{s}_i, \mathbf{s}_{-i}) - u_i(s_i, \mathbf{s}_{-i}). \tag{4.9}$$

We call the function ψ a **potential function** associated with the game.

CLAIM 4.4.3. *Every potential game has a Nash equilibrium in pure strategies.*

PROOF. The set $S_1 \times \cdots \times S_k$ is finite, so there exists some $\mathbf{s}$ that maximizes $\psi(\mathbf{s})$. Note that for this $\mathbf{s}$, the expression on the right-hand side in (4.9) is at most zero for any $i \in \{1, \ldots, k\}$ and any choice of $\tilde{\mathbf{s}}_i$. This implies that $\mathbf{s}$ is a Nash equilibrium. □

REMARK 4.4.4. Clearly, it suffices that $\mathbf{s}$ is a *local maximum*; i.e., for all i and s_i',

$$\psi(\mathbf{s}) \geq \psi(s_i', \mathbf{s}_{-i}).$$

In Example 4.4.1 the game was more naturally described in terms of agents trying to minimize their costs; i.e.

$$u_i(\mathbf{s}) = -\mathrm{cost}_i(\mathbf{s}).$$

In potential games with costs, as we saw above, it is more convenient to have the potential function decrease as the cost of each agent decreases. Thus, $\phi = -\psi$ is a potential for a game with given cost functions if ψ is a potential for the corresponding utilities u_i.

REMARK 4.4.5. If a function ψ is defined for strategy profiles of $k-1$ as well as k players and satisfies

$$\psi(\mathbf{s}) = \psi(\mathbf{s}_{-i}) + u_i(\mathbf{s}) \qquad \forall \mathbf{s} \quad \forall i, \tag{4.10}$$

then ψ is a potential function; i.e., (4.9) holds. Note that this held in congestion games. In fact, every potential function has such an extension. See Exercise 4.19. This suggests a recipe for constructing potential functions in games which are well-defined for any number of players: Let players join the game one at a time and add the utility of each player when he joins.

4.4.1.1. *Repeated play dynamics.* Consider a set of k players repeatedly playing a game starting with some initial strategy profile. In each step, exactly one player changes strategy and (strictly) improves his payoff. When no such improvement is possible, the process stops and the strategy profile reached must be a Nash equilibrium. In general, such a process might continue indefinitely, e.g., Rock, Paper, Scissors.

PROPOSITION 4.4.6. *In a potential game, the repeated play dynamics above terminate in a Nash equilibrium.*

PROOF. Equation (4.9) implies that in each improving move, the utility of the player that switches actions and the potential ψ increase by the same amount. Since there are finitely many strategy profiles $\mathbf{s}$, and in each improving move $\psi(\mathbf{s})$ increases, at some point a (local) maximum is reached and no player can increase his utility by switching strategies. In other words, a Nash equilibrium is reached.

$\square$

In the important special case where each player chooses his *best* improving move, the repeated play process above is called **best-response dynamics**.

4.4.2. Additional examples.

EXAMPLE 4.4.7 (**Consensus**). Consider a finite undirected graph $G = (V, E)$. In this game, each vertex $\{1, \ldots, n\} \in V$ is a player, and her action consists of choosing a bit in $\{0, 1\}$. We represent vertex i's choice by $b_i \in \{0, 1\}$. Let $N(i)$ be the set of neighbors of i in G and write $\mathbf{b} = (b_1, \ldots, b_n)$. The loss $D_i(\mathbf{b})$ for player i is the number of neighbors that she disagrees with; i.e.,

$$D_i(\mathbf{b}) = \sum_{j \in N(i)} |b_i - b_j|.$$

For example, the graph could represent a social network for a set of people, each deciding whether to go to Roxy or Hush (two nightclubs); each person wants to go to the club where most of his or her friends will be partying.

We define $\phi(\mathbf{b}) = \frac{1}{2} \sum_i D_i(\mathbf{b})$ and observe that this counts precisely the number of edges on which there is a disagreement. This implies that $\phi(\cdot)$ is a potential function for this game. Indeed, if we define $\phi(\mathbf{b}_{-i})$ to be the number of disagreements on edges excluding i, then $\phi(\mathbf{b}) = \phi(\mathbf{b}_{-i}) + D_i(\mathbf{b})$. Therefore, a series of improving moves, where exactly one player moves in each round, terminates in a pure Nash equilibrium. Two of the Nash equilibria are when all players are in agreement, but in some graphs there are other equilibria.

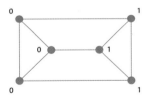

FIGURE 4.11. This is a Nash equilibrium in the Consensus game.

Now consider what happens when all players that would improve their payoff by switching their bit, do so *simultaneously*. In this case, the process might continue indefinitely. However, it will converge to a cycle of period at most two; i.e., it either stabilizes or alternates between two bit vectors. To see this, suppose that the strategy profile at time t is $\mathbf{b}^t = (b_1^t, b_2^t, \ldots, b_n^t)$. Let

$$f_t = \sum_i \sum_{j \in N(i)} |b_i^t - b_j^{t-1}|.$$

Observe that

$$\sum_{j \in N(i)} |b_i^{t+1} - b_j^t| \leq \sum_{j \in N(i)} |b_i^{t-1} - b_j^t|$$

since b_i^{t+1} is chosen at time $t+1$ to minimize the left hand side. Moreover, equality holds only if $b_i^{t+1} = b_i^{t-1}$. Summing over i shows that $f_{t+1} \leq f_t$. Thus, when f_t reaches its minimum, we must have $b_i^{t+1} = b_i^{t-1}$ for all i.

REMARK 4.4.8. To prove that a game has a pure Nash equilibrium that is reached by a finite sequence of improving moves, it suffices to find a *generalized potential function*, i.e., a function $\psi : S_1 \times \cdots \times S_k \to \mathbb{R}$ such that for each i and $\mathbf{s}_{-i} \in S_{-i}$

$$\mathrm{sgn}\left(\psi(\tilde{s}_i, \mathbf{s}_{-i}) - \psi(s_i, \mathbf{s}_{-i})\right) = \mathrm{sgn}\left(u_i(\tilde{s}_i, \mathbf{s}_{-i}) - u_i(s_i, \mathbf{s}_{-i})\right) \tag{4.11}$$

for all $s_i, \tilde{s}_i \in S_i$, where $\mathrm{sgn}(x) = x/|x|$ if $x \neq 0$ and $\mathrm{sgn}(0) = 0$.

EXAMPLE 4.4.9 (**Graph Coloring**). Consider an arbitrary undirected graph $G = (V, E)$ on n vertices. In this game, each vertex $v_i \in V$ is a player, and its possible actions consist of choosing a color s_i from the set $[n] := \{1, \ldots, n\}$. For any color c, define

$$n_c(\mathbf{s}) = \text{number of vertices with color } c \text{ when players color according to } \mathbf{s}.$$

The payoff of a vertex v_j (with color s_j) is equal to the number of other vertices with the same color if v_j's color is different from that of its neighbors, and it is 0 otherwise; i.e.,

$$u_j(\mathbf{s}) = \begin{cases} n_{s_j}(\mathbf{s}) & \text{if no neighbor of } v_j \text{ has the same color as } v_j, \\ 0 & \text{otherwise.} \end{cases}$$

For example, the graph could represent a social network, where each girl wants to wear the most popular dress (color) that none of her friends have.

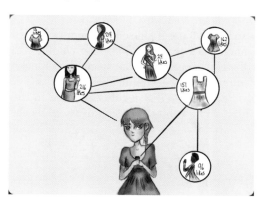

FIGURE 4.12. A social network.

Consider a series of moves in which one player at a time makes a best-response move. Then as soon as every player who has an improving move to make has done so, the graph will be **properly colored**; that is, no neighbors will have the same color. This is because a node's payoff is positive if it doesn't share its color with any neighbor and it is nonpositive otherwise. Moreover, once the graph is properly colored, it will never become improperly colored by a best-response move. Thus, we can restrict our attention to strategy profiles $\mathbf{s}$ in which the graph is properly colored.

LEMMA 4.4.10. *Graph Coloring has a pure Nash equilibrium.*

PROOF. We claim that, restricted to proper colorings, the function

$$\psi(\mathbf{s}) = \sum_{c=1}^{n} \sum_{\ell=1}^{n_c(\mathbf{s})} \ell$$

is a potential function: For any proper coloring $\mathbf{s}$ and any player i,

$$\psi(\mathbf{s}) = \psi(\mathbf{s}_{-i}) + n_{s_i}(\mathbf{s}) = \psi(\mathbf{s}_{-i}) + u_i(\mathbf{s});$$

i.e., (4.10) holds for proper colorings.   □

COROLLARY 4.4.11. *Let $\chi(G)$ be the **chromatic number** of the graph G, that is, the minimum number of colors in any proper coloring of G. Then the graph coloring game has a pure Nash equilibrium with $\chi(G)$ colors.*

PROOF. Suppose that $\mathbf{s}$ is a proper coloring with $\chi(G)$ colors. Then in a series of single-player improving moves starting at $\mathbf{s}$, no player will ever introduce an additional color, and the coloring will remain proper always. In addition, since the game is a potential game, the series of moves will end in a pure Nash equilibrium. Thus, this Nash equilibrium will have $\chi(G)$ colors.   □

4.5. Games with infinite strategy spaces

In some cases, a player's strategy space S_i is infinite.

EXAMPLE 4.5.1 (**Tragedy of the Commons**). Consider a set of k players that each want to send data along a shared channel of maximum capacity 1. Each player decides how much data to send along the channel, measured as a fraction of the capacity. Ideally, a player would like to send as much data as possible. The problem is that the quality of the channel degrades as a larger fraction of it is utilized, and if it is over-utilized, no data gets through. In this setting, each agent's strategy space S_i is $[0, 1]$. The utility function of each player i is

$$u_i(s_i, \mathbf{s}_{-i}) := s_i\left(1 - \sum_j s_j\right)$$

if $\sum_j s_j \leq 1$, and it is 0 otherwise.

We check that there is a pure Nash equilibrium in this game. Fix a player i and suppose that the other players select strategies $\mathbf{s}_{-i}$. Then player i's best response consists of choosing $s_i \in [0, 1]$ to maximize $s_i(1 - \sum_j s_j)$, which results in

$$s_i = \left(1 - \sum_{j \neq i} s_j\right)/2. \tag{4.12}$$

To be in Nash equilibrium, (4.12) must hold for all i. The unique solution to this system of equations has $s_i = 1/(k + 1)$ for all i.

This is a "tragedy" because each player's resulting utility is $1/(k+1)^2$, whereas if $s_i = 1/2k$ for all i, then each player would have utility $1/4k$. However, the latter choice is not an equilibrium.

EXAMPLE 4.5.2 (**Nightclub Pricing**). Three neighboring colleges have n students each that hit the nightclubs on weekends. Each of the two clubs, Roxy and Hush, chooses a price (cover charge) in $[0, 1]$. College A students go to Roxy, College C students go to Hush, and College B students choose whichever of Roxy or Hush has the lower cover charge that weekend, breaking ties in favor of Roxy. (See Figure 4.13.)

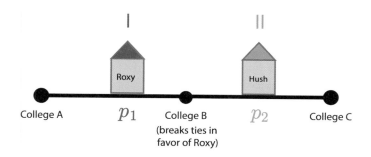

FIGURE 4.13. The Nightclub Pricing game.

Thus, if Roxy sets the price at p_1 and Hush sets the price at p_2, with $p_1 \leq p_2$, then Roxy's utility is $2np_1$ (n students from each of college A and college B pay the price p_1) and Hush's utility is np_2, whereas if $p_1 > p_2$, then Roxy's utility is np_1 and Hush's utility is $2np_2$.

In this game, there is no pure Nash equilibrium. To see this, suppose that Hush chooses a price $p_2 > 1/2$. Then Roxy's best response is to choose $p_1 = p_2$. But then p_2 is no longer a best response to p_1. If $p_2 = 1/2$, then Roxy's best response is either $p_1 = 1/2$ or $p_1 = 1$, but in either case $p_2 = 1/2$ is not a best-response. Finally, Hush will never set $p_2 < 1/2$ since this yields a payoff less than n, whereas a payoff of n is always achievable.

There is, however, a symmetric mixed Nash equilibrium. Any pure strategy with $p_1 < 1/2$ is dominated by the strategy $p_1 = 1$, and thus we can restrict attention to mixed strategies supported on $[1/2, 1]$. Suppose that Roxy draws its price from distribution F and Hush from distribution G, where both distributions are continuous and supported on all of $[1/2, 1]$. Then the expected payoff to Hush for any price p it might choose is

$$n\Big(pF(p) + 2p\big(1 - F(p)\big)\Big) = np(2 - F(p)), \tag{4.13}$$

which must[8] equal cn for some constant c and all p in $[1/2, 1]$. Setting $p = 1$ shows that $c = 1$, so $F(x) = 2 - 1/x$ in $[1/2, 1]$ (corresponding to density $f(x) = 1/x^2$ on that interval). Setting $G = F$ yields a Nash equilibrium. Note that the continuous distributions ensure that the chance of a tie is zero.

EXERCISE 4.b. Consider two nightclubs with cover charges $p_1 > 0$ and $p_2 > 0$ respectively. Students at the nearby college always go to the cheaper club, breaking

---

[8]We are using a continuous version of Lemma 4.3.7. If $\tilde{p}(2 - F(\tilde{p})) > p(2 - F(p))$ for some $p, \tilde{p} \in [1/2, 1]$, then the same inequality would hold if $p, \tilde{p}$ are slightly perturbed (by the continuity of F), and Hush would benefit by moving mass from a neighborhood of p to a neighborhood of $\tilde{p}$.

ties for club 1. The revenues per student will be $(p_1, 0)$ if $p_1 \leq p_2$ and $(0, p_2)$ if $p_1 > p_2$. Show that for any $c > 0$, there is a mixed Nash equilibrium that yields an expected revenue of c to each club.

4.6. The market for lemons

Economist George Akerlof won the Nobel Prize for analyzing how a used car market can break down in the presence of asymmetric information. Here is an extremely simplified version of his model. Suppose that there are cars of only two types: good cars (G) and lemons (L). A good car is worth \$9,000 to all sellers and \$12,000 to all buyers, while a lemon is worth only \$4,000 to sellers and \$6,000 to buyers. Obviously a seller knows what kind of car he is selling. If a buyer knew the type of the car being offered, he could split the difference in values with the seller and gain \$1,500 for a good car and \$1,000 for a lemon. However, a buyer doesn't know what kind of car he bought: lemons and good cars are indistinguishable at first, and a buyer only discovers what kind of car he bought after a few weeks, when the lemons break down. What a buyer does know is the fraction p of cars on the market that are lemons. Thus, the maximum amount that a rational buyer will pay for a car is $6,000p + 12,000(1 - p) = f(p)$, and a seller who advertises a car at $f(p) - \varepsilon$ will sell it.

However, if $p > \frac{1}{2}$, then $f(p) < \$9,000$, and sellers with good cars won't sell them. Thus, p will increase, $f(p)$ will decrease, and soon only lemons will be left on the market. In this case, asymmetric information hurts sellers with good cars, as well as buyers.

FIGURE 4.14. The seller, who knows the type of the car, may misrepresent it to the buyer, who doesn't know the type. (Drawing courtesy of Ranjit Samra; see http://rojaysoriginalart.com.)

Notes

The Prisoner's Dilemma game was invented in 1950 by Flood and Dresher [Axe84, Pou11]. The game is most relevant when it is repeated. See §6.4. Although Prisoner's Dilemma is most famous, Skyrms [Sky04] makes the case that Stag Hunt is more representative of real-life interactions. Stag Hunt and Chicken were both used as models for nuclear deterrence, e.g., by Schelling, who won a Nobel Prize in 2005 for "having enhanced our understanding of conflict and cooperation through game-theory analysis." See [O'N94] for a survey of game theory models of peace and war. Cheetahs and Antelopes and the Pollution Game are taken from [Gin00]. One Hundred Gnus and a Lioness is from [Sha17]. Finding Nash equilibria in multiplayer games often involves solving systems of polynomial equations as, for example, in the Pollution Game. An excellent survey of this topic is in [Stu02]. The Driver and Parking Inspector Game (Example 4.1.4) is from [TvS02].

Congestion games and best-response dynamics therein were introduced and analyzed by Rosenthal [Ros73]. Monderer and Shapley [MS96] studied the more general concept of potential games. The Consensus game (Example 4.4.7) is from [GO80], and the Graph Coloring game (Example 4.4.9) is from [PS12b]. Although the best-response dynamics in potential games is guaranteed to reach an equilibrium, this process might not be the most efficient method. Fabrikant et al. [FPT04] showed that there are congestion games in which best responses can be computed efficiently (in time polynomial in N, the sum of the number of players and the number of resources), but the minimum number of improving moves needed to reach a pure Nash equilibrium is exponential in N.

Tragedy of the Commons was introduced in a classic paper by Hardin [Har68a], who attributes the idea to Lloyd. Example 4.5.1 and Example 4.5.2 are from Chapter 1 of [NRTV07].

The market for lemons (§4.6) is due to Akerlof [Ake70]. George Akerlof won the 2001 Nobel Prize in Economics, together with A. Michael Spence and Joseph Stiglitz, "for their analyses of markets with asymmetric information."

John Nash

George Akerlof

The notion of Nash equilibrium, and Nash's Theorem showing that every finite game has a Nash equilibrium, are from [Nas50a].

While the Nash equilibrium concept is a cornerstone of game theory, the practical relevance of Nash equilibria is sometimes criticized for the following reasons. First, the ability of a player to play their Nash equilibrium strategy depends on knowledge and, indeed, common knowledge of the game payoffs. Second, Nash equilibria are viewed as representing a selfish point of view: A player would switch actions if it improves his utility, even if the damage to other players' utilities is much greater. Third, in situations where there are several Nash equilibria, it is unclear how players would agree on one. Finally, it may be difficult to find a Nash equilibrium, even in small games, when players have bounded rationality (see, e.g., [Rub98]). In large games, a series of recent results in computational complexity show that the problem of finding Nash equilibria is likely to be intractable (see, e.g., [DGP09, CD06, CDT06, Das13]). To quote Kamal Jain (personal communication), "If your laptop can't find an equilibrium, neither can the market."

A number of refinements have been proposed to address multiplicity of Nash equilbria. These include evolutionary stability (discussed in §7.1), focal points [Sch60], and trembling hand perfect equilibria [Sel75]. The latter are limits of equilibria in perturbed games, where each player must assign positive probability to all possible actions. In a trembling hand perfect equilibrium, every (weakly) dominated strategy is played with probability 0. See Section 7.3 in [MSZ13].

Regarding mixed Nash equilibria, critics sometimes doubt that players will explicitly randomize. There are several responses to this. First, in some contexts, e.g., the Penalty Kicks example from the Preface, randomness represents the uncertainty one player has about the selection process of the other. Second, sometimes players do explicitly randomize; for example, randomness is used to decide which entrances to an airport are patrolled, which passengers receive extra screening, and which days discounted airline tickets will be sold. Finally, probabilities in a mixed Nash equilibrium may represent population proportions as in §7.1.

Exercises

4.1. Modify the game of Chicken as follows. There is $p \in (0,1)$ such that, when a player swerves (plays S), the move is changed to drive (D) with probability p. Write the matrix for the modified game, and show that, in this case, the effect of increasing the value of M changes from the original version.

4.2. Two smart students form a study group in some math class where homework is handed in jointly by each group. In the last homework of the semester, each of the two students can choose to either work ("W") or party ("P"). If at least one of them solves the homework that week (chooses "W"), then they will both receive 10 points. But solving the homework incurs a substantial effort, worth -7 points for a student doing it alone, and an effort worth -2 points for each student, if both students work together. Partying involves no effort, and if both students party, they both receive 0 points. Assume that the students do not communicate prior to deciding whether they will work or party. Write this situation as a matrix game and determine all Nash equilibria.

4.3. Consider the following game:

		player II	
		C	D
player I	A	$(6,-10)$	$(0,10)$
	B	$(4,1)$	$(1,0)$

- Show that this game has a unique mixed Nash equilibrium.
- Show that if player I can commit to playing strategy A with probability slightly more than x^* (the probability she plays A in the mixed Nash equilibrium), then (a) player I can increase her payoff and (b) player II also benefits, obtaining a greater payoff than he did in the Nash equilibrium.

- Show similarly that if player II can commit to playing strategy C with probability slightly less than y^* (the probability he plays C in the mixed Nash equilibrium), then (a) player II can increase his payoff and (b) player I also benefits, obtaining a greater payoff than she did in the Nash equilibrium.

4.4.   Show that if the Prisoner's Dilemma game is played k times, where a player's payoff is the sum of his payoffs in the k rounds, then it is a dominant strategy to confess in every round. *Hint:* backwards induction.

4.5.   Two cheetahs and three antelopes: Two cheetahs each chase one of three antelopes. If they catch the same one, they have to share. The antelopes are Large, Small, and Tiny, and their values to the cheetahs are ℓ, s and t. Write the 3×3 matrix for this game. Assume that $t < s < \ell < 2s$ and that

$$\frac{\ell}{2}\left(\frac{2l - s}{s + \ell}\right) + s\left(\frac{2s - \ell}{s + \ell}\right) < t.$$

Find the pure equilibria and the symmetric mixed equilibria.

S 4.6.   Consider the game below and show that there is no pure Nash equilibrium, only a unique mixed one. Also, show that both commitment strategy pairs have the property that the player who did not make the commitment still gets the Nash equilibrium payoff.

		player II	
		C	D
player I	A	$(6, -10)$	$(0, 10)$
	B	$(4, 1)$	$(1, 0)$

4.7.   **Volunteering dilemma:** There are n players in a game show. Each player is put in a separate room. If some of the players volunteer to help the others, then each volunteer will receive \$1000 and each of the remaining players will receive \$1500. If no player volunteers, then they all get zero. Show that this game has a unique symmetric (mixed) Nash equilibrium. Let p_n denote the probability that player 1 volunteers in this equilibrium. Find p_2 and show that $\lim_{n \to \infty} np_n = \log(3)$.

4.8.   Three firms (players I, II, and III) put three items on the market and advertise them either on morning or evening TV. A firm advertises exactly once per day. If more than one firm advertises at the same time, their profits are zero. If exactly one firm advertises in the morning, its profit is \$200K. If exactly one firm advertises in the evening, its profit is \$300K. Firms must make their advertising decisions simultaneously. Find a symmetric mixed Nash equilibrium.

4.9.   Consider any two-player game of the following type:

player II

	A	B
A	(a,a)	(b,c)
B	(c,b)	(d,d)

player I

- Compute optimal safety strategies and show that they are not a Nash equilibrium.
- Compute the mixed Nash equilibrium and show that it results in the same player payoffs as the optimal safety strategies.

4.10.   Consider Example 4.1.3. Assume that with some probability p each country will be overtaken by extremists and will attack. Write down the game matrix and find Nash equilibria.

4.11.   The Welfare Game: John has no job and might try to get one. Or he might prefer to take it easy. The government would like to aid John if he is looking for a job but not if he stays idle. The payoffs are

jobless John

	try	not try
aid	(3,2)	$(-1,3)$
no aid	$(-1,1)$	(0,0)

government

Find the Nash equilibria.

*4.12.   Use Lemma 4.3.7 to derive an exponential time algorithm for finding a Nash equilibrium in two-player general-sum games using linear programming.

4.13.   The game of Hawks and Doves: Find the Nash equilibria in the game of Hawks and Doves whose payoffs are given by the matrix:

player II

	D	H
D	(1,1)	(0,3)
H	(3,0)	$(-4,-4)$

player I

4.14.   Consider the following n-person game. Each person writes down an integer in the range 1 to 100. A reward is given to the person whose number is closest to the mean. (In the case of ties, a winner is selected at random from among those whose number is closest to the mean.) What is a Nash equilibrium in this game?

4.15. Suppose that firm III shares a lake with two randomly selected firms from a population of firms of which proportion p purify. Show that firm III is better off purifying if $|p - 1/2| < \sqrt{3}/6$, whereas firm III is better off polluting if $|p - 1/2| > \sqrt{3}/6$.

4.16. Consider a k player game, where each player has the same set S of pure strategies. A permutation π of the set of players $\{1, \ldots, k\}$ is an **automorphism** of the game if for every i, and every strategy profile $\mathbf{s}$, we have

$$u_i(s_1, \ldots, s_k) = u_{\pi(i)}(s_{\pi^{-1}(1)}, \ldots, s_{\pi^{-1}(k)}).$$

Show that if all players have the same set of pure strategies, and for any two players $i_0, j_0 \in \{1, \ldots, k\}$, there is an automorphism π such that $\pi(i_0) = j_0$, then the game is symmetric in the sense of Definition 4.3.10.

4.17. A simultaneous congestion game: There are two drivers, one who will travel from A to C, the other from B to D. Each road is labelled (x, y), where x is the cost to any driver who travels the road alone, and y is the cost to each driver if both drivers use this road. Write the game in matrix form, and find all of the pure Nash equilibria.

4.18. Consider the following market sharing game discussed in §8.3. There are k NBA teams, and each of them must decide in which city to locate. Let v_j be the profit potential, i.e., number of basketball fans, of city j. If ℓ teams select city j, they each obtain a utility of v_j/ℓ. Let $\mathbf{c} = (c_1, \ldots, c_k)$ denote a strategy profile where c_i is the city selected by team i, and let $n_{c_i}(\mathbf{c})$ be the number of teams that select city c_i in this profile. Show that the market sharing game is a potential game with potential function

$$\Phi(\mathbf{c}) = \sum_{j \in \mathcal{C}} \sum_{\ell=1}^{n_{c_i}(\mathbf{c})} \frac{v_j}{\ell}$$

and hence has a pure Nash equilibrium.

4.19. Show that if ψ is a potential function for a game of k players, then ψ can be extended to strategy profiles of $k - 1$ players to satisfy (4.10).

4.20. Consider the following variant of the Consensus game (Example 4.4.7). Again, consider an arbitrary undirected graph $G = (V, E)$. In this game, each vertex $\{1, \ldots, n\} \in V$ is a player, and her action consists of choosing a bit in $\{0, 1\}$. We represent vertex i's choice by $b_i \in \{0, 1\}$. Let $N(i)$ be the set of neighbors of i in G and write $\mathbf{b} = (b_1, \ldots, b_n)$. The difference now is that there is a weight w_{ij} on each edge (i, j) that measures how much the two players i and j care about agreeing with each other. (Assume that $w_{ij} = w_{ji}$.) In this case, the loss $D_i(\mathbf{b})$ for player i is the total weight of neighbors that she disagrees with; i.e.,

$$D_i(\mathbf{b}) = \sum_{j \in N(i)} |b_i - b_j| w_{ij}.$$

Show that this is a potential game and that simultaneous improving moves converge to a cycle of period at most 2.

4.21.   Consider the setting of the previous exercise and show that if the weights w_{ij} and w_{ji} are different, then the game is not a potential game.

4.22.   Construct an example showing that the Graph Coloring game of Example 4.4.9 has a Nash equilibrium with more than $\chi(G)$ colors.

4.23.   The definition of a potential game extends to infinite strategy spaces S_i: Call $\psi : \prod_i S_i \to \mathbb{R}$ a potential function if for all i, the function $s_i \to \psi(s_i, \mathbf{s}_{-i}) - u_i(s_i, \mathbf{s}_{-i})$ is constant on S_i. Show that Example 4.5.1 is a potential game. *Hint:* Consider the case of two players with strategies $x, y \in [0, 1]$. It must be that

$$\psi(x, y) = c_y + x(1 - x - y) = c_x + y(1 - x - y);$$

i.e., $c_y + x(1 - x) = c_x + y(1 - y)$.

CHAPTER 5

Existence of Nash equilibria and fixed points

5.1. The proof of Nash's Theorem

Recall Nash's Theorem:

THEOREM 5.1.1. *For any general-sum finite game with $k \geq 2$ players, there exists at least one Nash equilibrium.*

We will use the following theorem that is proved in the next section.

THEOREM 5.1.2 (**Brouwer's Fixed-Point Theorem**). *Suppose that $K \subseteq \mathbb{R}^d$ is closed, convex, and bounded. If $T : K \to K$ is continuous, then there exists $\mathbf{x} \in K$ such that $T(\mathbf{x}) = \mathbf{x}$.*

PROOF OF NASH'S THEOREM VIA BROUWER'S THEOREM. First, consider the case of two players. Suppose that the game is specified by payoff matrices $A_{m \times n}$ and $B_{m \times n}$ for players I and II. Let $K = \Delta_m \times \Delta_n$. We will define a continuous map $T : K \to K$ that takes a pair of strategies $(\mathbf{x}, \mathbf{y})$ to a new pair $(\hat{\mathbf{x}}, \hat{\mathbf{y}})$ with the following properties:

(i) $\hat{\mathbf{x}}$ is a better response to $\mathbf{y}$ than $\mathbf{x}$ is, if such a response exists; otherwise $\hat{\mathbf{x}} = \mathbf{x}$.

(ii) $\hat{\mathbf{y}}$ is a better response to $\mathbf{x}$ than $\mathbf{y}$ is, if such a response exists; otherwise $\hat{\mathbf{y}} = \mathbf{y}$.

A fixed point of T will then be a Nash equilibrium.

Fix the strategy $\mathbf{y}$ of player II. Define c_i to be the maximum of zero and the gain player I obtains by switching from strategy $\mathbf{x}$ to pure strategy i. Formally, for $\mathbf{x} \in \Delta_m$

$$c_i := c_i(\mathbf{x}, \mathbf{y}) := \max \left\{ A_i \mathbf{y} - \mathbf{x}^T A \mathbf{y} , 0 \right\},$$

where A_i denotes the ith row of the matrix A. Define $\hat{\mathbf{x}} \in \Delta_m$ by

$$\hat{x}_i := \frac{x_i + c_i}{1 + \sum_{k=1}^m c_k} \,;$$

i.e., the weight of each action for player I is increased according to its performance against the mixed strategy $\mathbf{y}$.

Similarly, let

$$d_j := d_j(\mathbf{x}, \mathbf{y}) := \max \left\{ \mathbf{x}^T B^j - \mathbf{x}^T B \mathbf{y} , 0 \right\},$$

where B^j denotes the jth column of B, and define $\hat{\mathbf{y}} \in \Delta_n$ by

$$\hat{y}_j := \frac{y_j + d_j}{1 + \sum_{k=1}^n d_k} \,.$$

Finally, let $T(\mathbf{x}, \mathbf{y}) = (\hat{\mathbf{x}}, \hat{\mathbf{y}})$.

89

We claim that property (i) holds for this mapping. If $c_i = 0$ (i.e., $\mathbf{x}^T A \mathbf{y} \geq A_i \mathbf{y}$) for all i, then $\hat{\mathbf{x}} = \mathbf{x}$ is a best response to $\mathbf{y}$. Otherwise, $S := \sum_{i=1}^m c_i > 0$. We need to show that

$$\sum_{i=1}^m \hat{x}_i A_i \mathbf{y} > \mathbf{x}^T A \mathbf{y}. \tag{5.1}$$

Multiplying both sides by $1 + S$, this is equivalent to

$$\sum_{i=1}^m (x_i + c_i) A_i \mathbf{y} > (1 + S)\mathbf{x}^T A \mathbf{y},$$

which holds since

$$\sum_{i=1}^m c_i A_i \mathbf{y} > S \mathbf{x}^T A \mathbf{y} = \sum_i c_i \mathbf{x}^T A \mathbf{y}.$$

Similarly, property (ii) is satisfied.

Finally, we observe that K is convex, closed, and bounded and that T is continuous since c_i and d_j are. Thus, an application of Brouwer's theorem shows that there exists $(\mathbf{x}, \mathbf{y}) \in K$ for which $T(\mathbf{x}, \mathbf{y}) = (\mathbf{x}, \mathbf{y})$; by properties (i) and (ii), $(\mathbf{x}, \mathbf{y})$ is a Nash equilibrium.

For $k > 2$ players, we define for each player j and pure strategy ℓ of that player the quantity $c_\ell^{(j)}$ which is the gain player j gets by switching from his current strategy $\mathbf{x}^{(j)}$ to pure strategy ℓ, if positive, given the current strategies of all the other players. The rest of the argument follows as before. $\square$

Proposition 4.3.11 claimed that in a symmetric game, there is always a **symmetric Nash equilibrium**.

PROOF OF PROPOSITION 4.3.11: The map T, defined in the preceding proof from the k-fold product $\Delta_n \times \cdots \times \Delta_n$ to itself, can be restricted to the diagonal

$$D = \{(\mathbf{x}, \ldots, \mathbf{x}) \in \Delta_n^k : \mathbf{x} \in \Delta_n\}.$$

The image of D under T is a subset of D, because, in a symmetric game,

$$c_i^{(1)}(\mathbf{x}, \ldots, \mathbf{x}) = \cdots = c_i^{(k)}(\mathbf{x}, \ldots, \mathbf{x})$$

for all pure strategies i and $\mathbf{x} \in \Delta_n$. Brouwer's Fixed-Point Theorem yields a fixed point within D, which is a symmetric Nash equilibrium. $\square$

5.2. Fixed-point theorems*

Brouwer's Theorem is straightforward in dimension 1. Given $T : [a, b] \to [a, b]$, define $f(x) := T(x) - x$. Clearly, $f(a) \geq 0$, while $f(b) \leq 0$. By the Intermediate Value Theorem, there is $x \in [a, b]$ for which $f(x) = 0$, so $T(x) = x$.

In higher dimensions, Brouwer's Theorem is rather subtle; in particular, there is no generally applicable recipe to find or approximate a fixed point, and there may be many fixed points. Thus, before we turn to a proof of Theorem 5.1.2, we discuss some easier fixed point theorems, where iteration of the mapping from any starting point converges to the fixed point.

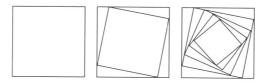

FIGURE 5.1. Under the transformation T a square is mapped to a smaller square, rotated with respect to the original. When iterated repeatedly, the map produces a sequence of nested squares. If we were to continue this process indefinitely, a single point (fixed by T) would emerge.

5.2.1. Easier fixed-point theorems.

Banach's Fixed-Point Theorem applies when the mapping T contracts distances, as in Figure 5.1

Recall that a metric space is **complete** if each Cauchy sequence therein converges to a point in the space. For example, any closed subset of $\mathbb{R}^n$ endowed with the Euclidean metric is complete.

THEOREM 5.2.1 (**Banach's Fixed-Point Theorem**). *Let K be a complete metric space. Suppose that $T : K \to K$ satisfies $d(T\mathbf{x}, T\mathbf{y}) \leq \lambda d(\mathbf{x}, \mathbf{y})$ for all $\mathbf{x}, \mathbf{y} \in K$, with $0 < \lambda < 1$ fixed. Then T has a unique fixed point $\mathbf{z} \in K$. Moreover, for any $\mathbf{x} \in K$, we have*

$$d(T^n\mathbf{x}, z) \leq \frac{d(\mathbf{x}, T\mathbf{x})\lambda^n}{1 - \lambda}.$$

PROOF. Uniqueness of fixed points: If $T\mathbf{x} = \mathbf{x}$ and $T\mathbf{y} = \mathbf{y}$, then

$$d(\mathbf{x}, \mathbf{y}) = d(T\mathbf{x}, T\mathbf{y}) \leq \lambda d(\mathbf{x}, \mathbf{y}).$$

Thus, $d(\mathbf{x}, \mathbf{y}) = 0$, so $\mathbf{x} = \mathbf{y}$.

As for existence, given any $\mathbf{x} \in K$, we define $\mathbf{x}_n = T\mathbf{x}_{n-1}$ for each $n \geq 1$, setting $\mathbf{x}_0 = \mathbf{x}$. Set $a = d(\mathbf{x}_0, \mathbf{x}_1)$, and note that $d(\mathbf{x}_n, \mathbf{x}_{n+1}) \leq \lambda^n a$. If $k > n$, then by the triangle inequality,

$$d(\mathbf{x}_n, \mathbf{x}_k) \leq d(\mathbf{x}_n, \mathbf{x}_{n+1}) + \cdots + d(\mathbf{x}_{k-1}, \mathbf{x}_k)$$

$$\leq a(\lambda^n + \cdots + \lambda^{k-1}) \leq \frac{a\lambda^n}{1 - \lambda}. \tag{5.2}$$

This implies that $\{\mathbf{x}_n : n \in \mathbb{N}\}$ is a Cauchy sequence. The metric space K is complete, whence $\mathbf{x}_n \to \mathbf{z}$ as $n \to \infty$. Note that

$$d(\mathbf{z}, T\mathbf{z}) \leq d(\mathbf{z}, \mathbf{x}_n) + d(\mathbf{x}_n, \mathbf{x}_{n+1}) + d(\mathbf{x}_{n+1}, T\mathbf{z}) \leq (1 + \lambda)d(\mathbf{z}, \mathbf{x}_n) + \lambda^n a \to 0$$

as $n \to \infty$. Hence, $d(\mathbf{z}, T\mathbf{z}) = 0$, and $T\mathbf{z} = \mathbf{z}$.

Thus, letting $k \to \infty$ in (5.2) yields

$$d(T^n\mathbf{x}, \mathbf{z}) = d(\mathbf{x}_n, \mathbf{z}) \leq \frac{a\lambda^n}{1 - \lambda}. \qquad \square$$

As the next theorem shows, the strong contraction assumption in Banach's Fixed-Point Theorem can be relaxed to decreasing distances if the space is compact. Recall that a metric space is **compact** if each sequence therein has a subsequence that converges to a point in the space. A subset of the Euclidean space $\mathbb{R}^d$ is compact if and only if it is closed and bounded.

See Exercise 5.2 for an example of a map $T : \mathbb{R} \to \mathbb{R}$ that decreases distances but has no fixed points.

THEOREM 5.2.2 (**Compact Fixed-Point Theorem**). *If K is a compact metric space and $T : K \to K$ satisfies $d(T(\mathbf{x}), T(\mathbf{y})) < d(\mathbf{x}, \mathbf{y})$ for all $\mathbf{x} \neq \mathbf{y} \in K$, then T has a unique fixed point $\mathbf{z} \in K$. Moreover, for any $x \in K$, we have $T^n(x) \to z$.*

PROOF. Let $f : K \to \mathbb{R}$ be given by $f(\mathbf{x}) := d(\mathbf{x}, T\mathbf{x})$. We first show that f is continuous. By the triangle inequality we have

$$d(\mathbf{x}, T\mathbf{x}) \leq d(\mathbf{x}, \mathbf{y}) + d(\mathbf{y}, T\mathbf{y}) + d(T\mathbf{y}, T\mathbf{x}),$$

so

$$f(\mathbf{x}) - f(\mathbf{y}) \leq d(\mathbf{x}, \mathbf{y}) + d(T\mathbf{y}, T\mathbf{x}) \leq 2d(\mathbf{x}, \mathbf{y}).$$

By symmetry, we also have $f(\mathbf{y}) - f(\mathbf{x}) \leq 2d(\mathbf{x}, \mathbf{y})$, and hence f is continuous.

Since K is compact, there exists $\mathbf{z} \in K$ such that

$$f(\mathbf{z}) = \min_{\mathbf{x} \in K} f(\mathbf{x}). \tag{5.3}$$

If $T\mathbf{z} \neq \mathbf{z}$, then $f(T(\mathbf{z})) = d(T\mathbf{z}, T^2\mathbf{z}) < d(\mathbf{z}, T\mathbf{z}) = f(\mathbf{z})$, and we have a contradiction to the minimizing property (5.3) of $\mathbf{z}$. Thus $T\mathbf{z} = \mathbf{z}$. Uniqueness is obvious.

Finally, we observe that iteration converges from any starting point x. Let $x_n = T^n x$, and suppose that x_n does not converge to z. Then for some $\epsilon > 0$, the set $S = \{n | d(x_n, z) \geq \epsilon\}$ is infinite. Let $\{n_k\} \subset S$ be an increasing sequence such that $y_k := x_{n_k} \to y \neq z$. Now

$$d(Ty_k, z) \to d(Ty, z) < d(y, z). \tag{5.4}$$

But $T^{n_{k+1} - n_k - 1}(Ty_k) = y_{k+1}$, so

$$d(Ty_k, z) \geq d(y_{k+1}, z) \to d(y, z),$$

contradicting (5.4). $\square$

EXERCISE 5.a. Prove that the convergence in the Compact Fixed-Point Theorem can be arbitrarily slow by showing that for any decreasing sequence $\{a_n\}_{n \geq 0}$ tending to 0, there is a distance decreasing $T : [0, a_0] \to [0, a_0]$ such that $T(0) = 0$ and $d(T^n a_0, 0) \geq a_n$ for all n.

5.2.2. Sperner's Lemma. In this section, we establish a combinatorial lemma that is key to proving Brouwer's Fixed-Point Theorem.

LEMMA 5.2.3 (**Sperner's Lemma**). *In $d = 1$: Suppose that the unit interval is subdivided $0 = t_0 < t_1 < \cdots < t_n = 1$, with each t_i being marked zero or one. If t_0 is marked zero and t_n is marked one, then the number of adjacent pairs (t_j, t_{j+1}) with different markings is odd.*

*In $d = 2$: Subdivide a triangle into smaller triangles in such a way that a vertex of any of the small triangles may not lie in the interior of an edge of another. Label the vertices of the small triangles 0, 1 or 2: the three vertices of the big triangle must be labeled 0, 1, and 2; vertices of the small triangles that lie on an edge of the big triangle must receive the label of one of the endpoints of that edge. Then the number of **properly labeled**[1] small triangles is odd; in particular, it is non-zero.*

---

[1] All three vertices have different labels.

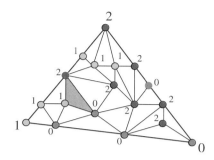

FIGURE 5.2. Sperner's lemma when $d = 2$.

PROOF. For $d = 1$, this is obvious: In a string of bits that starts with 0 and ends with 1, the number of bit flips is odd.

For $d = 2$, we will count in two ways the set Q of pairs consisting of a small triangle and an edge labeled 12 on that triangle. Let A_{12} denote the number of 12-labeled edges of small triangles that lie on the boundary of the big triangle. Let B_{12} be the number of such edges in the interior. Let N_{abc} denote the number of small triangles where the three labels are a, b and c. Note that

$$N_{012} + 2N_{112} + 2N_{122} = |Q| = A_{12} + 2B_{12},$$

because the left-hand side counts the contribution to Q from each small triangle and the right-hand side counts the contribution to Q from each 12-labeled edge. From the case $d = 1$, we know that A_{12} is odd, and hence N_{012} is odd too. □

For another proof, see Figure 5.3.

REMARK 5.2.4. Sperner's Lemma can be generalized to higher dimensions. See §5.4.

5.2.3. Brouwer's Fixed-Point Theorem.

DEFINITION 5.2.5. A set $S \subseteq \mathbb{R}^d$ has the **fixed-point property** (abbreviated **f.p.p.**) if for any continuous function $T : S \to S$, there exists $\mathbf{x} \in S$ such that $T(\mathbf{x}) = \mathbf{x}$.

Brouwer's Theorem asserts that every closed, bounded, convex set $K \subset \mathbb{R}^d$ has the f.p.p. Each of the hypotheses on K in the theorem is needed, as the following examples show:

(1) $K = \mathbb{R}$ (closed, convex, not bounded) with $T(x) = x + 1$.
(2) $K = (0, 1)$ (bounded, convex, not closed) with $T(x) = x/2$.
(3) $K = \{x \in \mathbb{R} : |x| \in [1, 2]\}$ (bounded, closed, not convex) with $T(x) = -x$.

REMARK 5.2.6. On first reading of the following proof, the reader should take $n = 2$. In two dimensions, simplices are triangles. To understand the proof for $n > 2$, §5.4 should be read first.

THEOREM 5.2.7 (**Brouwer's Fixed-Point Theorem for the simplex**). *The standard n-simplex $\Delta = \{\mathbf{x} \mid \sum_{i=0}^{n} x_i = 1, \forall i \ x_i \geq 0\}$ has the fixed-point property.*

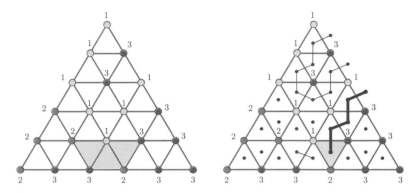

FIGURE 5.3. Sperner's Lemma: The left side of the figure shows a labeling and the three fully labeled subtriangles it induces. The right side of the figure illustrates an alternative proof of the case $d = 2$: Construct a graph G with a node inside each small triangle, as well as a vertex outside each 1-3 labeled edge on the outer right side of the big triangle. Put an edge in G between each pair of vertices separated only by a 1-3 labeled edge. In the resulting graph G (whose edges are shown in purple), each vertex has degree either 0, 1 or 2, so the graph consists of paths and cycles. Moreover, each vertex outside the big triangle has degree 0 or 1, and an odd number of these vertices have degree 1. Therefore, at least one (in fact an odd number) of the paths starting at these degree 1 vertices must end at a vertex interior to the large triangle. Each of the latter vertices lies inside a properly labeled small triangle. This is the highlighted subtriangle.

PROOF. Let Γ be a subdivision (as in Sperner's Lemma) of Δ where all triangles (or, in higher dimension, simplices) have diameter at most ϵ. Given a continuous mapping $T : \Delta \to \Delta$, write $T(\mathbf{x}) = (T_0(\mathbf{x}), \dots, T_n(\mathbf{x}))$. For any vertex $\mathbf{x}$ of Γ, let

$$\ell(\mathbf{x}) = \min\{i \; : \; T_i(\mathbf{x}) < x_i\}.$$

(Note that since $\sum_{i=0}^n x_i = 1$ and $\sum_{i=0}^n T_i(\mathbf{x}) = 1$, if there is no i with $T_i(\mathbf{x}) < x_i$, then $\mathbf{x}$ is a fixed point.)

By Sperner's Lemma, there is a properly labeled simplex Δ_1 in Γ, and this can already be used to produce an approximate fixed point of T; see the remark below.

To get a fixed point, find, for each k, a simplex with vertices $\{z^i(k)\}_{i=0}^n$ in Δ and diameter at most $\frac{1}{k}$, satisfying

$$T_i(\mathbf{z}^i(k)) < \mathbf{z}_i^i(k) \quad \text{for all } i \in [0, n]. \tag{5.5}$$

Find a convergent subsequence $\mathbf{z}^0(k_j) \to \mathbf{z}$ and observe that $\mathbf{z}^i(k_j) \to \mathbf{z}$ for all i. Thus, $T_i(\mathbf{z}) \leq z_i$ for all i, so $T(\mathbf{z}) = \mathbf{z}$. $\qquad\square$

REMARK 5.2.8. Let Δ_1 be a properly labeled simplex of diameter at most ϵ as in the proof above. Denote by $\mathbf{z}^0, \mathbf{z}^1, \dots, \mathbf{z}^n$ the vertices of Δ_1, where $\ell(\mathbf{z}^i) = i$. Let $\omega(\epsilon) := \max_{|\mathbf{x}-\mathbf{y}|\leq\epsilon} |T(\mathbf{x}) - T(\mathbf{y})|$. Then

$$T_i(\mathbf{z}^0) \leq T_i(\mathbf{z}^i) + \omega(\epsilon) < \mathbf{z}_i^i + \omega(\epsilon) \leq \mathbf{z}_i^0 + \epsilon + \omega(\epsilon).$$

On the other hand,

$$T_i(\mathbf{z}^0) = 1 - \sum_{j\neq i} T_j(\mathbf{z}^0) \geq 1 - \sum_{j\neq i} (\mathbf{z}_j^0 + \epsilon + \omega(\epsilon)) = \mathbf{z}_i^0 - n(\epsilon + \omega(\epsilon)).$$

Thus,

$$|T(\mathbf{z}^0) - \mathbf{z}^0| \leq n(n+1)(\epsilon + \omega(\epsilon)),$$

so $\mathbf{z}^0$ is an approximate fixed point.

DEFINITION 5.2.9. Let $S \subseteq \mathbb{R}^d$ and $\tilde{S} \subseteq \mathbb{R}^n$. A **homeomorphism** $h : S \to \tilde{S}$ is a one-to-one continuous map with a continuous inverse.

DEFINITION 5.2.10. Let $S \subseteq A \subseteq \mathbb{R}^d$. A **retraction** $g : A \to S$ is a continuous map where g restricted to S is the identity map.

LEMMA 5.2.11. *Let* $S \subseteq A \subseteq \mathbb{R}^d$ *and* $\tilde{S} \subseteq \mathbb{R}^n$.

(i) *If S has the f.p.p. and $h : S \to \tilde{S}$ is a homeomorphism, then $\tilde{S}$ has the f.p.p.*

(ii) *If $g : A \to S$ is a retraction and A has the f.p.p., then S has the f.p.p.*

PROOF. (i) Given $T : \tilde{S} \to \tilde{S}$ continuous, let $\mathbf{x} \in S$ be a fixed point of $h^{-1} \circ T \circ h : S \to S$. Then $h(\mathbf{x})$ is a fixed point of T.
(ii) Given $T : S \to S$, any fixed point of $T \circ g : A \to S$ is a fixed point of T. $\qquad\square$

LEMMA 5.2.12. *For $K \subset \mathbb{R}^d$ closed and convex, the nearest-point map $\Psi : \mathbb{R}^d \to K$ where*

$$\|\mathbf{x} - \Psi(\mathbf{x})\| = d(x, K) := \min_{y \in K} \|\mathbf{x} - \mathbf{y}\|$$

is uniquely defined and continuous.

PROOF. For uniqueness, suppose that $\|\mathbf{x} - \mathbf{y}\| = \|\mathbf{x} - \mathbf{z}\| = d(x, K)$ with $\mathbf{y}, \mathbf{z} \in K$. Assume by translation that $\mathbf{x} = 0$. Since $(\mathbf{y} + \mathbf{z})/2 \in K$, we have

$$d(0, K)^2 + \frac{\|\mathbf{y} - \mathbf{z}\|^2}{4} \leq \frac{\|\mathbf{y} + \mathbf{z}\|^2}{4} + \frac{\|\mathbf{y} - \mathbf{z}\|^2}{4} = \frac{\|\mathbf{y}\|^2 + \|\mathbf{z}\|^2}{2} = d(0, K)^2,$$

so $\mathbf{y} = \mathbf{z}$.

To show continuity, let $\Psi(\mathbf{x}) = \mathbf{y}$ and $\Psi(\mathbf{x} + \mathbf{u}) = \mathbf{y} + \mathbf{v}$. We show that $\|\mathbf{v}\| \leq \|\mathbf{u}\|$.

We know from the proof of the Separating Hyperplane Theorem that

$$\mathbf{v}^T(\mathbf{y} - \mathbf{x}) \geq 0$$

and

$$\mathbf{v}^T(\mathbf{x} + \mathbf{u} - \mathbf{y} - \mathbf{v}) \geq 0.$$

Adding these gives $\mathbf{v}^T(\mathbf{u} - \mathbf{v}) \geq 0$, so

$$\|\mathbf{v}\|^2 = \mathbf{v}^T \mathbf{v} \leq \mathbf{v}^T \mathbf{u} \leq \|\mathbf{v}\| \cdot \|\mathbf{u}\|$$

by the Cauchy-Schwarz inequality. Thus $\|\mathbf{v}\| \leq \|\mathbf{u}\|$. $\qquad\square$

Proof of Brouwer's Theorem (THEOREM 5.1.2). Let $K \subset \mathbb{R}^d$ be compact and convex. There is a simplex Δ_0 that contains K. Clearly Δ_0 is homeomorphic to a standard simplex, so it has the f.p.p. by Lemma 5.2.11(i). Then by Lemma 5.2.12, the nearest point map $\Psi : \Delta_0 \to K$ is a retraction. Thus, Lemma 5.2.11(ii) implies that K has the f.p.p. $\qquad\square$

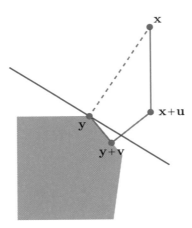

FIGURE 5.4. Illustration of the continuity argument in Lemma 5.2.12.

5.3. Brouwer's Fixed-Point Theorem via Hex*

In this section, we present a proof of Theorem 5.1.2 via Hex. Thinking of a Hex board as a hexagonal lattice, we can construct what is known as a **dual lattice** in the following way: The nodes of the dual are the centers of the hexagons and the edges link every two neighboring nodes (those are a unit distance apart).

Coloring the hexagons is now equivalent to coloring the nodes.

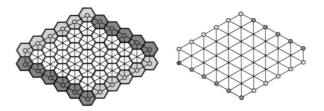

FIGURE 5.5. Hexagonal lattice and its dual triangular lattice.

This lattice is generated by two vectors $u, v \in \mathbb{R}^2$ as shown on the left side of Figure 5.6. The set of nodes can be described as $\{au + bv : a, b \in \mathbb{Z}\}$. Let's put $u = (0, 1)$ and $v = (\frac{\sqrt{3}}{2}, \frac{1}{2})$. Two nodes x and y are neighbors if $\|x - y\| = 1$.

We can obtain a more convenient representation of this lattice by applying a linear transformation G defined by

$$G(u) = \left(-\frac{\sqrt{2}}{2}, \frac{\sqrt{2}}{2}\right); \quad G(v) = (0, 1).$$

The game of Hex can be thought of as a game on the corresponding graph (see Figure 5.7). There, a Hex move corresponds to coloring one of the nodes. A player wins if she manages to create a connected subgraph consisting of nodes in her assigned color, which also includes at least one node from each of the two sets of her boundary nodes.

FIGURE 5.6. Action of G on the generators of the lattice.

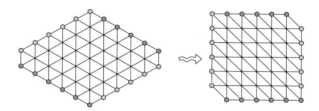

FIGURE 5.7. Under G an equilateral triangular lattice is transformed to an equivalent lattice.

The fact that any colored graph contains one and only one such subgraph is inherited from the corresponding theorem for the original Hex board.

PROOF OF BROUWER'S THEOREM USING HEX. As noted in §1.2.1, the fact that there is a winner in any play of Hex is the discrete analogue of the two-dimensional Brouwer fixed-point theorem. We now use this fact about Hex (proved as Theorem 1.2.6) to prove Brouwer's theorem, at least in two dimensions.

By Lemma 5.2.11, we may restrict our attention to a unit square. Consider a continuous map $T : [0,1]^2 \to [0,1]^2$. Componentwise we write $T(\mathbf{x}) = (T_1(\mathbf{x}), T_2(\mathbf{x}))$. Suppose it has no fixed points. Then define a function $f(\mathbf{x}) = T(\mathbf{x}) - \mathbf{x}$. The function f is never zero and continuous on a compact set; hence $\|f\|$ has a positive minimum $\varepsilon > 0$. In addition, as a continuous map on a compact set, T is uniformly continuous; hence $\exists\, \delta > 0$ such that $\|\mathbf{x} - \mathbf{y}\| < \delta$ implies $\|T(\mathbf{x}) - T(\mathbf{y})\| < \varepsilon$. Take such a δ with a further requirement $\delta < (\sqrt{2} - 1)\varepsilon$. (In particular, $\delta < \frac{\varepsilon}{\sqrt{2}}$.)

Consider a Hex board drawn in $[0,1]^2$ such that the distance between neighboring vertices is at most δ, as shown in Figure 5.8. Color a vertex $\mathbf{v}$ on the board yellow if $|f_1(\mathbf{v})|$ is at least $\varepsilon/\sqrt{2}$. If a vertex $\mathbf{v}$ is not yellow, then $\|f(\mathbf{v})\| \geq \varepsilon$ implies that $|f_2(\mathbf{v})|$ is at least $\varepsilon/\sqrt{2}$; in this case, color $\mathbf{v}$ blue. We know from Hex that in this coloring, there is a winning path, say, in yellow, between certain boundary vertices $\mathbf{a}$ and $\mathbf{b}$. For the vertex $\mathbf{a}^*$ neighboring $\mathbf{a}$ on this yellow path, we have $0 < a_1^* \leq \delta$. Also, the range of T is in $[0,1]^2$. Since $\mathbf{a}^*$ is yellow, $|T_1(\mathbf{a}^*) - a_1^*| \geq \varepsilon/\sqrt{2}$, and by the requirement on δ, we necessarily have $T_1(\mathbf{a}^*) - a_1^* \geq \varepsilon/\sqrt{2}$. Similarly, for the vertex $\mathbf{b}^*$ neighboring $\mathbf{b}$, we have $T_1(\mathbf{b}^*) - b_1^* \leq -\varepsilon/\sqrt{2}$. Examining the vertices on this yellow path one-by-one from $\mathbf{a}^*$ to $\mathbf{b}^*$, we must find neighboring vertices $\mathbf{u}$ and $\mathbf{v}$ such that $T_1(\mathbf{u}) - u_1 \geq \varepsilon/\sqrt{2}$ and $T_1(\mathbf{v}) - v_1 \leq -\varepsilon/\sqrt{2}$. Therefore,

$$T_1(\mathbf{u}) - T_1(\mathbf{v}) \geq 2\frac{\varepsilon}{\sqrt{2}} - (v_1 - u_1) \geq \sqrt{2}\varepsilon - \delta > \varepsilon.$$

However, $\|\mathbf{u} - \mathbf{v}\| \leq \delta$ should also imply $\|T(\mathbf{u}) - T(\mathbf{v})\| < \varepsilon$, a contradiction.   □

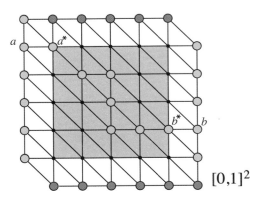

FIGURE 5.8. Proving Brouwer via Hex.

5.4. Sperner's Lemma in higher dimensions*

DEFINITION 5.4.1 (**Simplex**). An n-simplex $\Delta(v_0, v_1, \ldots, v_n)$ is the convex hull of a set of $n + 1$ points $v_0, v_1, \ldots, v_n \in \mathbb{R}^n$ that are affinely independent; i.e., the n vectors $v_i - v_0$, for $1 \leq i \leq n$, are linearly independent.

DEFINITION 5.4.2 (**Face**). A k-face of an n-simplex $\Delta(v_0, v_1, \ldots, v_n)$ is the convex hull of any $k + 1$ of the points $v_0, v_1, \ldots, v_n$. (See the left side of Figure 5.9.)

EXERCISE 5.b.
  (1) Show that $n + 1$ points $v_0, v_1, \ldots, v_n \in \mathbb{R}^d$ are affinely independent if and only if for every non-zero vector $(\alpha_0, \ldots, \alpha_n)$ for which $\sum_{0 \leq i \leq n} \alpha_i = 0$, it must be that $\sum_{0 \leq i \leq n} \alpha_i v_i \neq 0$. Thus, affine independence is a symmetric notion.
  (2) Show that a k-face of an n-simplex is a k-simplex.

DEFINITION 5.4.3 (**Subdivision of a simplex**). A **subdivision** of a simplex $\Delta(v_0, v_1, \ldots, v_n)$ is a collection Γ of n-simplices such that for every two simplices in Γ, either they are disjoint or their intersection is a face of both.

REMARK 5.4.4. Call an $(n-1)$-face of $\Delta_1 \in \Gamma$ an **outer face** if it lies on an $(n-1)$-face of $\Delta(v_0, v_1, \ldots, v_n)$; otherwise, call it an **inner face**. (See the right side of Figure 5.9.) It follows from the definition of subdivision that each inner face of $\Delta_1 \in \Gamma$ is an $(n-1)$-face of exactly one other simplex in Γ. Moreover, if F is an $(n-1)$-face of $\Delta(v_0, v_1, \ldots, v_n)$, then

$$\Gamma(F) := \{\Delta_1 \cap F\}_{\Delta_1 \in \Gamma}$$

is a subdivision of F. (See Figure 5.10.)

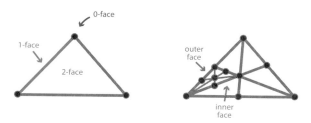

FIGURE 5.9. The left side shows a 2-simplex and its faces. The right side shows an inner face and an outer face in a subdivision.

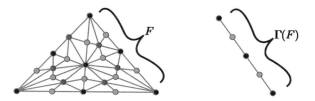

FIGURE 5.10. The figure shows a subdivision and its restriction to face F.

LEMMA 5.4.5. *For any simplex $\Delta(v_0, v_1, \ldots, v_n)$ and $\epsilon > 0$, there is a subdivision Γ such that all simplices in Γ have diameter less than ϵ.*

The case of $n = 2$ is immediate. See Figure 5.12. To prove the lemma in higher dimensions, we introduce **barycentric subdivision** defined as follows: Each subset $S \subset \{0, \ldots, n\}$ defines a face of the simplex of dimension $|S| - 1$, the convex hull of v_i, for $i \in S$. The average

$$v_S := \frac{1}{|S|} \sum_{i \in S} v_i$$

is called the **barycenter** of the face. Define a graph G_Δ on the vertices v_S with an edge (v_S, v_T) if and only if $S \subset T$. Each simplex in the subdivision is the convex hull of the vertices in a maximum clique in G_Δ.

Such a maximum clique corresponds to a collection of subsets $S_0 \subset S_1 \subset S_2 \cdots \subset S_n$ with $|S_i| = i + 1$. Thus there is a permutation on $\{0, \ldots, n\}$ with $\pi(0) = S_0$ and $\pi(i) = S_i \setminus S_{i-1}$ for all $i \geq 1$. If we write $w_i = v_{\pi(i)}$, then

$$v_{S_k} = \frac{w_0 + w_1 + \ldots + w_k}{k + 1}.$$

Thus the vertices of this clique are

$$w_0, \quad \frac{w_0 + w_1}{2}, \quad \frac{w_0 + w_1 + w_2}{3}, \quad \ldots, \quad \frac{1}{n+1} \sum_{i=0}^{n} w_i.$$

The convex hull of these vertices, which we denote by Δ_π, is

$$\Delta_\pi := \{ \sum_{0 \leq i \leq n} \alpha_i v_i \mid \alpha_{\pi(0)} \geq \cdots \geq \alpha_{\pi(n)} \geq 0 \text{ and } \sum_{0 \leq i \leq n} \alpha_i = 1 \}.$$

The full subdivision is $\Gamma_1 = \{\Delta_\pi \mid \pi \text{ a permutation of } \{0, \ldots, n\}\}$. See Figures 5.11 and 5.12.

EXERCISE 5.c.

(1) Verify the affine independence of $v_{S_0}, \ldots, v_{S_n}$.
(2) Verify that Δ_π is the convex hull of $v_{S_0}, \ldots, v_{S_n}$, where $\pi(i) = S_i \setminus S_{i-1}$.

FIGURE 5.11. This figure shows two steps of barycentric subdivision in two dimensions.

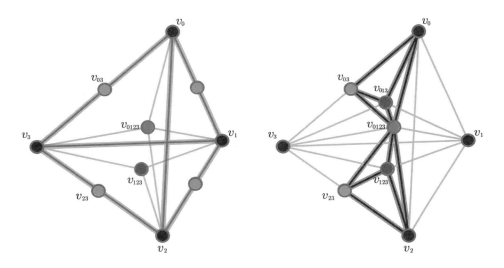

FIGURE 5.12. The left-hand side of this figure shows (most of) the vertices resulting from one step of barycentric subdivision in 3 dimensions. The green vertices are barycenters of simplices of dimension 1, the purple vertices (not all shown) are barycenters of simplices of dimension 2, and the pink vertex is the barycenter of the full simplex. The right-hand side of this figure shows two of the subsimplices that would result from barycentric subdivision. The upper subsimplex outlined corresponds to the permutation $\{0, 3, 1, 2\}$ and the bottom subsimplex corresponds to the permutation $\{2, 3, 1, 0\}$.

PROOF OF LEMMA 5.4.5: The diameter of each simplex Δ_π in Γ_1 is the maximum distance between any two vertices in Δ_π. We claim this diameter is at most $\frac{n}{n+1} D$, where D is the diameter of $\Delta(v_0, \ldots, v_n)$. Indeed, for any k, r

in $\{1, \ldots, n+1\}$,

$$\left| \frac{1}{k} \sum_{i=0}^{k-1} w_i - \frac{1}{r} \sum_{j=0}^{r-1} w_j \right| = \frac{1}{kr} \left| \sum_{i=0}^{k-1} \sum_{j=0}^{r-1} (w_i - w_j) \right|$$

$$\leq \frac{kr - r}{kr} D$$

$$= \left(\frac{k-1}{k} \right) D.$$

Iterating the barycentric subdivision m times yields a subdivision Γ_m in which the maximum diameter of any simplex is at most $\left(\frac{n}{n+1} \right)^m D$.

See Exercise 5.d below for the verification that this subdivision has the required intersection property.   □

The following corollary will be useful in Chapter 11.

COROLLARY 5.4.6. *Let Δ be a simplex on k vertices. Let Γ be any subdivision of Δ obtained by iterative barycentric subdivision. Let G_Γ be the graph whose edges are the 1-faces of Γ. Then G_Γ can be properly colored with k colors; i.e., all vertices in each subsimplex have different colors.*

PROOF. To see that such a proper coloring exists, suppose that Γ is constructed by iterating barycentric subdivision m times. Then color each vertex in the $m - 1^{st}$ barycentric subdivision with c_k. Within each of the subsimplices in this level, color each vertex that is a barycenter of a face of dimension i with c_i. Since every edge connects barycenters of faces of different dimension, this is a proper coloring.   □

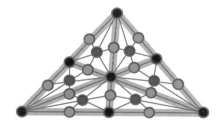

FIGURE 5.13. This picture shows the coloring of Corollary 5.4.6 for a simplex on three vertices to which iterative barycentric subdivision has been applied twice. The subdivision after one step is highlighted in gray. The corresponding vertices after the first subdivision are colored black (c_3). All vertices that are barycenters of dimension 1 are colored green (c_1), and all vertices that are barycenters of dimension 2 are colored purple (c_2).

EXERCISE 5.d. (1) Verify that Δ_π has one outer face determined by the equation $\alpha_{\pi(n)} = 0$ and n inner faces determined by the equations $\alpha_{\pi(k)} = \alpha_{\pi(k+1)}$ for $0 \leq k \leq n - 1$. (2) Verify that Γ_1 is indeed a subdivision. (3) Verify that for any $(n-1)$-face F of $\Delta(v_0, v_1, \ldots, v_n)$, the subdivision $\Gamma_1(F)$ is the barycentric subdivision of F.

DEFINITION 5.4.7 (**Proper labeling of a simplex**). A labeling ℓ of the vertices of an n-simplex $\Delta(v_0, v_1, \ldots, v_n)$ is proper if $\ell(v_0), \ell(v_1), \ldots, \ell(v_n)$ are all different.

DEFINITION 5.4.8 (**Sperner labeling of a subdivision**). A Sperner labeling ℓ of the vertices in a subdivision Γ of an n-simplex $\Delta(v_0, v_1, \ldots, v_n)$ is a labeling in which

- $\Delta(v_0, v_1, \ldots, v_n)$ is properly labeled,
- all vertices in Γ are assigned labels in $\{\ell(v_0), \ell(v_1), \ldots, \ell(v_n)\}$, and
- the labeling restricted to each face of $\Delta(v_0, \ldots, v_n)$ is a Sperner labeling there.

LEMMA 5.4.9 (**Sperner's Lemma for general** n). *Let ℓ be a Sperner labeling of the vertices in Γ, where Γ is a subdivision of the n-simplex $\Delta(v_0, v_1, \ldots, v_n)$. Then the number of properly labeled simplices in Γ is odd.*

PROOF. We prove the lemma by induction on n. The cases $n = 1, 2$ were proved in §5.2.2. For $n \geq 2$, consider a Sperner labeling of Γ. Call an $(n-1)$-face *good* if its vertex labels are $\ell(v_0), \ldots, \ell(v_{n-1})$.

Let g denote the number of good inner faces; let g_∂ be the number of good outer faces on $\Delta(v_0, \ldots, v_{n-1})$, and let N_j be the number of simplices in Γ with labels $\{\ell(v_i)\}_{i=0}^{n-1}$ and $\ell(v_j)$. Counting pairs

(simplex in Γ, good face of that simplex),

by the remark preceding Lemma 5.4.5 we obtain

$$2 \sum_{j=0}^{n-1} N_j + N_n = 2g + g_\partial.$$

Since g_∂ is odd by the inductive hypothesis, so is N_n. □

Notes

In his 1950 Ph.D. thesis, John Nash proved the existence of an equilibrium using Brouwer's Fixed Point Theorem [Bro11]. In the journal publication [Nas50a], he used Kakutani's Fixed Point Theorem instead. According to [OR14], the proof of Brouwer's Theorem from Sperner's Lemma [Spe28] is due to Knaster, Kuratowski, and Mazurkiewicz. The proof of Brouwer's theorem via Hex is due to David Gale [Gal79]. Theorem 5.2.2 is due to [Ede62]. See the book by Border [Bor85] for a survey of fixed-point theorems and their applications.

See [Rud76] for a discussion of general metric spaces.

Exercises

5.1.   Fill in the details showing that, in a symmetric game, with $A = B^T$, there is a symmetric Nash equilibrium. As suggested in the text, use the set $D = \{(x, x) : x \in \Delta_n\}$ in place of K in the proof of Nash's Theorem.

John Nash

5.2.   Show that the map $T : \mathbb{R} \to \mathbb{R}$ given by
$$T(x) = x + \frac{1}{1 + \exp(x)}$$
decreases distances but has no fixed point.

5.3.   Use Lemma 5.2.11(ii) to show that there is no retraction from a ball to a sphere.

5.4.   Show that there is no retraction from a simplex to its boundary directly from Sperner's Lemma, and use this to give an alternative proof of Brouwer's Theorem. (This is equivalent to the previous exercise because a simplex is homeomorphic to a ball.)

5.5.   Use Brouwer's Theorem to show the following: Let $\overline{B} = \overline{B(0,1)}$ be the closed ball in $\mathbb{R}^d$. There is no retraction from $\overline{B}$ to its boundary ∂B.

S 5.6.   Show that any d-simplex in $\mathbb{R}^d$ contains a ball.

S 5.7.   Let $K \subset \mathbb{R}^d$ be a compact convex set which contains a d-simplex. Show that K is homeomorphic to a closed ball.

CHAPTER 6

Games in extensive form

One of the key features of real-life games is that they take place over time and involve interaction, often with players taking turns. Such games are called **extensive-form games**.

6.1. Introduction

We begin with an example from Chapter 1.

EXAMPLE 6.1.1 (**Subtraction**). Starting with a pile of four chips, two players alternate taking one or two chips. Player I goes first. The player who removes the last chip wins.

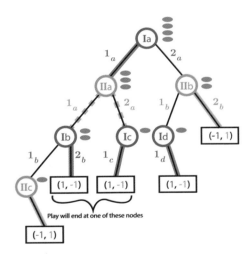

FIGURE 6.1. A game tree corresponding to the Subtraction game: Each leaf is labeled with the payoffs of the players. At nodes IIc, Ic, and Id, there is only one action that can be taken. At node Ib, player I loses if she removes one chip and wins if she removes two, so her choice, highlighted in the figure, is obvious. Proceeding up the tree, at node IIb, player II loses if he removes one chip and wins if he removes two, so again his choice is obvious. At node IIa, he loses either way, so in fact, his strategy at this node doesn't matter (indicated by the dots on the edges). Finally, at node Ia, player I wins if she removes one chip and loses if she removes two.

A natural way to find the best strategy for each player in a simple game like this is to consider the **game tree**, a representation of how the game unfolds, and then apply **backward induction** , that is, determine what action to play from the leaves up. At each node, the player will pick the action that leads to the highest payoff. Since we consider nodes in order of decreasing depth, when a node is considered, the payoffs that player will receive for each action he might take are already determined, and thus, the best response is determined. Figure 6.1 illustrates this process for the Subtraction game and shows that player I has a winning strategy.

Given an extensive-form game, we can in principle list the possible pure strategies of each of the players and the resulting payoffs. (This is called normal form.) In the Subtraction game, a strategy for player I specifies his action at node I_a and his action at node I_b. Similarly for player II. The resulting normal-form game is the following (where we show only the payoff to player I since this is a zero-sum game):

	player II			
player I	$1_a, 1_b$	$1_a, 2_b$	$2_a, 1_b$	$2_a, 2_b$
$1_a, 1_b$	-1	-1	1	1
$1_a, 2_b$	1	1	1	1
$2_a, 1_b$	1	-1	1	-1
$2_a, 2_b$	1	-1	1	-1

DEFINITION 6.1.2. A k-player finite **extensive-form game** is defined by a finite, rooted tree $\mathcal{T}$. Each node in $\mathcal{T}$ represents a possible state in the game, with leaves representing terminal states. Each internal (nonleaf) node v in $\mathcal{T}$ is associated with one of the players, indicating that it is his turn to play if/when v is reached. The edges from an internal node to its children are labeled with **actions**, the possible moves the corresponding player can choose from when the game reaches that state. Each leaf/terminal state results in a certain payoff for each player. We begin with games of **complete** information, where the rules of the game (the structure of the tree, the actions available at each node, and the payoffs at each leaf) are **common knowledge**[1] to all players.

A **pure strategy** for a player in an extensive-form game specifies an action to be taken at each of that player's nodes. A **mixed strategy** is a probability distribution over pure strategies.

The kind of equilibrium that is computed by backward induction is called a **subgame-perfect equilibrium** because the behavior in each **subgame**, is also an equilibrium. (Each node in the game tree defines a subgame, the game that would result if play started at that point.)

EXAMPLE 6.1.3 (**Line-Item Veto**). Congress and the President are at odds over spending. Congress prefers to increase military spending (M), whereas the President prefers a jobs package (J). However, both prefer a package that includes military spending and jobs to a package that includes neither. The following table gives their payoffs:

	Military	Jobs	Both	Neither
Congress	4	1	3	2
President	1	4	3	2

---

[1]That is, each player knows the rules, he knows the other players know the rules, he knows that they know that he knows the rules, etc.

A line-item veto gives the President the power to delete those portions of a spending bill he dislikes. Surprisingly though, as Figure 6.2 shows, having this power can lead to a less favorable outcome for the President.

In the games we've just discussed, we focused on subgame-perfect equilibria. However, as the following example shows, not all equilibria have this property.

EXAMPLE 6.1.4 (**Mutual Assured Destruction (MAD)**). Two countries, say A and B, each possess nuclear weapons. A is aggressive and B is benign. Country A chooses between two options. The first is to escalate the arms race, e.g., by firing test missiles, attacking a neighboring country, etc. The second is to do nothing and simply maintain the peace. If A escalates, then B has two options: retaliate, or back down. Figure 6.3 shows how the game might evolve.

If A believes that B will retaliate if she escalates, then her best action is to maintain the peace. Thus (maintain the peace, retaliate), resulting in payoffs of $(0,0)$, is a Nash equilibrium in this game. However, it is not a subgame-perfect equilibrium since in the subgame rooted at B's node, B's payoff is maximized by backing down, rather than retaliating — the subgame-perfect equilibrium is (escalate, back down).

Which equilibrium makes more sense? The threat by B to retaliate in the event that A escalates may or may not be credible since it will result in a significantly worse outcome to B than if he responds by backing down. Thus, A may not believe that B will, in fact, respond this way. On the other hand, nuclear systems are sometimes set up to automatically respond in the event of an attack, precisely to ensure that the threat is credible. This example illustrates the importance of being able to *commit* to a strategy.

REMARK 6.1.5. The structure of the extensive game shown in Figure 6.3 comes up in many settings. For example, consider a small and efficient airline (player A) trying to decide whether to offer a new route that encroaches on the territory of a big airline (player B). Offering this route corresponds to "escalating". Player B can then decide whether or not to offer a discount on its corresponding flights (retaliate) or simply cede this portion of the market (back down).

EXAMPLE 6.1.6 (**Centipede**). There is a pot of money that starts out with $4 and increases by a dollar each round the game continues. Two players take turns. When it is a player's turn and the pot has p, that player can either split the pot in his favor by taking $\$\lfloor \frac{p+4}{2} \rfloor$ (the "greedy" strategy), or allow the game to continue (the "continue" strategy) enabling the pot to increase.

Figure 6.4 shows that the unique subgame-perfect equilibrium is for the players to be greedy at each step. If, indeed, they play according to the subgame-perfect equilibrium, then player I receives $4 and player II gets nothing, whereas if they cooperate, they each end up with $50. (See Exercise 6.1.)

This equilibrium is counterintuitive. Indeed, laboratory experiments have shown that this equilibrium rarely arises when "typical" humans play this game. On the other hand, when the experimental subjects were chess players, the subgame-perfect outcome did indeed arise. Perhaps this is because chess players are more adept at backward induction. (See the notes for some of the relevant references.)

Regardless, a possible explanation for the fact that the subgame-perfect equilibrium does not arise in typical play is that the game is simply unnatural. It is not necessarily reasonable that the game goes on for a very long, but fixed, number n

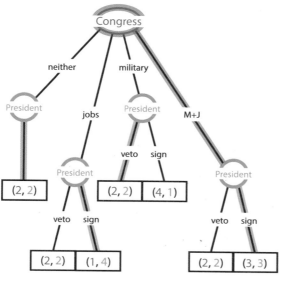

without line item veto

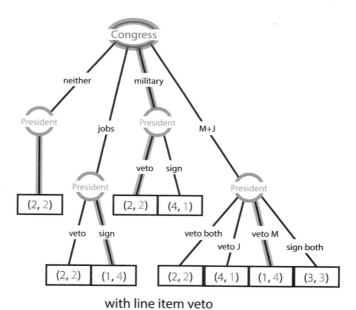

with line item veto

FIGURE 6.2. The top part of the figure shows the game from Example 6.1.3 without the line-item veto. The bottom part of the figure shows the game with the line-item veto. The highlighted edges show the actions taken in the subgame-perfect equilibrium: With the line-item veto, the result will be military spending, whereas without the line-item veto, the result will be a military and jobs bill.

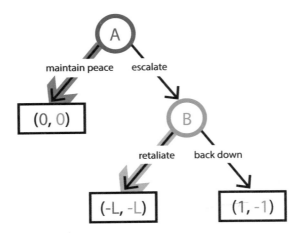

FIGURE 6.3. In the MAD game, (maintain peace, escalate) is a Nash equilibrium which is not subgame-perfect.

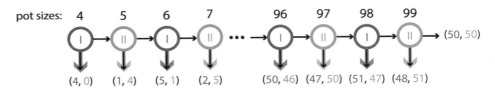

FIGURE 6.4. The top part of the figure shows the game and the resulting payoffs at each leaf. At each node, the "greedy" strategy consists of following the downward arrow, and the "continue" strategy is represented by the arrow to the right. Backward induction from the node with pot-size 99 shows that at each step the player is better off being greedy.

of rounds, and that it is common knowledge to all players that n is the number of rounds.

The extensive-form games we have seen so far are games of **perfect information**. At all times during play, a player knows the history of previous moves and hence which node of the tree represents the current state of play. In particular, she knows, for each possible sequence of actions that players take, exactly what payoffs each player will obtain.

In such games, the method of backward induction applies. Since this method leads to a strategy in which play at each node in the game tree is a best response to previous moves of the other players, we obtain the following:

THEOREM 6.1.7. *Every finite extensive-form game of perfect information has a subgame-perfect pure Nash equilibrium which can be computed by backward induction.*

EXERCISE 6.a. Prove Theorem 6.1.7.

6.2. Games of imperfect information

EXAMPLE 6.2.1. Recall the game of Chicken with the following payoff matrix:

	player II	
	Swerve (S)	Drive (D)
Swerve (S)	(1, 1)	(−1, 2)
Drive (D)	(2, −1)	(−M, −M)

player I

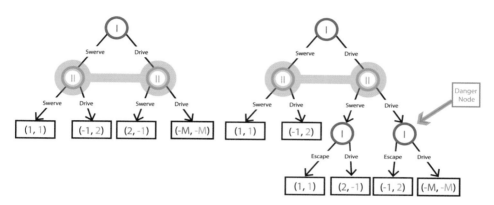

FIGURE 6.5. The figure shows an extensive-form version of the Chicken game. Player II's nodes are in the same **information set** because player I's and player II's actions occur simultaneously. Therefore, player II cannot distinguish which of two states he is in and must choose the same action at both. The figure on the right shows the version of Chicken in which player I's car is so easy to maneuver that she can escape at the very last minute after seeing player II's choice.

This, and any normal-form game, can be represented as an extensive-form game, as shown in Figure 6.5. When player II moves, he doesn't know what his opponent's move was. We capture this by defining an **information set**, consisting of the two player II nodes, and insisting that player II's action is the same at both nodes.

Consider now a variant of the game in which player I has a sports car that can escape collision in the last second, whereas player II's car cannot. This leads to the game shown on the right-hand side of Figure 6.5.

If we reduce to normal form, we obtain the following matrix, which is just the original matrix shown above, with one row repeated, and thus has the same equilibria:

	player II	
	Swerve (S)	Drive (D)
Swerve (S)	(1, 1)	(−1, 2)
Drive/Drive (DD)	(2, −1)	(−M, −M)
Drive/Escape (DE)	(1, 1)	(−1, 2)

player I

However, at the danger node, after players I and II both choose Drive, the Escape strategy is dominant for player I. Thus, the subgame-perfect equilibria are strategy pairs where player I escapes at the danger node and player II always drives. It is still a Nash equilibrium for player I to always drive and player II to swerve, but this equilibrium includes the "incredible threat" that player I will drive at the danger node.[2] The reduction from extensive form to normal form suppresses crucial timing of player decisions.

In an extensive game of **imperfect information** (but still complete information), each player knows the payoffs that all players will get for each possible action sequence, but does not know all actions that have been taken. This happens either because an opponents' action occurs simultaneously or simply because the player is not privy to information about what an opponent is doing.

Information sets are used to model the uncertainty a player has about which node of the tree the game is at when it's his turn to choose an action:

DEFINITION 6.2.2. In an extensive-form game, a player's nodes are partitioned into **information sets**. (In games of perfect information, each information set is a single node.) For any player i and any two nodes v, w in an information set S of that player, the set of actions available at v is identical to the set of actions available at w, and the same action must be selected at both nodes.

For any node v associated with player i, let $\mathcal{I}_1(v), \mathcal{I}_2(v), \ldots, \mathcal{I}_{T_v}(v)$ be the information sets of i along the path to v, and let $a_t(v)$ be the action i took at $\mathcal{I}_t(v)$. The information available to player i at v is the history of information sets and actions he took, which we denote by $H(v) := \{\mathcal{I}_t(v)\}_{t=1}^{T_v} \cup \{a_t(v)\}_{t=1}^{T_v-1}$.

A **pure strategy** for a player in a game with information sets defines an action for that player at each of his information sets, and, as always, a **mixed strategy** is a probability distribution over pure strategies.

Note that in a game of imperfect information, backward induction usually can not be employed to find an equilibrium. For example, the optimal strategy for player I depends on which node in the information set she is at, which depends on player II's strategy, and that decision is made at parents of player I's nodes in the tree. Indeed, once nonsingleton information sets are present, the notion of a subgame has to be defined more carefully: Subgames can't split up nodes in the same information set.

REMARK 6.2.3. Only a forgetful player would consider two nodes with different histories to be in the same information set. We restrict attention to players with **perfect recall**. For such a player, if two nodes v and w are in the same information set, then $H(v) = H(w)$.

6.2.1. Behavioral strategies. The use of the normal form version of an extensive game as a method for finding a Nash equilibrium is computationally complex for games that involve multiple rounds of play – the number of pure strategies a player i has is exponential in the number of information sets associated to i. A more natural way to construct a mixed strategy is to define the behavior at each information set independently. This is called a behavioral strategy.

---

[2]So she is worse off for having a better car.

DEFINITION 6.2.4. A **behavioral strategies** b_i for a player i in an extensive-form game is a map that associates to each information set I of i a probability distribution $b_i(\text{I})$ over the actions available to i at I. (For $v \in \text{I}$, we write $b_i(v) = b_i(\text{I})$.)

Every behavioral strategy b_i induces a corresponding mixed strategy, obtained by choosing an action independently at every information set S of player i according to the distribution $b_i(S)$.

REMARK 6.2.5. Some mixed strategies are not induced by a behavioral strategy because of dependence between the choice of actions at different information sets. (See Figure 6.6.) Thus, while a Nash equilibrium in mixed strategies always exists via reduction to the normal form case, it is not obvious that a Nash equilibrium in behavioral strategies exists.

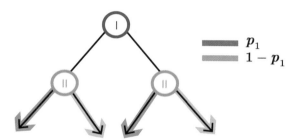

FIGURE 6.6. In this example, player II's mixed strategy puts probability p_1 on the two left actions and probability $1 - p_1$ on the two right actions. This mixed strategy is not induced by any behavioral strategy because the action player II takes at its two nodes is correlated. Notice though that if player II's strategy was to play the left action with probability p_1 and the right action with probability $1 - p_1$ at each node *independently* instead, then for any fixed player I strategy, her expected payoff would be the same as it is under the correlated strategy.

To show that Nash equilibria in behavioral strategies exist, we will need the following definition:

DEFINITION 6.2.6 (**Realization-equivalence**). Two strategies s_i and s_i' for player i in an extensive-form game are *realization-equivalent* if for each strategy $\mathbf{s}_{-i}$ of the opponents and every node v in the game tree, the probability of reaching v when strategy profile $(s_i, \mathbf{s}_{-i})$ is employed is the same as the probability of reaching v when $(s_i', \mathbf{s}_{-i})$ is employed.

REMARK 6.2.7. It is enough to verify realization-equivalence for opponent strategy profiles $\mathbf{s}_{-i}$ that are pure.

THEOREM 6.2.8. *Consider an extensive game of perfect recall. Then for any player i and every mixed strategy s_i, there is a realization-equivalent s_i' that is induced by a behavioral strategy b_i. Hence, for every possible strategy of the opponents $\mathbf{s}_{-i}$, every player i's expected utility under $(s_i, \mathbf{s}_{-i})$ is the same as his expected utility under $(s_i', \mathbf{s}_{-i})$.*

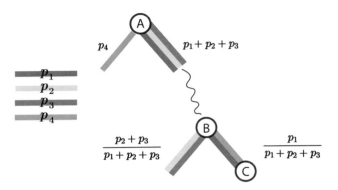

FIGURE 6.7. This figure gives a simple example of the construction of the behavioral strategy at nodes A and B. The labels on the edges represent the transition probabilities in the behavioral strategy.

PROOF. Let $s_i(A)$ denote the probability the mixed strategy s_i places on a set of pure strategies A. If v is a player i node, let $v_1, \ldots, v_{t-1}$ be the player i nodes on the path to v, and let a_j be the action at v_j leading towards v. Let $\Omega(v)$ be the set of pure strategies of player i where he plays a_j at v_j for $j < t$. The strategies in $\Omega(v)$ where he also plays action a at v are denoted $\Omega(v, a)$. The behavioral strategy $\mathbf{b}_i(v)$ is defined to be the conditional distribution over actions at v, given that v is reached. Thus the probability $\mathbf{b}_i(v)_a$ of taking action a is

$$\mathbf{b}_i(v)_a = \frac{s_i(\Omega(v, a))}{s_i(\Omega(v))}$$

if the denominator is nonzero. Otherwise, let $\mathbf{b}_i(v)$ be the uniform distribution over the actions at v. (See Figure 6.7.)

The key observation is that, by the assumption of perfect recall, $\mathbf{b}_i(v)_a = \mathbf{b}_i(w)_a$ if v and w are in the same information set $\mathcal{I}$, and therefore this is a valid behavioral strategy.

Finally, for a fixed pure strategy $\mathbf{s}_{-i}$, it follows by induction on the depth of the node v that the probability of reaching that node using the behavioral strategy $\mathbf{b}_i$ is the same as the probability of reaching that node using s_i. $\qquad \square$

COROLLARY 6.2.9. *In a finite extensive game of perfect recall, there is a Nash equilibrium in behavioral strategies.*

6.3. Games of incomplete information

Sometimes a player does not know exactly what game he is playing, e.g., how many players there are, which moves are available to the players, and what the payoffs at terminal nodes are. For example, in a game of poker, each player doesn't know which cards his opponents received, and therefore doesn't know the payoffs in each terminal state; in an eBay auction, a player doesn't know how many competing bidders there are or how much they value the object being auctioned. These are games of **incomplete information**.

FIGURE 6.8. Poker is a game of incomplete information.

In such a game, there is not much a player can do except guard against the worst case. Thus, a natural strategy in such a game is a *safety strategy*, wherein a player chooses a strategy which maximizes his payoff in the worst case.

6.3.1. Bayesian games. In many situations, however, the players have probabilistic prior information about which game is being played. Under this assumption, a game of incomplete information can be converted to a game of complete but imperfect information using **moves by nature** and information sets.

EXAMPLE 6.3.1 (**Fish-Selling Game**). Fish being sold at the market is fresh with probability 2/3 and old otherwise, and the customer knows this. The seller knows whether the particular fish on sale now is fresh or old. The customer asks the fish-seller whether the fish is fresh, the seller answers, and then the customer decides to buy the fish or to leave without buying it. The price asked for the fish is $12. It is worth $15 to the customer if fresh and nothing if it is old. Thus, if the customer buys a fresh fish, her gain is $3. The seller bought the fish for $6, and if it remains unsold, then he can sell it to another seller for the same $6 if it is fresh, and he has to throw it out if it is old. On the other hand, if the fish is old, the seller claims it to be fresh, and the customer buys it, then the seller loses $R in reputation. The game tree is depicted in Figure 6.10.

The seller clearly should not say "old" if the fish is fresh. Hence we should examine two possible pure strategies for him: FF means he always says "fresh"; OF means he always tells the truth. For the customer, there are four ways to react to what he might hear. Hearing "old" means that the fish is indeed old, so it is clear that she should leave in this case. Thus two rational strategies remain: BL means she buys the fish if she hears "fresh" and leaves if she hears "old"; LL means she always leaves. Here are the expected payoffs for the two players, averaged over the randomness coming from the actual condition of the fish:

FIGURE 6.9. The seller knows whether the fish is fresh; the customer only knows the probability.

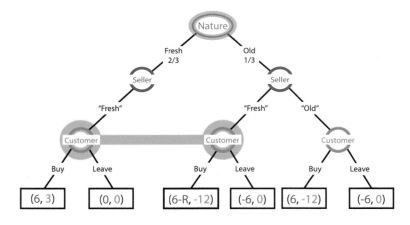

FIGURE 6.10. The game tree for the Fish-Selling game. The top node in the tree is a move by nature, with the outcome being fresh with probability 2/3 or old with probability 1/3.

		customer	
		BL	LL
seller	FF	$(6 - R/3, -2)$	$(-2, 0)$
	OF	$(2, 2)$	$(-2, 0)$

We see that if losing reputation does not cost too much in dollars, i.e., if $R < 12$, then there is only one pure Nash equilibrium: FF against LL. However, if $R \geq 12$, then the (OF, BL) pair also becomes a pure equilibrium, and the payoffs to both players from this equilibrium are much higher than the payoffs from the other equilibrium.

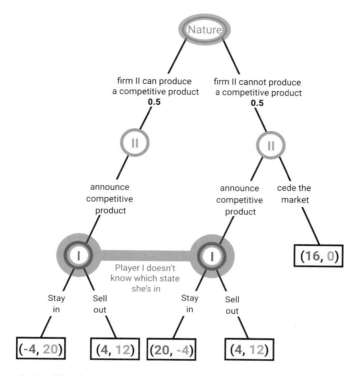

FIGURE 6.11. The figure shows the Large Company vs Startup game. Prior to the beginning of the game, player I announces her new technology. At the beginning of the game, there is a move by nature, which determines whether or not II actually can pull together a competitive product. Only player II is privy to the outcome of this move by nature. The two nodes at which player I makes a move form a single information set: player I does not know which of these states she is in. All she knows is that II has announced a competitive product, and knowing only that, she has to decide between competing with the giant or letting the giant buy her out. Thus, her strategy is the same at both nodes in the information set.

EXAMPLE 6.3.2 (**Large Company versus Startup**). A startup (player I) announces an important technology threatening a portion of the business that a very large company (player II) engages in. Given the resources available to II, e.g., a very large research and development group, it is possible that II will be able to pull together a competitive product in short order. One way or another, II may want to announce that a competitive product is in the works *regardless* of its existence, simply to intimidate the startup and motivate it to accept a buyout offer. The resulting game tree, which has a move by nature and an information set is shown in Figure 6.11. We can reduce the game to normal form by averaging over the randomness:

		player II	
		announce/cede	announce/announce
player I	stay in (I)	$(6, 10)$	$(8, 8)$
	sell out (O)	$(10, 6)$	$(4, 12)$

For example, if player I's strategy is to stay in and player II's strategy is announce/cede (i.e., announces a competitive strategy only if he can produce a competitive product), then the payoffs are the average of $(-4, 20)$ and $(16, 0)$.

A **Bayesian game** is an extensive-form game of imperfect information, with a first move by nature and probabilities that are in common knowledge to all the players. Different players may have different information about the outcome of the move by nature. This is captured by their information sets.

We summarize the procedure for constructing a normal-form game G_N associated to a two-player Bayesian game G: The actions available to player I in G_N are her pure strategies in G, and similarly for player II. The payoff matrices A and B for players I and II have entries

$$A(s_I, s_{II}) = \mathbb{E}\left[u_I(s_I, s_{II}, \mathcal{M})\right] \quad \text{and} \quad B(s_I, s_{II}) = \mathbb{E}\left[u_{II}(s_I, s_{II}, \mathcal{M})\right]$$

where $u_i(s_I, s_{II}, \mathcal{M})$ is the payoff to player i when player I plays pure strategy s_I, player II plays pure strategy s_{II}, and $\mathcal{M}$ is the move by nature in G.

6.3.2. Signaling.

EXAMPLE 6.3.3 (**Lions and Antelopes**). Antelopes have been observed to jump energetically when they notice a lion. Why do they expend energy in this way? One theory is that the antelopes are signaling danger to others at some distance, in a community-spirited gesture. However, the antelopes have been observed doing this even when there are no other antelopes nearby. The currently accepted theory is that the signal is intended for the lion, to indicate that the antelope is in good health and is unlikely to be caught in a chase. This is the idea behind **signaling**.

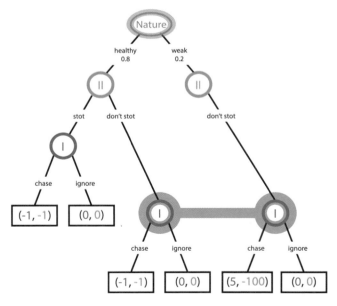

FIGURE 6.12. Lions and antelopes. Given that an antelope doesn't stot, chasing yields the lioness a positive payoff. Therefore, a healthy antelope is motivated to stot.

FIGURE 6.13. An antelope stotting to indicate its good health.

Consider the situation of an antelope catching sight of a lioness in the distance. Suppose there are two kinds of antelope, healthy (H) and weak (W). A lioness can catch a weak antelope but has no chance of catching a healthy antelope (and would expend a lot of energy if he tried).

This can be modeled as a combination of two simple games (A^H and A^W), depending on whether the antelope is healthy or weak, in which case the antelope has only one strategy (to run if chased), but the lioness has the choice of chasing (C) or ignoring (I):

$$A^H = \quad \begin{array}{c|c} & \text{antelope} \\ \hline & \text{run if chased} \\ \hline \text{chase} & (-1,-1) \\ \hline \text{ignore} & (0,0) \end{array} \quad \text{and} \quad A^W = \quad \begin{array}{c|c} & \text{antelope} \\ \hline & \text{run if chased} \\ \hline \text{chase} & (5,-100) \\ \hline \text{ignore} & (0,0) \end{array}$$

The lioness does not know which game she is playing — and if 20% of the antelopes are weak, then the lioness can expect a payoff of $(.8)(-1)+(.2)(5) = .2$ by chasing. However, the antelope does know, and if a healthy antelope can credibly[3] convey that information to the lioness by jumping very high, both will be better off — the antelope much more than the lioness!

6.3.3. Zero-sum games of incomplete information.

EXAMPLE 6.3.4 (**A simultaneous randomized game**). A zero-sum game is chosen by a fair coin toss. The players then make simultaneous moves. These moves are revealed and then they play a second round of the same game before any payoffs are revealed:

$$A_H = \quad \begin{array}{c|cc} & \multicolumn{2}{c}{\text{player II}} \\ & L & R \\ \hline U & -1 & 0 \\ D & 0 & 0 \end{array} \quad \text{and} \quad A_T = \quad \begin{array}{c|cc} & \multicolumn{2}{c}{\text{player II}} \\ & L & R \\ \hline U & 0 & 0 \\ D & 0 & -1 \end{array}$$

---

[3]A weak antelope cannot jump that high.

If neither player knows the result of the initial coin toss, each player will use the mixed strategy $(\frac{1}{2}, \frac{1}{2})$, and the value of the game to player I (for the two rounds) is $-\frac{1}{2}$. Now suppose that player I learns the result of the coin toss before playing the game. Then she can simply choose the row with all zeros and lose nothing, regardless of whether player II knows the coin toss as well.

Next consider the same story, but with matrices

$$A^H = \begin{array}{c} \\ \text{player I} \end{array} \begin{array}{c} \text{player II} \\ \begin{array}{c|cc} & L & R \\ \hline U & 1 & 0 \\ D & 0 & 0 \end{array} \end{array} \qquad \text{and} \qquad A^T = \begin{array}{c} \\ \text{player I} \end{array} \begin{array}{c} \text{player II} \\ \begin{array}{c|cc} & L & R \\ \hline U & 0 & 0 \\ D & 0 & 1 \end{array} \end{array}$$

Again, without knowing the result of the coin toss, the value to player I in each round is $\frac{1}{4}$. If player I is informed of the coin toss at the start, then in the second round, she will be *greedy*, i.e., choose the row with the 1 in it. The question remains of what she should do in the first round.

Player I has a simple strategy that will get her $\frac{3}{4}$ — this is to ignore the coin flip on the first round (and choose U with probability $\frac{1}{2}$), but then, on the second round, to be greedy.

In fact, $\frac{3}{4}$ is the value of the game. A strategy for player II that shows this is the following: In the first round, he plays L with probability $\frac{1}{2}$. In the second round, he flips a fair coin. If it comes up heads, then he assumes that player I played greedily in the first round[4] and he responds accordingly; if it comes up tails, then he chooses L with probability $\frac{1}{2}$. If player I plays greedily in the first round, then she gets $\frac{1}{2}$ in the first round and $\frac{1}{4}$ in the second round. If player I is sneaky (plays D in A^H and U in A^T), then she gets 0 in the first round and $\frac{3}{4}$ in the second round. Finally, if player I plays the same action in round 1 for both A^H and A^T, then she will receive $\frac{1}{4}$ in that round and $\frac{1}{2}$ in the second round.

It is surprising that sometimes the best use of information is to ignore it.

6.3.4. Summary: Comparing imperfect and incomplete information.

Recall that in a game of *perfect information*, each player knows the entire game tree and, whenever it is his turn, he knows the history of all previous moves (including any moves by nature). Thus, all information sets are of size one.

In a game of *imperfect information*, each player knows the entire game tree (including the probabilities associated with any move by nature). A player also knows the information set he is in whenever it is his turn. However, there is at least one information set of size greater than one.

In a game of *incomplete information*, players do not know the entire game tree or exactly which game they are playing. This is, in general, an intractable setting without further assumptions.

One way to handle this is to extend the game tree by adding an initial move by nature, with a commonly known prior on this move. This approach converts the game of incomplete information to a *Bayesian game*, which is a game of complete but imperfect information.

---

[4]That is, played U in A^H and D in A^T.

6.4. Repeated games

A special kind of extensive-form game arises when a regular one-round game of simultaneous moves is played repeatedly for some number of rounds.

For example, recall Prisoner's Dilemma.[5] We saw that the unique Nash equilibrium, indeed dominant strategy, in this game is for both players to defect:

		player II	
		cooperate (C)	defect (D)
player I	cooperate (C)	$(6,6)$	$(0,8)$
	defect (D)	$(8,0)$	$(2,2)$

What if the game is played n times? We assume that each player is trying to maximize the sum of his payoffs over the rounds. Both players' actions in each round are revealed simultaneously, and they know the actions taken on previous rounds when deciding how to play in the current round.

As in the one-shot game, in the final round, it will be a dominant strategy for each player to defect. Therefore, it is also a dominant strategy for each player to defect in the previous round, etc. Backward induction implies that the unique Nash equilibrium is to defect in each round.

It is crucial for the analysis we just gave that the number of rounds of the game is common knowledge. But for very large n, this is not necessarily realistic. Rather, we would like to model the fact that the number of times the game will be played is not known in advance.

One possibility is to let the game run forever and consider the **limsup average payoff**:

$$\limsup_{T \to \infty} \frac{1}{T} \sum_{t=1}^{T} (\text{the player's payoff in round } t). \tag{6.1}$$

(When the limit exists, we will refer to it as the **average payoff**.)

We emphasize that this is very different from a limit of fixed horizon games. In the latter case, a player can select a strategy that depends on the horizon T, while if the goal is to maximize the (limiting) average payoff, then the player must select *one* strategy independently of T.

Another way to assign utility to a player in an infinitely repeated game is to use a discount factor $\beta < 1$ and consider the **discounted payoff**:

$$\sum_{t=1}^{\infty} \beta^t (\text{the player's payoff in round } t). \tag{6.2}$$

There are two common interpretations for this:

- For each $t \geq 1$, given that the game has lasted for $t-1$ rounds, it continues to the next round with probability β. Then the probability of still playing at time t is β^t and equation (6.2) represents the player's expected payoff.
- A dollar earned today is better than a dollar earned next year since it can be enjoyed in the intervening year or invested to earn interest.

The strategies we consider in repeated games are analogous to behavioral strategies in extensive-form games:

---

[5] This version differs from the one in Chapter 4 in that a constant has been added to all payoffs to make them nonnegative.

DEFINITION 6.4.1. Let G be a k-player normal-form game, where player i's action set is A_i. Let $\mathbf{A} := \prod_{i=1}^{k} A_i$ be the set of action profiles. A **(behavioral) strategy** s_i for player i in the **infinitely repeated game** G^∞ is a mapping that for each t assigns to every possible history of actions $H_{t-1} \in \mathbf{A}^{t-1}$ a mixed strategy $s_i(H_{t-1})$ for player i in G (i.e., a distribution over A_i) to be played in round t.

6.4.1. Repetition with discounting. Consider Iterated Prisoner's Dilemma[6] with discount factor β.

DEFINITION 6.4.2. The **Tit-for-Tat** strategy in Iterated Prisoner's Dilemma is the following:

- Cooperate in round 1.
- For every round $k > 1$, play what the opponent played in round $k - 1$.

This strategy fares surprisingly well against a broad range of competing strategies. See the notes.

LEMMA 6.4.3. *For $\beta > 1/3$, it is a Nash equilibrium in Iterated Prisoner's Dilemma for both players to play Tit-for-Tat.*

REMARK 6.4.4. The threshold of $1/3$ for β depends on the specific payoff matrix used, but the principle applies more broadly.

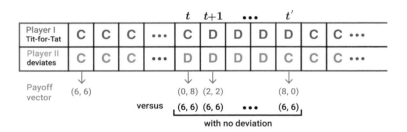

FIGURE 6.14. Illustration of deviation in Tit-for-Tat strategies

PROOF. If both players play Tit-for-Tat, then the payoff to each player in every round is 6. Suppose though that the first player plays Tit-for-Tat and the second player deviates. Consider the first round t at which he defects. Suppose first that he never switches back to cooperating. Then his payoff from t on is $8 \cdot \beta^t + 2 \sum_{j>t} \beta^j$, versus $6 \sum_{j \geq t} \beta^j$ if he had kept on cooperating. The latter is larger for $\beta > 1/3$.

If he does switch back to cooperating at some round $t' > t$ then his payoff in rounds t through t' is

$$8 \cdot \beta^t + 2 \sum_{t<j<t'} \beta^j, \quad \text{versus} \quad 6 \sum_{t \leq j \leq t'} \beta^j$$

if he doesn't defect during this period. The latter is also greater when $\beta > 1/3$ (and even for β slightly smaller).

Applying this argument to each interval where player II defected proves the theorem.   □

---

[6]This is the game G^∞, where G is the version of Prisoner's Dilemma shown at the beginning of §6.4.

The following strategy constitutes a more extreme form of punishment for an opponent who doesn't cooperate.

DEFINITION 6.4.5. The **Grim** strategy in Iterated Prisoner's Dilemma is the following: Cooperate until a round in which the other player defects, and then defect from that point on.

EXERCISE 6.b. Determine for which values of β it is a Nash equilibrium in Iterated Prisoner's Dilemma for both players to use the Grim strategy.

The previous exercise shows that (Grim, Grim) is a Nash equilibrium in Iterated Prisoner's Dilemma if β is sufficiently large. In fact, (Grim, Tit-for-Tat) is also a Nash equilibrium. But these are far from the only equilibria.

In the next section we characterize the payoffs achievable in a Nash equilibrium. To simplify the discussion, we consider average, rather than discounted, payoffs.

6.4.2. The Folk Theorem for average payoffs. Consider two infinite sequences of actions $\mathbf{a}_{II}$ and $\mathbf{a}_{I}$. One way for player II to try to force player I to stick with $\mathbf{a}_{I}$ is to "punish" player I if she deviates from that sequence. In order for this threat to work, any gain from deviating must be outweighed by the punishment.

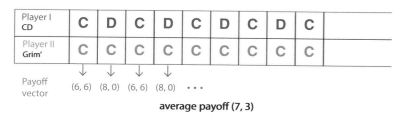

FIGURE 6.15. (CD, Grim′) strategy pair without deviation.

EXAMPLE 6.4.6. Consider the **Cooperate-Defect (CD)** strategy for player I in Iterated Prisoner's Dilemma defined as follows: Alternate between cooperating and defecting as long as the other player cooperates. If the other player ever defects, defect from that point on. Let **Grim′** be the player II strategy that cooperates as long as player I alternates between cooperate and defect, but if player I ever defects on an odd round, then player II defects henceforth.

We claim that the strategy pair (CD, Grim′) is a Nash equilibrium that yields an average payoff of 7 to player I and 3 to player II.

To see that these are the average payoffs, observe that if neither player deviates, then in alternate rounds, they play (cooperate, cooperate) yielding payoffs of (6, 6), and (defect, cooperate), yielding payoffs of (8,0) for average payoffs of (7, 3).

Figures 6.15 and 6.16 illustrate why this is a Nash equilibrium.

The previous example gives a special case of one of the famous folk theorems, characterizing the payoffs achievable by Nash equilibria in repeated games.

We will need two definitions:

DEFINITION 6.4.7 (**Payoff polytope**). Let G be a finite k-person normal-form game, where player i's action set is A_i. Let $\mathbf{A} := A_1 \times A_2 \times \cdots \times A_k$, the set of action profiles, and $u_i : \mathbf{A} \to \mathbb{R}$ the utility of player i. For $\mathbf{a} \in \mathbf{A}$, write

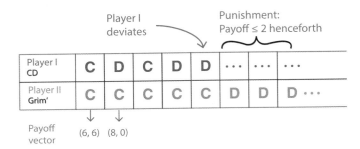

FIGURE 6.16. (CD, Grim$'$) strategy pair with deviation. The figure shows that the Cooperate-Defect (CD) player's average payoff drops by deviating. Similar analysis shows that Grim$'$ is a best response to CD.

$\mathbf{u}(\mathbf{a}) = (u_1(\mathbf{a}), \ldots, u_k(\mathbf{a}))$ for the utility vector in G corresponding to the action profile $\mathbf{a}$. The convex hull of $\{\mathbf{u}(\mathbf{a}) \mid \mathbf{a} \in \mathbf{A}\}$ is called the *payoff polytope* of the game.

DEFINITION 6.4.8 (**Individually-rational payoff profiles**). Let G be a finite k-person game with action sets A_i. We say a payoff vector $\mathbf{g} = (g_1, \ldots, g_k)$ is *individually rational* if each player's payoff is at least his minmax value, the lowest payoff his opponents can limit him to. That is, for all i

$$g_i \geq \min_{\mathbf{x}_{-i}} \max_{a_i} u_i(a_i, \mathbf{x}_{-i}).$$

Note that the strategies $\mathbf{x}_j$ in $\mathbf{x}_{-i}$ could be randomized. (Recall that $\mathbf{x}_j$ is a mixed strategy for player j in a single round of the game G.)

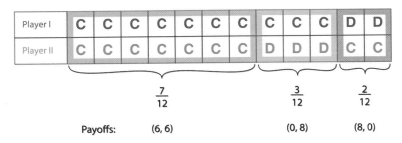

FIGURE 6.17. This figure shows the cycle that would be used in Iterated Prisoner's Dilemma when the probability distribution is $p_{C,C} = \frac{7}{12}$, $p_{C,D} = \frac{1}{4}$, and $p_{D,C} = \frac{1}{6}$. In this case, the cycle is of length 12, and the average payoffs are $\frac{7}{12} \cdot 6 + \frac{3}{12} \cdot 0 + \frac{2}{12} \cdot 8 = 4\frac{5}{6}$ for player I and $5\frac{1}{2}$ for player II.

DEFINITION 6.4.9 (**Nash equilibrium in a repeated game with average payoffs**). A strategy profile $\mathbf{s} = (s_1, \ldots, s_k)$ in the infinitely repeated game G^{∞} yields a payoff $u_i^t := u_i(\mathbf{s}(H_{t-1}))$ to player i in round t. (Notice that u_i^t is a random variable since each $s_j(H_{t-1})$ is a mixed strategy; see Definition 6.4.1.) The profile

s is a Nash equilibrium (average payoffs) if the following two conditions hold:

- The limit of payoffs exists; i.e., for each player j, with probability 1,

$$\lim_{T \to \infty} \frac{1}{T} \sum_{t=1}^{\infty} u_j^t$$

  exists.

- There is a vector (the average payoff vector) $\mathbf{g} = (g_1, \ldots, g_k)$ such that for each player j and deviation s_j',

$$\mathbb{E}\left[\lim_{T \to \infty} \frac{1}{T} \sum_{t=1}^{\infty} u_j^t \right] = g_j \geq \mathbb{E}\left[\limsup_{T \to \infty} \frac{1}{T} \sum_{t=1}^{T} u_j\big(s_j'(H_{t-1}), \mathbf{s}_{-j}(H_{t-1})\big) \right].$$

THEOREM 6.4.10 (**The Folk Theorem for Average Payoffs:**). *Let G be a finite k-person game.*

(1) *If $\mathbf{s}^* = (s_1^*, \ldots, s_k^*)$ is a Nash equilibrium (average payoffs) in the infinitely repeated game G^∞, then the resulting average payoff vector $(g_1, \ldots, g_k)$ is in the payoff polytope and is individually rational.*

(2) *If $\mathbf{g} = (g_1, \ldots, g_k)$ is individually rational and is in the payoff polytope, then there is a Nash equilibrium in G^∞ for which the players obtain these average payoffs.*

6.4.3. Proof of Theorem 6.4.10*. Part (1): First, since the payoff limit exists for $\mathbf{s}^*$, the strategies are in the payoff polytope. Second, if the strategies $\mathbf{s}^*$ yield an average payoff g_i to player i that is not individually rational, then player i has a better response. Specifically, for each round t, have her play a best response to whatever strategy $\mathbf{s}^*$ prescribes for the other agents in round t given the history up to and including $t - 1$. By construction, this yields her at least her minmax utility in each round, showing that $\mathbf{s}^*$ is not an equilibrium.

Part (2): Let $\mathbf{p}$ be a probability distribution over action profiles for which $\mathbf{g} = \sum_{\mathbf{a}} p_{\mathbf{a}} \mathbf{u}(\mathbf{a})$. We first prove the theorem assuming that the entries in $\mathbf{p}$ are rational. Let D be a common denominator of the numbers in $\mathbf{p}$. Construct a cycle of action tuples $\mathbf{a} = (a_1, a_2, \ldots, a_k)$ of length D consisting of $D \cdot p_{\mathbf{a}}$ occurrences of tuple $\mathbf{a}$ for each possible action profile. The equilibrium strategies are then defined as follows: Each player plays the strategy specified by the cycle just described. (See Figure 6.17 for an example.) If some player j ever deviates from this strategy, then from that point on, the rest of the players punish him by switching to the strategy that yields the minmax payoff to j. This is a Nash equilibrium because if player i deviates in any way, his payoff from the next round on is the minmax payoff which is at most g_i.

Next we show how to extend this to a Nash equilibrium for an irrational payoff vector $\mathbf{g}$. Let $\mathbf{g}(1), \mathbf{g}(2), \ldots$ be a sequence of rational payoff vectors in the payoff polytope, that converges to $\mathbf{g}$ and satisfies $\|\mathbf{g}(j) - \mathbf{g}\| \geq \|\mathbf{g}(j+1) - \mathbf{g}\|$ for all j. Let D_j be the common denominator of the action tuple probabilities corresponding to $\mathbf{g}(j)$ (as in the previous paragraph). The Nash equilibrium we construct will have the following form: For $j = 1, 2, \ldots$ play the strategy profile cycle achieving $\mathbf{g}(j)$ as described above for n_j rounds, where n_j is selected so that

$$n_j D_j > 2^j \left(D_{j+1} + \sum_{k<j} n_k D_k \right). \tag{6.3}$$

We refer to these n_j rounds as the jth stage. By construction, the jth stage lasts longer than 2^j times all earlier stages plus a single round of stage $j + 1$.

We now argue that the limiting payoff the players obtain is $\mathbf{g}$. Without loss of generality assume that $\|u(\mathbf{a})\| \leq 1$ for all $\mathbf{a}$. Also, let $\mathbf{a}_t$ be the action vector prescribed for step t of the game. Now suppose that at some time T, the players are in stage $\ell + 1$; that is,

$$0 < T - \left(\sum_{j=1}^{\ell} n_j D_j + m D_{\ell+1} \right) \leq D_{\ell+1},$$

for some nonnegative integer $m < n_{\ell+1}$. Then by (6.3), the current stage $\ell + 1$ round plus all stages $1, \ldots, \ell - 1$ last for no more than $T2^{-\ell}$ steps. Therefore,

$$\left\| \sum_{t=1}^{T} \mathbf{u}(\mathbf{a}_t) - T\mathbf{g} \right\| < T\|\mathbf{g}(\ell) - \mathbf{g}\| + T2^{1-\ell},$$

and therefore as $\ell \to \infty$, the average payoff vector converges to $\mathbf{g}$.

If a player ever deviates from the plan above, then from that point on, the rest of the players punish him so he receives his minmax payoff. Since $\mathbf{g}$ is individually rational, this strategy profile is a Nash equilibrium.

Notes

The notion of subgame-perfect equilibrium was formalized by Selten [Sel65]. The proof that every finite game of perfect information has a pure Nash equilibrium, indeed a subgame-perfect equilibrium, is due to Zermelo [Zer13, SW01] and Kuhn [Kuh53]. In the same paper, Kuhn [Kuh53] proved Theorem 6.2.8 showing that in games of perfect recall, every mixed strategy has a realization-equivalent behavior strategy. The Line-Item Veto game is from [DN08]; it represents the conflict that arose in 1987 between Reagan and Congress, though with the preferences reversed from our example. The Large Company versus Startup game is from [TvS02]. The Centipede Game is due to Rosenthal [Ros81]. See [PHV09, MP92] for the results of behavioral experiments and related literature on the Centipede Game.

The mathematical approach to the analysis of games of incomplete information was initiated by John Harsanyi [Har67, Har68b, Har68c] and was the major factor in his winning of the 1994 Nobel Prize in Economics. The prize was shared with Nash and Selten "for their pioneering analysis of equilibria in the theory of non-cooperative games."

In §6.3.1, we found equilibria in Bayesian games by reducing the game to normal form with payoffs averaged over the moves by nature. We know that such equilibria have realization-equivalent equilibria in behavioral strategies. These equilibria have the property that for each player, given the information he has on the move by nature, his strategy is a best response to the strategies of the other players. This is called a *Bayesian equilibrium*. The way Harsanyi made these ideas precise was by referring to the information a player has about the move by nature as his type.[7] Thus, we can think of the move by nature as assigning types to the different players. The interpretation of an equilibrium in behavioral strategies, in a game with moves by nature, as a Bayesian equilibrium is due to Harsanyi [Har67]. For more details, see [MSZ13, Theorem 9.53].

Influential early works on signaling and asymmetric information were the book by Spence [Spe74] on signaling in economics and society and the paper of Zahavi [Zah75] on the handicap principle that emphasized the role of costly signaling. For a broad introduction to signaling theory, see [Ber06]. A. Michael Spence won the 2001 Nobel Prize

---

[7]We take this perspective in Chapter 14.

in Economics, together with George Akerlof and Joseph Stiglitz, "for their analyses of markets with asymmetric information."

Repeated games have been the subject of intense study. In a famous experiment (see [AH81, Axe84]), Axelrod asked people to send him computer programs that play Iterated Prisoner's Dilemma and pitted them against each other. Tit-for-Tat, a four-line program sent by Anatol Rapoport, won the competition.

Robert Aumann

Thomas Schelling

The Folk Theorem was known in the game theory community before it appeared in journals. A version of Theorem 6.4.10 for discounted payoffs is also known. See [MSZ13]. Some relevant references are [Fri71, Rub79, Aum81, Aum85, AS94]. For a broader look at the topic of repeated games, see [MSZ15].

In 2005, Robert Aumann won the Nobel Prize in Economics for his work on repeated games, and more generally "for having enhanced our understanding of conflict and cooperation through game-theory analysis." The prize was shared with Thomas Schelling.

A theory of repeated games with incomplete information was initiated by Aumann and Maschler in the 1960s but only published in 1995 [AM95]. In particular, if the games in §6.3.3 are repeated T times, then the gain to player I from knowing which game is being played is $T/2$ in the first example, but only $o(T)$ in the second example.

This chapter provides only a brief introduction to the subject of extensive-form games. The reader is encouraged to consult one of the many books that cover the topic in depth and analyze other equilibrium notions, e.g., Rasmusen [Ras07], Maschler, Solan, and Zamir [MSZ13], and Fudenberg and Tirole [FT91].

Exercises

6.1.   Find all pure equilibria of the Centipede Game (Example 6.1.6).

6.2.   In the Fish Seller Game (Example 6.3.1), suppose that the seller only knows with probability 0.9 the true status of his fish (fresh or old). Draw the game tree for this Bayesian game and determine the normal-form representation of the game.

S 6.3.   Consider the zero-sum two-player game in which the game to be played is randomized by a fair coin toss. (This example was discussed in §2.5.1.) If the toss comes up heads, the payoff matrix is given by A^H, and if tails, it is given by A^T:

$$A^H = \begin{array}{c} \\ \text{player I} \end{array} \begin{array}{c} \text{player II} \\ \begin{array}{c|cc} & L & R \\ \hline U & 8 & 2 \\ D & 6 & 0 \end{array} \end{array} \quad \text{and} \quad A^T = \begin{array}{c} \\ \text{player I} \end{array} \begin{array}{c} \text{player II} \\ \begin{array}{c|cc} & L & R \\ \hline U & 2 & 6 \\ D & 4 & 10 \end{array} \end{array}$$

For each of the settings below, draw the Bayesian game tree, convert to normal form, and find the value of the game.

- Suppose that player I is told the result of the coin toss and both players play simultaneously.
- Suppose that player I is told the result of the coin toss but she must reveal her move first.

6.4.   Kuhn Poker: Consider a simplified form of poker in which the deck has only three cards: a Jack, a Queen and a King (ranked from lowest to highest), and there are two players, I and II. The game proceeds as follows:

- The game starts with each player anteing $1.
- Each player is dealt one of the cards.
- Player I can either *pass* (P) or *bet* (B) $1.
  - If player I bets, then player II can either *fold* (F) or *call* (C) (adding $1 to the pot).
  - if player I passes, then player II can *pass* (P) or *bet* $1 (B).
    * If player II raises, then player I can either *fold* or *call*.
- If one of the players folds, the other player takes the pot. If neither folds, the player with the high card wins what's in the pot.

Find a Nash equilibrium in this game via reduction to the normal form. Observe that in this equilibrium, there is bluffing and overbidding.

6.5.   Consider an extensive-form game consisting of a series of sequential auctions for three different laptops. In round 1, there is an auction for laptop 1, with participants A and B. In round 2, there is an auction for laptop 2, with participants C and D. In round 3, there is an auction for laptop 3, with participants B and C. Each auction is a second-price auction: Each participant submits a bid, and the person with the higher bid wins but pays the bid of the loser. Suppose also that each participant has a value for a laptop: Assume that $v_A = 1$, $v_B = 100$, $v_C = 100$, and $v_D = 99$. The utility of a participant is 0 if he loses in all auctions he participates in, and it is his value for a laptop minus the sum of all payments he makes otherwise.

A strategy for each player specifies, given the history, what bid that player submits in each auction he participates in. Show that there is an equilibrium in which players A, B, and C win.

CHAPTER 7

Evolutionary and correlated equilibria

7.1. Evolutionary game theory

Biology has brought a kind of thuggish brutality to the refined intellectual world of game theory.
— Alan Grafen

Most of the games we have considered so far involve rational players optimizing their strategies. A new perspective was proposed by John Maynard Smith and George Price in 1973: Each player could be an organism whose pure strategy is encoded in its genes[1]. A strategy that yields higher payoffs enables greater reproductive success.[2] Thus, genes coding for such strategies increase in frequency in the next generation.

Interactions in the population are modeled by randomly selecting two individuals, who then play a game. Thus, each player faces a mixed strategy with probabilities corresponding to population frequencies.

We begin with an example, a variant of our old nemesis, the game of Chicken.

7.1.1. Hawks and Doves. The game described in Figure 7.1 is a simple model for two behaviors – one bellicose, the other pacifist – within the population of a single species. This game has the following payoff matrix:

$$
\begin{array}{c|c|c|}
 & \multicolumn{2}{c}{\text{player II}} \\
 & H & D \\
\hline
H & (\frac{v}{2}-c, \frac{v}{2}-c) & (v, 0) \\
\hline
D & (0, v) & (\frac{v}{2}, \frac{v}{2}) \\
\hline
\end{array}
$$

(player I on the left, rows H, D)

Now imagine a large population, each of whose members are hardwired genetically either as hawks or as doves, and assume that those who do better at this game have more offspring. We will argue that if $(x, 1-x)$ is a symmetric Nash equilibrium in this game, then these will also be equilibrium proportions in the population.

Let's see what the Nash equilibria are. If $c < \frac{v}{2}$, the game is a version of Prisoner's Dilemma and (H, H) is the only equilibrium. When $c > \frac{v}{2}$, there are two pure Nash equilibria: (H, D) and (D, H); and since the game is symmetric, there is a symmetric mixed Nash equilibrium. Suppose each player plays H with probability $x \in (0, 1)$. For this to be player I's strategy in a Nash equilibrium, the payoffs to player II from playing H and D must be equal:

$$\text{(L)} \qquad x\left(\frac{v}{2} - c\right) + (1-x)v = (1-x)\frac{v}{2} \quad \text{(R)}. \qquad (7.1)$$

---

[1] A player may not be aware of his strategy.
[2] This is known as natural selection.

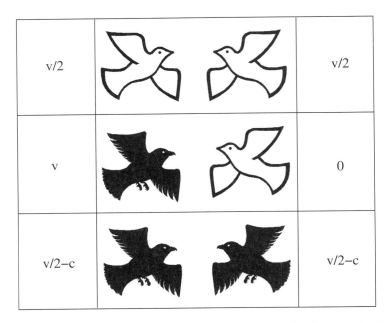

FIGURE 7.1. Two players play this game for a prize of value $v > 0$. They confront each other, and each chooses (simultaneously) to fight or to flee; these two strategies are called the "hawk" (H) and the "dove" (D) strategies, respectively. If they both choose to fight (two hawks), then each incurs a cost c, and the winner (either is equally likely) takes the prize. If a hawk faces a dove, the dove flees, and the hawk takes the prize. If two doves meet, they split the prize equally.

For this to hold, we need $x = \frac{v}{2c}$, which by the assumption is less than 1. By symmetry, player II will do the same thing.

Population dynamics for Hawks and Doves. Now suppose we have the following dynamics in the population: Throughout their lives, random members of the population pair off and play Hawks and Doves; at the end of each generation, members reproduce in numbers proportional to their winnings. Let x denote the fraction of hawks in the population.

If $x < \frac{v}{2c}$, then in equation (7.1), (L) > (R) – the expected payoff for a hawk is greater than that for a dove, and so in the next generation, x, the fraction of hawks, will increase.

On the other hand, if $x > \frac{v}{2c}$, then (L) < (R) – the expected payoff for a dove is higher than that of a hawk, and so in the next generation, x will decrease.

7.1.2. Evolutionarily stable strategies. Consider a symmetric, two-player game with n pure strategies each and payoff matrices A and B for players I and II, with $A_{i,j} = B_{j,i}$.

We take the point of view that a symmetric mixed strategy in this game corresponds to the proportions of each type within the population.

To motivate the formalism, suppose a population with strategy $\mathbf{x}$ is invaded by a small population of mutants of type $\mathbf{z}$ (that is, playing strategy $\mathbf{z}$), so the new composition is $\varepsilon\mathbf{z} + (1 - \varepsilon)\mathbf{x}$, where ε is small. The new payoffs will be

$$\varepsilon\mathbf{x}^T A\mathbf{z} + (1 - \varepsilon)\mathbf{x}^T A\mathbf{x} \quad \text{(for } \mathbf{x}\text{'s),} \tag{7.2}$$

$$\varepsilon\mathbf{z}^T A\mathbf{z} + (1 - \varepsilon)\mathbf{z}^T A\mathbf{x} \quad \text{(for } \mathbf{z}\text{'s).} \tag{7.3}$$

The criteria for $\mathbf{x}$ to be an evolutionary stable strategy will imply that, for small enough ε, the average payoff for $\mathbf{x}$'s will be strictly greater than that for $\mathbf{z}$'s, so the invaders will disappear. Formally:

DEFINITION 7.1.1. A mixed strategy $\mathbf{x}$ in Δ_n is an **evolutionarily stable strategy (ESS)** if for any pure "mutant" strategy $\mathbf{z}$:

(a) $\mathbf{z}^T A\mathbf{x} \leq \mathbf{x}^T A\mathbf{x}$.
(b) If $\mathbf{z}^T A\mathbf{x} = \mathbf{x}^T A\mathbf{x}$, then $\mathbf{z}^T A\mathbf{z} < \mathbf{x}^T A\mathbf{z}$.

Observe that criterion (a) is equivalent to $\mathbf{x}$ being a (symmetric) Nash equilibrium.[3] Thus, if $\mathbf{x}$ is a Nash equilibrium, criterion (a) holds with equality for any $\mathbf{z}$ in the support of $\mathbf{x}$.

Assuming (a), no mutant will fare strictly better against the current population strategy $\mathbf{x}$ than $\mathbf{x}$ itself. However, a mutant strategy $\mathbf{z}$ could still successfully invade if it does just as well as $\mathbf{x}$ when playing against $\mathbf{x}$ and when playing against another $\mathbf{z}$ mutant. Criterion (b) excludes this possibility.

EXAMPLE 7.1.2 (**Hawks and Doves**). We will verify that the mixed Nash equilibrium $\mathbf{x} = \left(\frac{v}{2c}, 1 - \frac{v}{2c}\right)$ (i.e., H is played with probability $\frac{v}{2c}$) is an ESS when $c > \frac{v}{2}$. First, we observe that both pure strategies satisfy criterion (a) with equality, so we check (b).

- If $\mathbf{z} = (1, 0)$ ("H"), then $\mathbf{z}^T A\mathbf{z} = \frac{v}{2} - c$, which is strictly less than $\mathbf{x}^T A\mathbf{z} = x(\frac{v}{2} - c) + (1 - x)0$.
- If $\mathbf{z} = (0, 1)$ ("D"), then $\mathbf{z}^T A\mathbf{z} = \frac{v}{2} < \mathbf{x}^T A\mathbf{z} = xv + (1 - x)\frac{v}{2}$.

Thus, the mixed Nash equilibrium for Hawks and Doves is an ESS.

EXAMPLE 7.1.3 (**Rock-Paper-Scissors**). The unique Nash equilibrium in Rock-Paper-Scissors, $\mathbf{x} = (\frac{1}{3}, \frac{1}{3}, \frac{1}{3})$, is **not** evolutionarily stable.

		player II		
		Rock	Paper	Scissors
player I	Rock	0	−1	1
	Paper	1	0	−1
	Scissors	−1	1	0

This is because the payoff of $\mathbf{x}$ against any strategy is 0, and the payoff of any pure strategy against itself is also 0, and thus, the expected payoff of $\mathbf{x}$ and $\mathbf{z}$ will be equal. This suggests that under appropriate notions of population dynamics, cycling will occur: A population with many Rocks will be taken over by Paper, which in turn will be invaded (bloodily, no doubt) by Scissors, and so forth. These dynamics have been observed in nature — in particular, in a California lizard[4].

---

[3]This is shorthand for $(\mathbf{x}, \mathbf{x})$ being a Nash equilibrium.
[4]The description of this example follows, almost verbatim, the exposition of Gintis [Gin00].

The side-blotched lizard *Uta stansburiana* has three distinct types of male: orange-throat, blue-throat, and yellow-striped. All females of the species are yellow-striped. The orange-throated males are violently aggressive, keep large harems of females, and defend large territories. The yellow-striped males are docile and look like receptive females. In fact, the orange-throats can't distinguish between the yellow-striped males and females. This enables the yellow-striped males to sneak into their territory and secretly copulate with the females. The blue-throats are less aggressive than the orange-throats, keep smaller harems (small enough to distinguish their females from yellow-striped males), and defend small territories.

Researchers have observed a six-year cycle starting with domination, say, by the orange-throats. Eventually, the orange-throats amass territories and harems so large that they can no longer be guarded effectively against the sneaky yellow-striped males, who are able to secure a majority of copulations and produce the largest number of offspring. When the yellow-striped lizards become very common, however, the males of the blue-throated variety get an edge: Since they have small harems, they can detect yellow-striped males and prevent them from invading their harems. Thus, a period when the blue-throats become dominant follows. However, the aggressive orange-throats do comparatively well against blue-throats since they can challenge them and acquire their harems and territories, thus propagating themselves. In this manner, the population frequencies eventually return to the original ones, and the cycle begins anew.

When John Maynard Smith learned that *Uta stansburia* were "playing" Rock-Paper-Scissors, he reportedly[5] exclaimed, "They have read my book!"

FIGURE 7.2. The three types of male lizard *Uta stansburiana*. Picture courtesy of Barry Sinervo; see http://bio.research.ucsc.edu/~barrylab.

---

[5]This story is reported in [Sig05].

EXAMPLE 7.1.4 (**Unstable mixed Nash equilibrium**). In this game,

player II

	A	B
A	$(10, 10)$	$(0, 0)$
B	$(0, 0)$	$(5, 5)$

(player I labels the rows A, B)

both pure strategies (A, A) and (B, B) are evolutionarily stable, while the symmetric mixed Nash equilibrium $\mathbf{x} = (\frac{1}{3}, \frac{2}{3})$ is not.

Although (B, B) is evolutionarily stable, if a sufficiently large population of A's invades, then the "stable" population will in fact shift to being entirely composed of A's. Specifically, if, after the A's invade, the new composition is α fraction A's and $1 - \alpha$ fraction B's, then using (7.2), the payoffs for each type are

$$5(1 - \alpha) \quad \text{(for } B\text{'s)}$$
$$10\alpha \quad \text{(for } A\text{'s).}$$

Thus if $\alpha > 1/3$, the payoffs of the A's will be higher and they will "take over".

EXERCISE 7.a (**Mixed population invasion**). Consider the following game:

player II

	A	B	C
A	$(0, 0)$	$(6, 2)$	$(-1, -1)$
B	$(2, 6)$	$(0, 0)$	$(3, 9)$
C	$(-1, -1)$	$(9, 3)$	$(0, 0)$

(player I labels the rows A, B, C)

Find two mixed Nash equilibria, one supported on $\{A, B\}$, the other supported on $\{B, C\}$. Show that they are both ESS, but the $\{A, B\}$ equilibrium is not stable when invaded by an arbitrarily small population composed of half B's and half C's.

EXAMPLE 7.1.5 (**Sex ratios**). Evolutionary stability can be used to explain sex ratios in nature. In mostly monogomous species, it seems natural that the birth rate of males and females should be roughly equal. But what about sea lions, in which a single male gathers a large harem of females, while many males never reproduce? Game theory helps explain why reproducing at a 1:1 ratio remains stable. To illustrate this, consider the following highly simplified model. Suppose that each harem consists of one male and ten females. If M is the number of males in the population and F the number of females, then the number of "lucky" males, that is, males with a harem, is $M_L = \min(M, F/10)$. Suppose also that each mating pair has b offspring on average. A random male has a harem with probability M_L/M, and if he does, he has $10b$ offspring on average. Thus, the expected number of offspring a random male has is $\mathbb{E}[C_m] = 10bM_L/M = b\min(10, F/M)$. On the other hand, the number of females that belong to a harem is $F_L = \min(F, 10M)$, and thus the expected number of offspring a female has is $\mathbb{E}[C_f] = bF_L/F = b\min(1, 10M/F)$.

If $M < F$, then $\mathbb{E}[C_m] > \mathbb{E}[C_f]$, and individuals with a higher propensity to have male offspring than females will tend to have more grandchildren, resulting in a higher proportion of genes in the population with a propensity for male offspring. In other words, the relative birthrate of males will increase. On the other hand, if

FIGURE 7.3. Sea lion life.

$M > F$, then $\mathbb{E}[C_m] < \mathbb{E}[C_f]$, and the relative birthrate of females increases. (Of course, when $M = F$, we have $\mathbb{E}[C_m] = \mathbb{E}[C_f]$, and the sex ratio is stable.)

7.2. Correlated equilibria

If there is intelligent life on other planets, in a majority of them, they would have discovered correlated equilibrium before Nash equilibrium.   – Roger Myerson

EXAMPLE 7.2.1 (**Battle of the Sexes**). The wife wants to head to the opera, but the husband yearns instead to spend an evening watching baseball. Neither is satisfied by an evening without the other. In numbers, player I being the wife and player II the husband, here is the scenario:

		husband	
		opera	baseball
wife	opera	(4,1)	(0,0)
	baseball	(0,0)	(1,4)

In this game, there are two pure Nash equilibria: Both go to the opera or both watch baseball. There is also a mixed Nash equilibrium which yields each player an expected payoff of $4/5$ (when the wife plays $(4/5, 1/5)$ and the husband plays $(1/5, 4/5)$). This mixed equilibrium hardly seems rational: The payoff a player gets is lower than what he or she would obtain by going along with his or her spouse's preference. How might this couple decide between the two pure Nash equilibria?

One way to do this would be to pick a joint action based on a flip of a single coin. For example, the two players could agree that if the coin lands heads, then both go to the opera; otherwise, both watch baseball. Observe that even after the coin toss, neither player has an incentive to unilaterally deviate from the agreement.

To motivate the concept of correlated equilibrium introduced below, observe that a mixed strategy pair in a two-player general-sum game with action spaces $[m]$ and $[n]$ can be described by a random pair of actions: $\mathcal{R}$ with distribution $\mathbf{x} \in \Delta_m$ and $\mathcal{C}$ with distribution $\mathbf{y} \in \Delta_n$, picked independently by players I and II. Thus,

$$\mathbb{P}[\mathcal{R} = i, \ \mathcal{C} = j] = x_i y_j.$$

It follows from Lemma 4.3.7 that $\mathbf{x}$, $\mathbf{y}$ is a Nash equilibrium if and only if

$$\mathbb{P}\left[\mathcal{R}=i\right] > 0 \Longrightarrow \mathbb{E}\left[a_{i,\mathcal{C}}\right] \geq \mathbb{E}\left[a_{\ell,\mathcal{C}}\right]$$

for all i and ℓ in $[n]$ and

$$\mathbb{P}\left[\mathcal{C}=j\right] > 0 \Longrightarrow \mathbb{E}\left[b_{\mathcal{R},j}\right] \geq \mathbb{E}\left[b_{\mathcal{R},k}\right]$$

for all j and k in $[m]$.

Player II

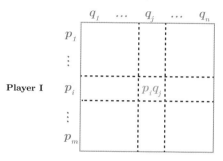
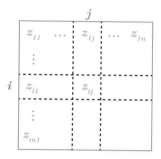

FIGURE 7.4. This figure illustrates the difference between a Nash equilibrium and a correlated equilibrium. In a Nash equilibrium, the probability that player I plays i and player II plays j is the product of the two corresponding probabilities (in this case $p_i q_j$), whereas a correlated equilibrium puts a probability, say z_{ij}, on each pair (i,j) of strategies.

DEFINITION 7.2.2. A **correlated strategy pair** is a pair of random actions $(\mathcal{R},\mathcal{C})$ with an arbitrary joint distribution

$$z_{ij} = \mathbb{P}\left[\mathcal{R}=i,\ \mathcal{C}=j\right].$$

The next definition formalizes the idea that, in a correlated equilibrium, if player I knows that the players' actions $(\mathcal{R},\mathcal{C})$ are picked according to the joint distribution $\mathbf{z}$ and player I is informed only that $\mathcal{R}=i$, then she has no incentive to switch to some other action ℓ.

DEFINITION 7.2.3. A correlated strategy pair in a two-player game with payoff matrices A and B is a **correlated equilibrium** if

$$\mathbb{P}\left[\mathcal{R}=i\right] > 0 \Longrightarrow \mathbb{E}\left[a_{i,\mathcal{C}} \mid \mathcal{R}=i\right] \geq \mathbb{E}\left[a_{\ell,\mathcal{C}} \mid \mathcal{R}=i\right] \qquad (7.4)$$

for all i and ℓ in $[n]$ and

$$\mathbb{P}\left[\mathcal{C}=j\right] > 0 \Longrightarrow \mathbb{E}\left[b_{\mathcal{R},j} \mid \mathcal{C}=j\right] \geq \mathbb{E}\left[b_{\mathcal{R},k} \mid \mathcal{C}=j\right]$$

for all j and k in $[m]$.

REMARK 7.2.4. In terms of the distribution $\mathbf{z}$, the inequality in condition (7.4) is

$$\sum_j \left(\frac{z_{ij}}{\sum_k z_{ik}}\right) a_{ij} \geq \sum_j \left(\frac{z_{ij}}{\sum_k z_{ik}}\right) a_{\ell j}.$$

Thus, $\mathbf{z}$ is a correlated equilibrium iff for all i and ℓ,

$$\sum_j z_{ij} a_{ij} \geq \sum_j z_{ij} a_{\ell j}$$

Player II strategy conditioned on i

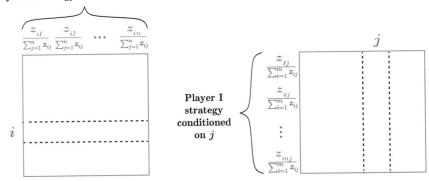

FIGURE 7.5. The left figure shows the distribution player I faces (the labels on the columns) when the correlated equilibrium indicates that she should play i. Given this distribution over columns, Definition 7.2.3 says that she has no incentive to switch to a different row strategy. The right figure shows the distribution player II faces when told to play j.

and for all j and k,

$$\sum_i z_{ij} b_{ij} \geq \sum_i z_{ij} b_{ik}.$$

The next example illustrates a more sophisticated correlated equilibrium that is not simply a mixture of Nash equilibria.

EXAMPLE 7.2.5 (**Chicken, revisited**). In this game, (S, D) and (D, S) are Nash equilibria with payoffs of $(2, 7)$ and $(7, 2)$, respectively. There is also a mixed Nash equilibrium in which each player plays S with probability $2/3$ and D with probability $1/3$ resulting in an expected payoff of $4\frac{2}{3}$.

		player II	
		Swerve (S)	Drive (D)
player I	Swerve (S)	(6, 6)	(2, 7)
	Drive (D)	(7, 2)	(0, 0)

The following probability distribution $\mathbf{z}$ is a correlated equilibrium which results in an expected payoff of $4\frac{1}{2}$ to each player, worse than the mixed Nash equilibrium:

		player II	
		Swerve (S)	Drive (D)
player I	Swerve(S)	0	1/2
	Drive (D)	1/2	0

A more interesting correlated equilibrium, that yields a payoff outside the convex hull of the Nash equilibrium payoffs, is the following:

		player II	
		Swerve (S)	Drive (D)
player I	Swerve (S)	1/3	1/3
	Drive (D)	1/3	0

For this correlated equilibrium, it is crucial that the row player only knows $\mathcal{R}$ and the column player only knows $\mathcal{C}$. Otherwise, in the case that the outcome is (C, C), each player would have an incentive to deviate (unilaterally).

Thus, to implement a correlated equilibrium, an external mediator is typically needed. Here, the external mediator chooses the pair of actions $(\mathcal{R}, \mathcal{C})$ according to this distribution $((S,D),(D,S),(S,S))$ with probability $\frac{1}{3}$ each) and then discloses to each player which action he or she should take (but not the action of the opponent). At this point, each player is free to follow or to reject the suggested action. It is in their best interest to follow the mediator's suggestion, and thus this distribution is a correlated equilibrium.

To see this, suppose the mediator tells player I to play D. In this case, she knows that player II was told to play S and player I does best by complying to collect the payoff of 7. She has no incentive to deviate.

On the other hand, if the mediator tells her to play S, she is uncertain about what player II was told, but conditioned on what she is told, she knows that (S,S) and (S,D) are equally likely. If she follows the mediator's suggestion and plays S, her payoff will be $6 \times \frac{1}{2} + 2 \times \frac{1}{2} = 4$, while her expected payoff from switching is $7 \times \frac{1}{2} = 3.5$, so player I is better off following the suggestion.

We emphasize that the random actions $(\mathcal{R}, \mathcal{C})$ used in this correlated equilibrium are dependent, so this is not a Nash equilibrium. Moreover, the expected payoff to each player when both follow the suggestion is $2 \times \frac{1}{3} + 6 \times \frac{1}{3} + 7 \times \frac{1}{3} = 5$. This is better than what they would obtain from the symmetric Nash equilibrium or from averaging the two asymmetric Nash equilibria.

Notes

The Alan Grafen quote at the beginning of the chapter is from [HH07]. Evolutionary stable strategies were introduced by John Maynard Smith and George Price [SP73, Smi82] (though Nash in his thesis [Nas50b] already discussed the interpretation of a mixed strategy in terms of population frequencies). For a detailed account of how game theory affected evolutionary biology, see the classic book by John Maynard Smith [Smi82]. The concept has found application in a number of fields including biology, ecology, psychology, and political science. For more information on evolutionary game theory, see Chapters 6, 11, and 13 in [YZ15]. The study of evolutionary stable strategies has led to the development of evolutionary game dynamics, usually studied via systems of differential equations. See, e.g., [HS98].

John Maynard Smith

The description of cycling in the frequencies of different types of *Uta stansburiana* males and the connection to Rock-Paper-Scissors is due to B. Sinervo and C. M. Lively [SL96].

The notion of correlated equilibrium was introduced in 1974 by Robert Aumann [Aum74, Aum87]. The fact that every finite game has at least one Nash equilibrium implies that every finite game has a correlated equilibrium. Hart and Schmeidler [HS89] provide an elementary direct proof (via the minimax theorem) of the existence of correlated equilibria in games with finitely many players and strategies. Note that while we have only defined correlated equilibrium here in two-player games, the notion extends naturally to more players. See, e.g., Chapter 8 of [MSZ13].

Surprisingly, finding a correlated equilibrium in large scale problems is computationally easier than finding a Nash equilibrium. In fact, there are no computationally efficient algorithms known for finding Nash equilibria, even in two-player games. However, correlated equilibria can be computed via linear programming. (See, e.g., [MG07] for an introduction to linear programming.) For a discussion of the complexity of computing Nash equilibria and correlated equilibria, see the survey by Papadimitriou in [YZ15].

Exercises

7.1.   Find all Nash equilibria and determine which of the symmetric equilibria are evolutionarily stable in the following games:

<table>
<tr><td></td><td colspan="2">player II</td></tr>
<tr><td></td><td>A</td><td>B</td></tr>
<tr><td>A</td><td>(4, 4)</td><td>(2, 5)</td></tr>
<tr><td>B</td><td>(5, 2)</td><td>(3, 3)</td></tr>
</table>

and

<table>
<tr><td></td><td colspan="2">player II</td></tr>
<tr><td></td><td>A</td><td>B</td></tr>
<tr><td>A</td><td>(4, 4)</td><td>(3, 2)</td></tr>
<tr><td>B</td><td>(2, 3)</td><td>(5, 5)</td></tr>
</table>

S 7.2.   Consider the following symmetric game as played by two drivers, both trying to get from Here to There (or two computers routing messages along cables of different bandwidths). There are two routes from Here to There; one is wider and therefore faster, but congestion will slow them down if both take the same route. Denote the wide route W and the narrower route N. The payoff matrix is

FIGURE 7.6. The leftmost image shows the payoffs when both drivers drive on the narrower route, the middle image shows the payoffs when both drivers drive on the wider route, and the rightmost image shows what happens when the red driver (player I) chooses the wide route and the yellow driver (player II) chooses the narrow route.

player I (red)	player II (yellow)	
	W	N
W	$(3,3)$	$(5,4)$
N	$(4,5)$	$(2,2)$

Find all Nash equilibria and determine which ones are evolutionarily stable.

7.3.   Argue that in a symmetric game, if $a_{ii} > b_{i,j}$ $(= a_{j,i})$ for all $j \neq i$, then pure strategy i is an evolutionarily stable strategy.

7.4.   Occasionally, two parties resolve a dispute (pick a "winner") by playing a variant of Rock-Paper-Scissors. In this version, the parties are penalized if there is a delay before a winner is declared; a delay occurs when both players choose the same strategy. The resulting payoff matrix is the following:

player I	player II		
	Rock	Paper	Scissors
Rock	$(-1,-1)$	$(0,1)$	$(1,0)$
Paper	$(1,0)$	$(-1,-1)$	$(0,1)$
Scissors	$(0,1)$	$(1,0)$	$(-1,-1)$

Show that this game has a unique Nash equilibrium that is fully mixed, and results in expected payoffs of 0 to both players. Then show that the following probability distribution is a correlated equilibrium in which the players obtain expected payoffs of 1/2:

player I	player II		
	Rock	Paper	Scissors
Rock	0	1/6	1/6
Paper	1/6	0	1/6
Scissors	1/6	1/6	0

CHAPTER 8

The price of anarchy

In this chapter, we study **the price of anarchy**, the worst-case ratio between the quality of a socially optimal outcome and the quality of a Nash equilibrium outcome.

8.1. Selfish routing

On Earth Day in 1990, the New York City traffic commissioner made the decision to close 42nd Street, one of the most congested streets in Manhattan. Many observers predicted that disastrous traffic conditions would ensue. Surprisingly, however, overall traffic and typical travel times actually improved. As we shall see next, phenomena like this, where reducing the capacity of a road network actually improves travel times, can be partially explained with game theory.

FIGURE 8.1. Each link in the left figure is labeled with a latency function $\ell(x)$ which describes the travel time on an edge as a function of the congestion x on that edge. (The congestion x is the fraction of traffic going from A to B that takes this edge.) In Nash equilibrium, each driver chooses the route that minimizes his own travel time, given the routes chosen by the other drivers. The unique Nash equilibrium in this network, shown on the right, is obtained by sending half the traffic to the top and half to the bottom. Thus, the latency each driver experiences is $3/2$. This is also the optimal routing; i.e., it minimizes the average latency experienced by the drivers.

EXAMPLE 8.1.1 (**Braess Paradox**). A large number of drivers head from point A to point B each morning. There are two routes, through C and through D. The travel time on each road may depend on the traffic on it as shown in Figure 8.1. Each driver, knowing the traffic on each route, will choose his own path selfishly, that is, to minimize his own travel time, given what everyone else is doing. In this example, in equilibrium, exactly half of the traffic will go on each route, yielding an average travel time of $3/2$. This setting, where the proportion of drivers taking a route can have any value in $[0,1]$, is called "infinitesimal drivers".

Now consider what can happen when a new, very fast highway is added between C and D. Indeed, we will assume that this new route is so fast that we can simply think of the travel time on it as being 0.

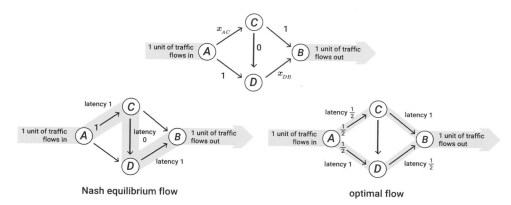

Nash equilibrium flow   **optimal flow**

FIGURE 8.2. The Braess Paradox: Each link in the top figure is labeled with a latency function $\ell(x)$ which describes the travel time on that edge as a function of the fraction x of traffic using that edge. These figures show the effect of adding a 0 latency road from C to D: The travel time on each of $\gamma_C = A - C - B$ and $\gamma_D = A - D - B$ is always at least the travel time on the new route $\gamma = A - C - D - B$. Moreover, if a positive fraction of the traffic takes route γ_C (resp. γ_D), then the travel time on γ is *strictly* lower than that of γ_C (resp. γ_D). Thus, the unique Nash equilibrium is for all the traffic to go on the path γ, as shown in the bottom left figure. In this equilibrium, the average travel time the drivers experience is 2, as shown on the bottom left. On the other hand, if the drivers could be forced to choose routes that would minimize the average travel time, it would be reduced to 3/2, the social optimum, as shown on the bottom right.

One would think that adding a fast road could never slow down traffic, but surprisingly in this case it does: As shown in Figure 8.2, average travel time in equilibrium increases from 3/2 to 2. This phenomenon, where capacity is added to a system and, in equilibrium, average driver travel time increases, is called the **Braess Paradox**.

We define the **socially optimal traffic flow** to be the partition of traffic that minimizes average latency. The crux of the Braess Paradox is that while the social optimum can only improve when roads are added to the network, the Nash equilibrium can get worse.

We use the term **price of anarchy** to measure the ratio between performance in equilibrium and the social optimum. In Example 8.1.1,

$$\text{price of anarchy} := \frac{\text{average travel time in worst Nash equilibrium}}{\text{average travel time in socially optimal outcome}} = \frac{2}{3/2} = \frac{4}{3}.$$

In fact, we will see that in **any** road network with affine latency functions and infinitesimal drivers, as in this example, the price of anarchy is at most 4/3! We will develop this result soon, but first, let's calculate the price of anarchy in a couple of simple scenarios.

EXAMPLE 8.1.2 (**Pigou-type examples**). In the example of Figure 8.3, with latency functions ax and bx (a and b are both constants greater than 0), the Nash equilibrium and optimal flow are the same: the solution to $ax = b(1 - x)$. Thus, the price of anarchy is 1.

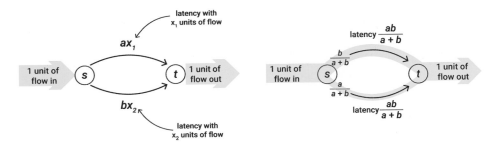

FIGURE 8.3. The figure on the right shows the equilibrium and optimal flows (which are identical in this case). Both have a fraction $b/(a + b)$ of the traffic on the upper link, resulting in the same latency of $ab/(a + b)$ on both the top and bottom links. Thus, the price of anarchy in this network is 1.

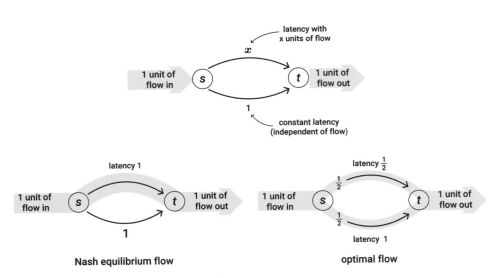

FIGURE 8.4. The top figure shows the latency functions on each edge. The bottom left figure shows the Nash equilibrium flow, which has an average latency of 1. The bottom right shows the optimal flow, which has an average latency of 3/4.

On the other hand, as shown in Figure 8.4, with latency functions of x and 1, the Nash equilibrium sends all traffic to the top link, whereas the optimal flow sends half the traffic to the top and half to the the bottom, for a price of anarchy of 4/3.

8.1.1. Bounding the price of anarchy. Consider an arbitrary road network, in which one unit of traffic flows from source s to destination t. Let $\mathcal{P}_{st}$ be the set of paths in G from s to t. Each driver chooses a path $P \in \mathcal{P}_{st}$. Let f_P be the fraction of drivers that take path P. Write $\mathbf{f} := (f_P)_{P \in \mathcal{P}_{st}}$ for the resulting traffic flow. The space of possible flows is

$$\Delta(\mathcal{P}_{st}) = \left\{ \mathbf{f} : f_P \geq 0 \ \forall P \text{ and } \sum_{P \in \mathcal{P}_{st}} f_P = 1 \right\}.$$

Given such a flow, the induced (traffic) flow on an edge e is

$$F_e := \sum_{P | e \in P} f_P. \tag{8.1}$$

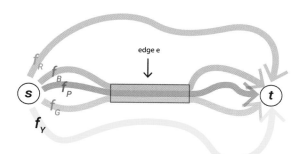

$$\boldsymbol{F_e = f_B + f_P + f_G} \qquad \boldsymbol{\ell_e(F_e) = \ell_e(f_B + f_P + f_G)}$$

FIGURE 8.5. This figure shows the relationship between the edge flow F_e and the path flows that contribute to it (in this example f_B, f_P, and f_G) and depicts the computation of $\ell_e(F_e)$. The contribution of edge e to $L(\mathbf{f}) = \sum_P f_P L_P(\mathbf{f})$ is precisely $F_e \ell_e(F_e)$. See (8.3).

Denote by $\ell_e(x)$, the latency on edge e as a function of x, the amount of traffic on the edge. Throughout this section, we assume that latency functions are weakly increasing and continuous. Notice that each driver that chooses path P experiences the same latency:

$$L_P(\mathbf{f}) = \sum_{e \in P} \ell_e(F_e). \tag{8.2}$$

We denote by $L(\mathbf{f})$ the total latency of all the traffic. Since a fraction f_P of the traffic has latency $L_P(\mathbf{f})$, this total latency is

$$L(\mathbf{f}) = \sum_P f_P L_P(\mathbf{f}).$$

In equilibrium, each driver will choose some lowest latency path with respect to the current choices of other drivers:

DEFINITION 8.1.3. A flow $\mathbf{f}$ is a **(Nash) equilibrium flow** if and only if whenever $f_P > 0$, the path P is a minimum latency path; that is

$$L_P(\mathbf{f}) = \sum_{e \in P} \ell_e(F_e) = \min_{P' \in \mathcal{P}_{st}} L_{P'}(\mathbf{f}).$$

REMARK 8.1.4. In §8.1.3, we show that equilibrium flows exist.

REMARK 8.1.5. We can equivalently calculate the latency $L(\mathbf{f})$ from the edge flows F_e. (See Figure 8.5.) Since the latency experienced by the flow F_e across edge e is $\ell_e(F_e)$, we can write $L(\mathbf{f})$ in two ways:

$$L(\mathbf{f}) = \sum_P f_P L_P(\mathbf{f}) = \sum_e F_e \ell_e(F_e). \tag{8.3}$$

The next proposition generalizes this equation to the setting where the edge latencies are determined by one flow ($\mathbf{f}$) and the routing is specified by a different flow ($\tilde{\mathbf{f}}$).

PROPOSITION 8.1.6. *Let $\mathbf{f}$ and $\tilde{\mathbf{f}}$ be two path flows with corresponding edge flows $\{F_e\}_{e \in E}$ and $\{\tilde{F}_e\}_{e \in E}$, respectively. Then*

$$\sum_P \tilde{f}_P L_P(\mathbf{f}) = \sum_e \tilde{F}_e \ell_e(F_e). \tag{8.4}$$

PROOF. We have

$$\sum_P \tilde{f}_P L_P(\mathbf{f}) = \sum_P \tilde{f}_P \sum_{e \in P} \ell_e(F_e) = \sum_e \ell_e(F_e) \sum_{P | e \in P} \tilde{f}_P = \sum_e \tilde{F}_e \ell_e(F_e),$$

using (8.2) for the first equality and (8.1) for the last.   □

The next lemma asserts that if the edge latencies are determined by an equilibrium flow and are *fixed* at these values, then any other flow has weakly higher latency.

LEMMA 8.1.7. *Let $\mathbf{f}$ be an equilibrium flow and let $\tilde{\mathbf{f}}$ be any other path flow, with corresponding edge flows $\{F_e\}_{e \in E}$ and $\{\tilde{F}_e\}_{e \in E}$, respectively. Then*

$$\sum_e (\tilde{F}_e - F_e) \ell_e(F_e) \geq 0. \tag{8.5}$$

PROOF. Let $L = \min_{P' \in \mathcal{P}_{st}} L_{P'}(\mathbf{f})$. By Definition 8.1.3, if $f_P > 0$, then $L_P(\mathbf{f}) = L$. Since $\sum_P f_P = \sum_P \tilde{f}_P = 1$, it follows that

$$\sum_P f_P L_P(\mathbf{f}) = L \quad \text{and} \quad \sum_P \tilde{f}_P L_P(\mathbf{f}) \geq L. \tag{8.6}$$

We combine these using (8.4) and (8.3) to get

$$\sum_e \tilde{F}_e \ell_e(F_e) \geq \sum_e F_e \ell_e(F_e).   □$$

8.1.2. Affine latency functions.

THEOREM 8.1.8. *Let G be a network where one unit of traffic is routed from a source s to a destination t. Suppose that the latency function on each edge e is affine; that is, $\ell_e(x) = a_e x + b_e$, for constants $a_e, b_e \geq 0$. Let $\mathbf{f}$ be an equilibrium flow in this network and let $\mathbf{f}^*$ be an optimal flow; that is,*

$$L(\mathbf{f}^*) = \min\{L(\tilde{\mathbf{f}}) : \tilde{\mathbf{f}} \in \Delta(\mathcal{P}_{st})\}.$$

Then the price of anarchy is at most $4/3$; i.e.,

$$L(\mathbf{f}) \leq \frac{4}{3} L(\mathbf{f}^*).$$

REMARK 8.1.9. When the latency functions are linear (i.e., when $b_e = 0$ for all links e), the price of anarchy is 1. See Exercise 8.3.

PROOF. Let $\{F_e^*\}_{e \in E}$ be the set of edge flows corresponding to the optimal (overall latency minimizing) path flow $\mathbf{f}^*$. By Lemma 8.1.7,

$$L(\mathbf{f}) = \sum_e F_e(a_e F_e + b_e) \leq \sum_e F_e^*(a_e F_e + b_e). \tag{8.7}$$

Thus,

$$L(\mathbf{f}) - L(\mathbf{f}^*) \leq \sum_e F_e^* a_e (F_e - F_e^*).$$

Using the inequality $x(y-x) \leq y^2/4$ (which follows from $(x - y/2)^2 \geq 0$), we deduce that

$$L(\mathbf{f}) - L(\mathbf{f}^*) \leq \frac{1}{4} \sum_e a_e F_e^2 \leq \frac{1}{4} L(\mathbf{f}). \qquad \square$$

A corollary of this theorem is that the Braess Paradox example we saw earlier is extremal.

COROLLARY 8.1.10. *If additional roads are added to a road network with affine latency functions, the latency at Nash equilibrium can increase by at most a factor of $4/3$.*

PROOF. Let G be a road network and H an augmented version of G. Let $\mathbf{f}_G$ denote an equilibrium flow in G and $\mathbf{f}_G^*$ an optimal flow in G. Similarly for H. Clearly $L(\mathbf{f}_H^*) \leq L(\mathbf{f}_G^*)$. It follows that

$$L(\mathbf{f}_H) \leq \frac{4}{3} L(\mathbf{f}_H^*) \leq \frac{4}{3} L(\mathbf{f}_G^*) \leq \frac{4}{3} L(\mathbf{f}_G). \qquad \square$$

8.1.3. Existence of equilibrium flows.

LEMMA 8.1.11. *Consider an s-t road network, where the latency function on edge e is $\ell_e(\cdot)$. If all latency functions are nonnegative, weakly increasing and continuous, then a Nash equilibrium exists.*

REMARK 8.1.12. This game is a continuous analogue of the congestion games discussed in §4.4.

PROOF. The set of possible path flows is the simplex $\Delta(\mathcal{P}_{st})$ of distributions over paths. The mapping (8.1) from path flows to edge flows is continuous and linear, so the set K of all possible edge flows is compact and convex. Define a potential function over edge flows:

$$\Phi(\mathbf{F}) := \sum_e \int_0^{F_e} \ell_e(x)dx.$$

Note that Φ is a convex function since the latency functions $\ell_e(\cdot)$ are weakly increasing.

We will show that an edge flow $\mathbf{F}$ that minimizes Φ is a Nash equilibrium flow. To see this, let $\mathbf{f}$ be any path flow corresponding to edge flow $\mathbf{F}$. (Note that this path flow is not necessarily unique.) Let P^* be any path of minimum latency under $\mathbf{F}$, and let L^* be the latency on this path. If P is a path of latency $L > L^*$, we claim that $f_P = 0$. If not, then the flow obtained by moving δ units of flow from P to P^* has a lower value of Φ: Doing this changes Φ by

$$\Delta\Phi = \sum_{e \in P^* \setminus P} \int_{F_e}^{F_e + \delta} \ell_e(x)dx - \sum_{e \in P \setminus P^*} \int_{F_e - \delta}^{F_e} \ell_e(x)dx$$

$$= \delta \sum_{e \in P^* \setminus P} \ell_e(F_e) - \delta \sum_{e \in P \setminus P^*} \ell_e(F_e) + o(\delta)$$

$$= \delta(L^* - L) + o(\delta),$$

which is negative for δ sufficiently small.   □

8.1.4. Beyond affine latency functions. Let $\mathcal{L}$ be a class of latency functions. In §8.1.2 we showed that if

$$\mathcal{L} = \{ax + b | a, b \geq 0\},$$

then the price of anarchy is 4/3, which is achieved in the Pigou network shown in Figure 8.4. When the set of latency functions is expanded, e.g.,

$$\mathcal{L}' = \{ax^2 + bx + c | a, b, c \geq 0\},$$

the price of anarchy is worse, as shown in Figure 8.6. However, as we shall see next, the price of anarchy for *any class of latency functions and any network* is maximized in a Pigou network (Figure 8.7)!

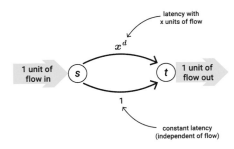

FIGURE 8.6. With the given latency functions, the optimal flow routes x units of flow on the upper link (and thus $1 - x$ on the lower link) so as to minimize the average latency, which is $x \cdot x^d + (1 - x)$. The Nash equilibrium flow routes all the flow on the upper link. The resulting price of anarchy is approximately 1.6 for $d = 2$, approximately 1.9 for $d = 3$, and is asymptotic to $d/\ln d$ as d tends to infinity.

Suppose that there are r units of flow from s to t. If $\ell(x)$ is the latency function on the upper link with x units of flow and if it is strictly increasing, then $\ell(r)$ is the smallest constant latency that can be assigned to the bottom edge that induces an equilibrium flow using only the top edge.

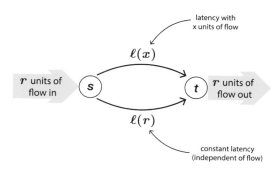

FIGURE 8.7. Here we are assuming that there are r units of flow from s to t. The Pigou price of anarchy $\alpha_r(\ell)$ is the price of anarchy in this network. Since latency functions are weakly increasing, $\ell(r) \geq \ell(x)$ for any $0 \leq x \leq r$ and thus in the worst Nash equilibrium, all flow is on the top edge. (There may be other Nash equilibria as well.)

DEFINITION 8.1.13. Let $\alpha_r(\ell)$ be the **Pigou price of anarchy** for latency function $\ell(\cdot)$ in the network shown in Figure 8.7 when the total flow from s to t is r; i.e.,

$$\alpha_r(\ell) = \frac{r\ell(r)}{\min_{0 \leq x \leq r}[x \cdot \ell(x) + (r - x) \cdot \ell(r)]}. \tag{8.8}$$

REMARK 8.1.14. It will be useful below to note that the minimum in the denominator is unchanged if you take it over $x \geq 0$ since $x(\ell(x) - \ell(r)) \geq 0$ for $x \geq r$.

THEOREM 8.1.15. *Let $\mathcal{L}$ be a class of latency functions, and define*

$$\mathcal{A}_r(\mathcal{L}) := \max_{0 \leq \tilde{r} \leq r} \sup_{\ell \in \mathcal{L}} \alpha_{\tilde{r}}(\ell).$$

Let G be a network with latency functions in $\mathcal{L}$ and total flow r from s to t. Then the price of anarchy in G is at most $\mathcal{A}_r(\mathcal{L})$.

PROOF. Let $\mathbf{f}$ and $\mathbf{f}^*$ be an equilibrium flow and an optimal flow in G, with corresponding edge flows $\mathbf{F}$ and $\mathbf{F}^*$. Fix an edge e in G and consider a Pigou network (as in Figure 8.7) with $\ell(x) := \ell_e(x)$ and total flow $r := F_e$. Then, by (8.8),

$$\alpha_{F_e}(\ell_e) = \frac{F_e \cdot \ell_e(F_e)}{\min_{0 \leq x \leq F_e}[x \cdot \ell_e(x) + (F_e - x) \cdot \ell_e(F_e)]}$$

$$\geq \frac{F_e \cdot \ell_e(F_e)}{F_e^* \cdot \ell_e(F_e^*) + (F_e - F_e^*) \cdot \ell_e(F_e)},$$

where the final inequality follows from Remark 8.1.14. Rearranging, we obtain

$$F_e^* \ell_e(F_e^*) \geq \frac{1}{\alpha_{F_e}(\ell_e)} F_e \cdot \ell_e(F_e) + (F_e^* - F_e) \cdot \ell_e(F_e),$$

and summing over e yields

$$L(\mathbf{f}^*) \geq \frac{1}{\mathcal{A}_r(\mathcal{L})} L(\mathbf{f}) + \sum_e (F_e^* - F_e) \cdot \ell_e(F_e).$$

Observe that Lemma 8.1.7 also applies to the case where the total flow is r. (In (8.6), replace L by rL.) Thus, applying (8.5) to the second sum yields

$$L(\mathbf{f}^*) \geq \frac{1}{\mathcal{A}_r(\mathcal{L})} L(\mathbf{f}). \qquad \square$$

8.1.5. A traffic-anarchy tradeoff. The next result shows that the effect of anarchy cannot be worse than doubling the total amount of traffic.

THEOREM 8.1.16. *Let G be a road network with a specified source s and sink t where r units of traffic are routed from s to t, and let $\mathbf{f}$ be a corresponding equilibrium flow with total latency $L(\mathbf{f})$. Let $\mathbf{f}^*$ be an optimal flow when $2r$ units of traffic are routed in the same network, resulting in total latency $L(\mathbf{f}^*)$. Then*

$$L(\mathbf{f}) \leq L(\mathbf{f}^*).$$

PROOF. Suppose that all the paths in use in the equilibrium flow $\mathbf{f}$ have latency L. As in the proof of Lemma 8.1.7, we have that

$$L(\mathbf{f}) = \sum_P f_P L_P(\mathbf{f}) = rL \quad \text{and} \quad \sum_P f_P^* L_P(\mathbf{f}) \geq 2rL. \qquad (8.9)$$

We will show that

$$\sum_P f_P^* L_P(\mathbf{f}) \leq L(\mathbf{f}) + L(\mathbf{f}^*) \qquad (8.10)$$

which together with (8.9) completes the proof.

We first rewrite (8.10) in terms of edges using (8.4):

$$\sum_e F_e^* \ell_e(F_e) \leq \sum_e F_e^* \ell_e(F_e^*) + \sum_e F_e \ell_e(F_e). \qquad (8.11)$$

We claim that this inequality holds for each edge; i.e.,

$$F_e^* \big(\ell_e(F_e) - \ell_e(F_e^*) \big) \leq F_e \ell_e(F_e). \qquad (8.12)$$

To verify this, consider separately the case where $F_e^* > F_e$ and $F_e^* \leq F_e$, and use the fact that $\ell_e(\cdot)$ is increasing. $\qquad \square$

REMARK 8.1.17. An alternative interpretation of this result is that doubling the capacity of every link can compensate for the lack of central control. See Exercise 8.6.

8.2. Network formation games

Consider a set of companies jointly constructing a communication network. Each company needs to connect a source to a sink. The cost of a link used by several companies is shared equally between them. How does this cost sharing rule affect their choices?

We model this via the following **fair network formation game:** There is a directed graph G whose edges E represent links that can be constructed. Associated with each link e is a cost c_e. There are k players; player i chooses a path P_i in G from node s_i to node t_i and pays his fair share of the cost of each link in path P_i. Thus, if link e is on the paths selected by r players, then each of them must

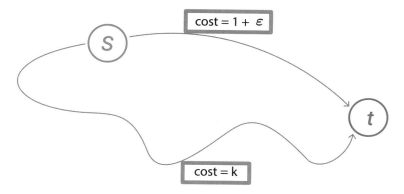

FIGURE 8.8. In this example, there are two Nash equilbria. It is a Nash equilibrium for all players to choose the upper path, resulting in a cost of $(1+\epsilon)/k$ to each player. However, there is also a bad Nash equilibrium: If all players choose the lower path, then no player has an incentive to deviate. In this case, each player's cost is 1, whereas if he switched to the upper path, his cost would be $1 + \epsilon$.

pay c_e/r to construct that link. The goal of each player is to minimize his total payment.

REMARK 8.2.1. A fair network formation game is a potential game via the potential function

$$\phi(\mathbf{s}) = \sum_{e \in E(\mathbf{s})} \sum_{j=1}^{n_e(\mathbf{s})} \frac{c_e}{j}.$$

(See Section 4.4.)

In the example shown in Figure 8.8, there are k players, each requiring the use of a path from a source s to a destination t. There is a bad Nash equilibrium in which the cost of constructing the network is approximately k times as large as the minimum possible cost, yielding a price of anarchy of about k. However, this equilibrium is extremely unstable: After a single player switches to the top path, all the others will follow.

The example of Figure 8.8 inspires us to ask if there always exists a Nash equilibrium which is close to optimal. The network in Figure 8.9 shows that the ratio can be as high as $H_k \sim \ln k$.

THEOREM 8.2.2. *In every fair network formation game with k players, there is a Nash equilibrium with cost at most $H_k = 1 + \frac{1}{2} + \frac{1}{3} + \cdots + \frac{1}{k} = \ln k + O(1)$ times the optimal cost.*

PROOF. Given a pure strategy profile $\mathbf{s}$ for the players, let $n_e(\mathbf{s})$ be the number of players that use link e. Let $E(\mathbf{s})$ be the set of links with $n_e(\mathbf{s}) \geq 1$. Since this game is a potential game, we know that best-response dynamics will lead to a pure Nash equilibrium. Observe that all strategy profiles $\mathbf{s}$ satisfy

$$\mathrm{cost}(\mathbf{s}) := \sum_{e \in E(\mathbf{s})} c_e \leq \phi(\mathbf{s}) = \sum_{e \in E(\mathbf{s})} \sum_{j=1}^{n_e(\mathbf{s})} \frac{c_e}{j}.$$

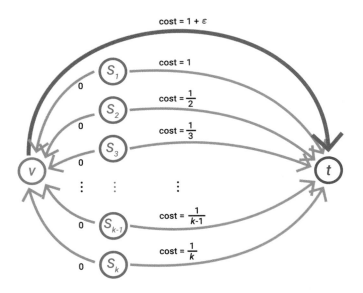

FIGURE 8.9. In this example, each of the k players needs to choose a path from s_i to t. The optimal cost network here is for all players i to choose the path from s_i to v to t, resulting in a network of total cost $1 + \epsilon$. However, this is not a Nash equilibrium. Indeed, for player k, it is a dominant strategy to use the path $s_k \to t$ since his alternative cost is at least $(1 + \epsilon)/k$. Given this choice of player k, it is dominant for player $k-1$ to choose path $s_{k-1} \to t$. Iterating yields an equilibrium where player i chooses the path $s_i \to t$. This equilibrium is unique since it arises by iterated removal of dominated strategies. (In fact, this is the only correlated equilibrium for this example.) The cost of the resulting network is approximately $H_k = 1 + \frac{1}{2} + \frac{1}{3} + \cdots + \frac{1}{k}$ times that of the cheapest network.

Let $\mathbf{s}_{\mathrm{opt}}$ be a strategy profile that minimizes $\mathrm{cost}(\mathbf{s})$. Best-response dynamics starting from $\mathbf{s}_{\mathrm{opt}}$ reduce the value of the potential function and terminate in a Nash equilibrium, say $\mathbf{s}_f$. It follows that

$$\mathrm{cost}(\mathbf{s}_f) \leq \phi(\mathbf{s}_f) \leq \phi(\mathbf{s}_{\mathrm{opt}}) \leq \mathrm{cost}(\mathbf{s}_{\mathrm{opt}}) H_k,$$

which completes the proof. $\qquad\qquad\qquad\qquad\qquad\qquad\qquad\qquad\qquad\qquad\qquad\square$

8.3. A market sharing game

There are k NBA teams, and each of them must decide in which city to locate.[1] Let v_j be the profit potential, i.e., the number of basketball fans, of city j. If ℓ teams select city j, they each obtain a utility of v_j/ℓ. See Figure 8.10.

PROPOSITION 8.3.1. *The market sharing game is a potential game and hence has a pure Nash equilibrium. (See Exercise 4.18.)*

---

[1]In 2014, Steve Ballmer, based in Seattle, bought the Clippers, an NBA team based in Los Angeles. He chose to keep the team in Los Angeles, even though there was already another NBA team there, but none in Seattle. Moving the team would increase the total number of fans with a hometown basketball team but would reduce the profit potential of the Clippers.

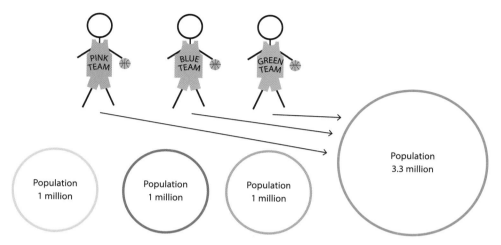

FIGURE 8.10. Three basketball teams are deciding which city to locate in when four choices are available. It is a Nash equilibrium for all of them to locate in the largest city where they will each have a utility of 1.1 million. If one of the teams were to switch to one of the smaller cities, that team's utility would drop to 1 million.

For any set S of cities, define the total value

$$V(S) := \sum_{j \in S} v_j.$$

Assume that $v_j \geq v_{j+1}$ for all j. Clearly $S^* = \{1, \ldots, k\}$ maximizes $V(S)$ over all sets of size k.

We use $\mathbf{c} = (c_1, \ldots, c_k)$ to denote a strategy profile in the market sharing game, where c_i represents the city chosen by team i. Let $S := \{c_1, \ldots, c_k\}$ be the set of cities selected.

LEMMA 8.3.2. *Let $\mathbf{c}$ and $\tilde{\mathbf{c}}$ be any two strategy profiles in the market sharing game, where the corresponding sets of cities selected are S and $\tilde{S}$. Denote by $u_i(c_i, \mathbf{c}_{-i})$ the utility obtained by team i if it chooses city c_i and the other teams choose cities $\mathbf{c}_{-i}$. Then*

$$\sum_i u_i(\tilde{c}_i, \mathbf{c}_{-i}) \geq V(\tilde{S}) - V(S).$$

PROOF. Let $\tilde{c}_i \in \tilde{S} \setminus S$. Then $u_i(\tilde{c}_i, \mathbf{c}_{-i}) = v_{\tilde{c}_i}$. Thus

$$\sum_i u_i(\tilde{c}_i, \mathbf{c}_{-i}) \geq V(\tilde{S} \setminus S) \geq V(\tilde{S}) - V(S). \qquad (8.13)$$

$\square$

REMARK 8.3.3. Lemma 8.3.2 is a typical step in many price of anarchy proofs. It disentangles the sum of utilities of the players when each separately deviates from c_i to $\tilde{c}_i$ in terms of the quantities $V(S)$ and $V(\tilde{S})$. This is sufficient to prove that the price of anarchy of this game is at most 2.

THEOREM 8.3.4. *Suppose that $\mathbf{c} = (c_1, \ldots, c_k)$ is a Nash equilibrium in the market sharing game and $S := S(\mathbf{c})$ is the corresponding set of cities selected. Then the price of anarchy is at most 2; i.e., $V(S^*) \leq 2V(S)$.*

PROOF. We claim that

$$V(S) = \sum_i u_i(c_i, \mathbf{c}_{-i}) \geq \sum_i u_i(c_i^*, \mathbf{c}_{-i}) \geq V(S^*) - V(S), \qquad (8.14)$$

which proves the theorem. The first equality in (8.14) is by definition. The inequality $u_i(c_i, \mathbf{c}_{-i}) \geq u_i(c_i^*, \mathbf{c}_{-i})$ follows from the fact that $\mathbf{c}$ is a Nash equilibrium. The final inequality follows from Lemma 8.3.2.   □

REMARK 8.3.5. In Exercise 8.7, we'll see that the price of anarchy bound of 2 can be replaced by $2 - 1/k$.

8.4. Atomic selfish routing

In §8.1.2, we considered a selfish routing game in which each driver was infinitesimal. We revisit selfish routing, but in a setting where there are few drivers and each one contributes significantly to the total travel time.

EXAMPLE 8.4.1. Consider a road network $G = (V, E)$ and a set of k drivers, with each driver i traveling from a starting node $s_i \in V$ to a destination $t_i \in V$. Associated with each edge $e \in E$ is a latency function $\ell_e(n) = a_e \cdot n + b_e$ representing the cost of traversing edge e if n drivers use it. Driver i's strategic decision is which path P_i to choose from s_i to t_i, and her objective is to choose a path with minimum latency.

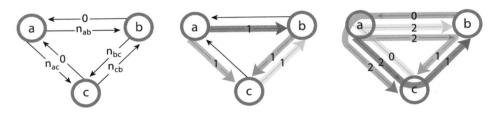

FIGURE 8.11. In the left graph, each directed edge is labeled with the travel time as a function of the number of drivers using it. Here the purple driver is traveling from a to b, the green driver is traveling from a to c, the blue driver is traveling from b to c, and the yellow driver is traveling from c to b. The figure in the middle shows the socially optimal routes (all single-hop), that is, the routes that minimize the average driver travel time. Each edge is labeled with the latency that driver experiences. This set of routes is also a Nash equilibrium. On the other hand, there is also a bad Nash equilibrium, in which each driver takes a 2-hop path, shown in the figure on the right. In the middle and right figures, the label on each colored path is the latency that particular driver experiences on that link. In this case, the price of anarchy is 5/2.

In the example shown in Figure 8.11, the socially optimal outcome is also a Nash equilibrium with a total latency of 4, but there is another Nash equilibrium with a total latency of 10. Next, we show that in any network with affine latency functions, the price of anarchy (the ratio between travel time in the *worst* Nash equilibrium and that in the socially optimal outcome) is at most 5/2.

Denote by $L_i(P_i, \mathbf{P}_{-i})$ the latency along the path P_i selected by driver i, given the paths $\mathbf{P}_{-i}$ selected by the other drivers, and let

$$L(\mathbf{P}) := \sum_i L_i(P_i, \mathbf{P}_{-i})$$

be the sum of these latencies.

We will need the following claim:

CLAIM 8.4.2. *Let n and m be any nonnegative integers. Then*

$$n(m+1) \le \frac{5}{3}n^2 + \frac{1}{3}m^2.$$

PROOF. We have to show that $f(m,n) := 5n^2 + m^2 - 3n(m+1) \ge 0$ for nonnegative integers m, n. The cases where $m = 0$ or $n = 0$ or $m = n = 1$ are clear. In all other cases $n + m \ge 3$, so $f(m,n) = (2n - m)^2 + n(n + m - 3) \ge 0$. $\square$

The next lemma is the key to our price of anarchy bounds.

LEMMA 8.4.3. *Let $\mathbf{P} = (P_1, \ldots, P_k)$ be any strategy profile (a path P_i from s_i to t_i for each i) in the atomic selfish routing game G. Let $\mathbf{P}^* = (P_1^*, \ldots, P_k^*)$ be the paths that minimize the total travel time $L(\mathbf{P}^*)$. Then*

$$\sum_i L_i(P_i^*, \mathbf{P}_{-i}) \le \frac{5}{3}L(\mathbf{P}^*) + \frac{1}{3}L(\mathbf{P}). \qquad (8.15)$$

PROOF. Let n_e be the number of paths in $\mathbf{P}$ that use edge e. Then the total travel time experienced by the drivers in this equilibrium is

$$L(\mathbf{P}) = \sum_i L_i(P_i, \mathbf{P}_{-i}) = \sum_i \sum_{e \in P_i} (a_e \cdot n_e + b_e) = \sum_e n_e (a_e \cdot n_e + b_e).$$

Let n_e^* be the number of paths among $P_1^*, \ldots, P_k^*$ that use edge e. Observe that

$$\sum_i L_i(P_i^*, \mathbf{P}_{-i}) \le \sum_i \sum_{e \in P_i^*} (a_e \cdot (n_e + 1) + b_e),$$

since switching from P_i to P_i^* can increase the number of drivers that use e by at most 1. Thus

$$\sum_i L_i(P_i^*, \mathbf{P}_{-i}) \le \sum_e n_e^* (a_e \cdot (n_e + 1) + b_e),$$

since the (upper bound on the) travel time on each edge is counted a number of times equal to the number of paths in $\mathbf{P}^*$ that use it. Finally, using Claim 8.4.2, we have $n_e^*(n_e + 1) \le \frac{5}{3}(n_e^*)^2 + \frac{1}{3}(n_e)^2$, so

$$\sum_i L_i(P_i^*, \mathbf{P}_{-i}) \le \sum_e \left(a_e \left(\frac{5}{3}(n_e^*)^2 + \frac{1}{3}n_e^2 \right) + b_e n_e^* \right)$$

$$= \frac{5}{3} \sum_e n_e^*(a_e n_e^* + b_e) + \frac{1}{3} \sum_e a_e n_e^2$$

$$\le \frac{5}{3}L(\mathbf{P}^*) + \frac{1}{3}L(\mathbf{P}). \qquad \square$$

THEOREM 8.4.4. *The price of anarchy of the atomic selfish routing game is $5/2$.*

PROOF. The example in Figure 8.11 shows that the price of anarchy is at least $5/2$. To see that it is at most $5/2$, let $\mathbf{P} = (P_1, \ldots, P_k)$ be a Nash equilibrium profile in the atomic selfish routing game. Then

$$L(\mathbf{P}) = \sum_i L_i(P_i, \mathbf{P}_{-i}) \leq \sum_i L_i(P_i^*, \mathbf{P}_{-i}),$$

where the second inequality follows from the fact that $\mathbf{P}$ is a Nash equilibrium. Thus, by (8.15),

$$L(\mathbf{P}) \leq \frac{5}{3}L(\mathbf{P}^*) + \frac{1}{3}L(\mathbf{P}).$$

Finally, rearranging, we get

$$L(\mathbf{P}) \leq \frac{5}{2}L(\mathbf{P}^*). \qquad \square$$

8.4.1. Extension theorems. The crucial step in the price of anarchy bound we just obtained was Lemma 8.4.3, which enabled us to disentangle the term $\sum_i L_i(P_i^*, \mathbf{P}_{-i})$ into a linear combination of $L(\mathbf{P}^*)$ and $L(\mathbf{P})$, the latencies associated with the two "parent" strategy profiles. Such a disentanglement enables us to prove a price of anarchy bound for a pure Nash equilibrium that extends automatically to certain scenarios in which players are not in Nash equilibrium. In this section, we prove this fact for a general cost minimization game.

Let G be a game in which players are trying to minimize their costs. For a strategy profile $\mathbf{s} \in S_1 \times S_2 \times \cdots \times S_k$, let $C_i(\mathbf{s})$ be the cost incurred by player i on strategy profile $\mathbf{s}$. As usual, strategy profile $\mathbf{s} = (s_1, \ldots, s_k)$ is a Nash equilibrium if for each player i and $s_i' \in S_i$,

$$C_i(s_i, \mathbf{s}_{-i}) \leq C_i(s_i', \mathbf{s}_{-i}).$$

Define the overall cost in profile $\mathbf{s}$ to be

$$\text{cost}(\mathbf{s}) := \sum_i C_i(\mathbf{s}).$$

DEFINITION 8.4.5. Let $\mathbf{s} = (s_1, \ldots, s_n)$ be a strategy profile in cost-minimization game G. Let $\mathbf{s}^*$ be the strategy profile that minimizes $\text{cost}(\mathbf{s})$. Then G is (λ, μ)-**smooth** if

$$\sum_i C_i(s_i^*, \mathbf{s}_{-i}) \leq \lambda \cdot \text{cost}(\mathbf{s}^*) + \mu \cdot \text{cost}(\mathbf{s}). \tag{8.16}$$

REMARK 8.4.6. Lemma 8.4.3 proves that the atomic selfish routing game is $(5/3, 1/3)$-smooth.

The following theorem shows that (λ, μ)-smooth cost minimization games have price of anarchy at most $\lambda/(1 - \mu)$ and enables extension theorems that yield the same bound with respect to other solution concepts.

THEOREM 8.4.7. *Let G be a cost-minimization game as discussed above. Let $\mathbf{s}^* = (s_1^*, \ldots, s_k^*)$ be a strategy profile which minimizes $\text{cost}(\mathbf{s}^*)$ and suppose that G is (λ, μ)-smooth. Then the following price of anarchy bounds hold:*

(1) *If $\mathbf{s}$ is a pure Nash equilibrium, then*

$$\text{cost}(\mathbf{s}) \leq \frac{\lambda}{1 - \mu}\text{cost}(\mathbf{s}^*).$$

(2) *Mixed Nash equilibria and (coarse) correlated equilibria: If* $\mathbf{p}$ *is a distribution over strategy profiles such that for all i and $\tilde{s}_i$*

$$\sum_{\mathbf{s}} p_{\mathbf{s}} C_i(\mathbf{s}) \leq \sum_{\mathbf{s}} p_{\mathbf{s}} C_i(\tilde{s}_i, \mathbf{s}_{-i}), \tag{8.17}$$

then[2]

$$\sum_{\mathbf{s}} p_{\mathbf{s}} \mathrm{cost}(\mathbf{s}) \leq \frac{\lambda}{1-\mu} \mathrm{cost}(\mathbf{s}^*). \tag{8.18}$$

(3) *Sublinear regret: If the game G is played T times, and each player uses a sublinear regret algorithm to minimize their cost [3], with player i using strategy s_i^t in the tth round and $\mathbf{s}^t = (s_1^t, \ldots, s_k^t)$, then*

$$\frac{1}{T} \sum_i C_i(\mathbf{s}^t) \leq \frac{\lambda}{1-\mu} \mathrm{cost}(\mathbf{s}^*) + o(1). \tag{8.19}$$

PROOF. The proof of part (1) is essentially the same as that of Theorem 8.4.4 and it is a special case of part (2).

Proof of (2):

$$\sum_{\mathbf{s}} p_{\mathbf{s}} \mathrm{cost}(\mathbf{s}) = \sum_{\mathbf{s}} p_{\mathbf{s}} \sum_i C_i(s_i, \mathbf{s}_{-i})$$

$$\leq \sum_{\mathbf{s}} p_{\mathbf{s}} \sum_i C_i(s_i^*, \mathbf{s}_{-i}) \qquad \text{by (8.17)}$$

$$\leq \sum_{\mathbf{s}} p_{\mathbf{s}} (\lambda \cdot \mathrm{cost}(\mathbf{s}^*) + \mu \cdot \mathrm{cost}(\mathbf{s})) \qquad \text{by (8.16)}$$

$$= \lambda \cdot \mathrm{cost}(\mathbf{s}^*) + \mu \cdot \sum_{\mathbf{s}} p_{\mathbf{s}} \mathrm{cost}(\mathbf{s}).$$

Rearranging yields (8.18).

Proof of (3): We have

$$\frac{1}{T} \sum_{t=1}^T \mathrm{cost}(\mathbf{s}^t) = \frac{1}{T} \sum_{t=1}^T \sum_i C_i(s_i^t, \mathbf{s}_{-i}^t)$$

$$\leq \frac{1}{T} \sum_{t=1}^T \sum_i C_i(s_i^*, \mathbf{s}_{-i}^t) + o(1)$$

where the second inequality is the guarantee from the sublinear regret learning algorithm. Next we use the smoothness inequality (8.16) to upper bound the right hand side yields:

$$\frac{1}{T} \sum_{t=1}^T \mathrm{cost}(\mathbf{s}^t) \leq \frac{1}{T} \sum_{t=1}^T \left(\lambda \cdot \mathrm{cost}(\mathbf{s}^*) + \mu \cdot \mathrm{cost}(\mathbf{s}^t) \right) + o(1).$$

---

[2]The condition for $\mathbf{p}$ to be a correlated equilibrium is that for all i, s_i, and s_i', we have $\sum_{\mathbf{s}_{-i}} p_{(s_i, \mathbf{s}_{-i})} C_i(s_i, \mathbf{s}_{-i}) \leq \sum_{\mathbf{s}_{-i}} p_{(s_i, \mathbf{s}_{-i})} C_i(s_i', \mathbf{s}_{-i})$. See §7.2. This condition implies (8.17) by taking $s_i' := s_i^*$ and summing over s_i. Note though that (8.17) is a weaker requirement, also known as a *coarse correlated equilibrium*.

[3]See Chapter 18.

Finally, rearranging, we get

$$\frac{1}{T}\sum_{t=1}^{T} \text{cost}(\mathbf{s}^t) \leq \frac{\lambda}{1-\mu} \cdot \text{cost}(\mathbf{s}^*) + o(1)$$

Rearranging yields (8.19).   □

8.4.2. Application to atomic selfish routing. Using the fact that the atomic selfish routing game is $(5/3, 1/3)$-smooth (by Lemma 8.4.3), we can apply Theorem 8.4.7, Part (3) and obtain the following corollary.

COROLLARY 8.4.8. *Let G be a road network with affine latency functions. Suppose that every day driver i travels from s_i to t_i, choosing his route P_i^t on day t using a sublinear regret algorithm (such as the Multiplicative Weights Algorithm from §18.3.2). Then*

$$\frac{1}{T}\sum_{t=1}^{T} L(\mathbf{P}^t) \leq \frac{5}{2}L(\mathbf{P}^*) + o(1).$$

REMARK 8.4.9. Since drivers are unlikely to know each other's strategies, this corollary seems more applicable than Theorem 8.4.4.

For other applications of Theorem 8.4.7, see the exercises.

Notes

In 2012, Elias Koutsoupias, Christos Papadimitriou, Noam Nisan, Amir Ronen, Tim Roughgarden, and Éva Tardos won the Gödel Prize[4] for "laying the foundations of algorithmic game theory." Two of the three papers [KP09, RT02] cited for the prize are concerned with the price of anarchy.

Elias Koutsoupias   Christos Papadimitriou

The "price of anarchy" concept was introduced by Koutsoupias and Papadimitriou in 1999 [KP99, KP09], though in the original paper the relevant terminology was "coordination ratio". The term "price of anarchy" is due to Papadimitriou [Pap01]. The first use of this style of analysis in congestion games was in 2000 [RT02].

An extensive discussion of the material on selfish routing discussed in §8.1, including numerous references, can be found in the book by Roughgarden [Rou05]. See also Chapters 17 and 18 of [NRTV07]. Pigou's example is discussed in his 1920 book [Pig20]. The

---

[4]The Gödel Prize is an annual prize for outstanding papers in theoretical computer science.

Tim Roughgarden   Èva Tardos

traffic model and definition of Nash flows are due to Wardrop [War52] and the proof that Nash flows exist is due to Beckmann, McGuire, and Winsten [BMW56]. The Braess Paradox is from [Bra68]. Theorem 8.1.8 and Theorem 8.1.16 are due to Roughgarden and Tardos [RT02]. A version of Theorem 8.1.15 under a convexity assumption was proved by Roughgarden [Rou03]; this assumption was removed by Correa, Schulz, and Stier-Moses [CSSM04]. The proofs of Theorem 8.1.8 and Theorem 8.1.15 presented here are due to Correa et al. [CSSM04]. With suitable tolling, the inefficiency of Nash equilibria in selfish routing can be eliminated. See, e.g., [FJM04]).

The network formation results of §8.2 are due to Anshelevich et al. [ADK+08]. In their paper, they introduce the notion of the *price of stability* of a game, in which the ratio between the optimal value of a global objective and the value of this objective in Nash equilibrium is compared. The difference between the price of stability and the price of anarchy is that the latter is concerned with the worst Nash equilibrium, whereas the former is concerned with the *best* Nash equilbrium. In fact, Theorem 8.2.2 is a price of stability result. See also the survey of network formation games by Jackson [Jac05] and Chapter 19 of [NRTV07].

The market sharing game of §8.3 is a special case of a class of games called *utility games* introduced by Vetta [Vet02]. In these games, players must choose among a set of locations and the social surplus is a function of the locations selected. For example, he considers a facility location game in which service providers choose locations at which they can locate their facilities, in response to customer demand that depends on the distribution of customer locations. All of the games in Vetta's class have price of anarchy 2.

The results on nonatomic selfish routing in §8.4 are due to Christodoulou and Kout-soupias [CK05b] and Awerbuch, Azar, and Epstein [AAE13]. The smoothness framework and extension theorems described in §8.4.1 are due to Roughgarden [Rou09]. These results synthesize a host of prior price of anarchy proofs and extensions that allow for weaker assumptions on player rationality (e.g. [BHLR08, BEDL10, CK05a, GMV05, MV04, Vet02]).

The price of anarchy and smoothness notions have been extended to settings of incomplete information and to more complex mechanisms. For a taste of this topic, see §14.7, and for a detailed treatment, see, e.g., [ST13, Rou12, Syr14, HHT14, Rou14, Har17].

For more on the price of anarchy and related topics, see Chapters 17–21 of [NRTV07] and the lecture notes by Roughgarden [Rou13, Rou14].

Exercise 8.6 is due to [RT02]. Exercises 8.11, 8.12, and 8.13 are due to [Rou09]. Exercise 8.14 is from [KP09] and Exercise 8.15 is from [FLM+03].

Exercises

8.1.   Show that Theorem 8.1.8 holds in the presence of multiple traffic flows. Specifically, let G be a network where $r_i > 0$ units of traffic are routed from source s_i to destination t_i, for each $i = 1, \ldots, k$. Suppose that the latency function on each edge e is affine; that is, $\ell_e(x) = a_e x + b_e$, for constants $a_e, b_e \geq 0$. Show that the price of anarchy is at most $4/3$; that is, the total latency in equilibrium is at most $4/3$ that of the optimal flow.

8.2.   Suppose that $\mathcal{L}$ is the set of all nonnegative, weakly increasing, concave functions. Show that for this class of functions
$$\mathcal{A}_r(\mathcal{L}) \leq \frac{4}{3}.$$

8.3.   Let G be a network where one unit of traffic is routed from a source s to a destination t. Suppose that the latency function on each edge e is linear; that is, $\ell_e(x) = a_e x$, for constants $a_e \geq 0$. Show that the price of anarchy in such a network is 1. Hint: In (8.7), use the inequality $xy \leq (x^2 + y^2)/2$.

8.4.   Extend Theorem 8.1.15 to the case where there are multiple flows: Let G be a network with latency functions in $\mathcal{L}$ and total flow r_i from s_i to t_i, for $i = 1, \ldots, k$. Then the price of anarchy in G is at most $\mathcal{A}_r(\mathcal{L})$, where $r = \sum_{i=1}^{k} r_i$.

8.5.   Let $c > 0$ and suppose that f_c is the function
$$f_c(x) := \begin{cases} \frac{1}{c-x}, & 0 \leq x < c, \\ \infty, & x \geq c. \end{cases}$$

Consider a selfish routing network where all latency functions are in $\mathcal{L}$ where
$$\mathcal{L} = \{f_c(\cdot) | c > 0\}.$$
Suppose that in equilibrium, for every edge e, the flow $F_e \leq (1 - \beta)c_e$, where the latency function on edge e is $f_{c_e}(x)$. Show that the price of anarchy is upper bounded by
$$\frac{1}{2}\left(1 - \sqrt{\frac{1}{\beta}}\right).$$

8.6.   Let G be a selfish routing network in which r units of traffic are routed from source s to destination t. Suppose that $\ell_e(\cdot)$ is the latency function associated with edge $e \in E$. Consider another network G' with exactly the same network topology and the same amount of traffic being routed, but where the latency function $\ell'_e(\cdot)$ on edge e satisfies
$$\ell'_e(x) := \frac{\ell_e(x/2)}{2} \quad \forall e \in E.$$

This corresponds to doubling the capacity of the link. Suppose that $\mathbf{f}^*$ is an optimal flow in G and $\mathbf{f}'$ is an equilibrium flow in G'. Use Theorem 8.1.16 to prove that
$$L_{G'}(\mathbf{f}') \leq L_G(\mathbf{f}^*).$$

S   8.7.   Show that the price of anarchy bound for the market sharing game from §8.3 can be improved to $2 - 1/k$ when there are k teams. Show that this bound is tight.

S*8.8.   Consider an auctioneer selling a single item via a **first-price auction**[5] Each of the n bidders submits a bid, say b_i for the ith bidder, and, given the bid vector $\mathbf{b} = (b_1, \ldots, b_n)$, the auctioneer allocates the item to the highest bidder at a price equal to her bid. (The auctioneer employs some deterministic tie-breaking rule.) Each bidder has a **value** v_i for the item. A bidder's **utility** from the auction when the bid vector is $\mathbf{b}$ and her value is v_i is

$$u_i[\mathbf{b}|v_i] := \begin{cases} v_i - b_i & i \text{ wins the auction,} \\ 0 & \text{otherwise.} \end{cases}$$

Each bidder will bid in the auction so as to maximize her (expected) utility. The expectation here is over any randomness in the bidder strategies. The **social surplus** $V(\mathbf{b})$ of the auction is the sum of the utilities of the bidders and the auctioneer revenue. Since the auctioneer revenue equals the winning bid, we have

$$V(\mathbf{b}) := \text{value of winning bidder.}$$

Show that the price of anarchy is at most $1 - 1/e$; that is, for $\mathbf{b}$ a Nash equilibrium,

$$\mathbb{E}\left[V(\mathbf{b})\right] \geq \left(1 - \frac{1}{e}\right) \max_i v_i.$$

Hint: Consider instead what happens when bidder i deviates from b_i to the distribution with density $f(x) = 1/(v_i - x)$, with support $[0, (1 - 1/e)v_i]$.

8.9.   Consider a two-bidder, two-item auction, where the values of the bidders are shown in Table 1.
         What is the optimal (i.e., social surplus maximizing) allocation of items to bidders? Suppose that the seller sells the items by asking the bidders to submit one bid for each item and then running separate second-price auctions for each item. (In a second-price auction, the item is allocated to the bidder that bid the highest at a price equal to the second highest bid.) Show that there is a pure Nash equilibrium in which the social surplus is maximized. Consider the following alternative set of bids $\mathbf{b}_1 = (0, 1)$ and $\mathbf{b}_2 = (1, 0)$. Show that these bids are a Nash equilibrium and that the price of anarchy in this example is 2.

---

[5]See Chapter 14 for a detailed introduction to auctions, and Theorem 14.7.1 for a generalization of this result.

TABLE 1. The valuations of the bidders for each combination of items.

Bidder:	1	2
Items received:		
No items	0	0
Item 1	2	1
Item 2	1	2
Both items	2	2

8.10.   Consider atomic selfish routing in a Pigou network with latency x on the top edge and latency 2 on the bottom edge shown in Figure 8.12. What is the total latency for the optimal routing? Show that there are two equilibria and that they have different costs.

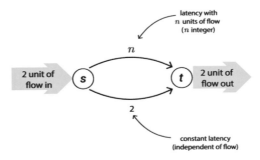

FIGURE 8.12. Figure for Exercise 8.10.

8.11.   Prove the analogue of Theorem 8.4.7, Part (1), for games in which we are interested in maximizing a global objective function such as social surplus: Consider a k-player game G. Let $V(\mathbf{s})$ be a global objective function such that

$$V(\mathbf{s}) \geq \sum_i u_i(\mathbf{s}). \qquad (8.20)$$

We say the game is (λ, μ)-smooth for strategy profile $\mathbf{s}' = (s_1', \ldots, s_k')$ if for all strategy profiles $\mathbf{s} = (s_1, \ldots, s_k)$

$$\sum_i u_i(s_i', \mathbf{s}_{-i}) \geq \lambda V(\mathbf{s}') - \mu V(\mathbf{s}). \qquad (8.21)$$

Let $\mathbf{s}^* = (s_1^*, \ldots, s_k^*)$ be a strategy profile which maximizes global objective function $V(\mathbf{s})$, and let $\mathbf{s}$ be a Nash equilibrium. Show that if G is (λ, μ)-smooth for $\mathbf{s}^*$, then

$$V(\mathbf{s}) \geq \frac{\lambda}{1 + \mu} V(\mathbf{s}^*).$$

For example, Lemma 8.3.2 shows that the market sharing game is $(1, 1)$-smooth for all strategy profiles $\mathbf{s}'$. From this, we derived Theorem 8.3.4.

8.12. Prove the analogue of Theorem 8.4.7, Part (2), for games in which we are interested in maximizing a global objective function such as social surplus: Suppose that $V(\cdot)$ is a global objective function that satisfies (8.20) and let G be a game that is (λ, μ)-smooth for $\mathbf{s}^*$, where $\mathbf{s}^*$ maximizes $V(\cdot)$. (See (8.21).) Also, let $\mathbf{p}$ be a distribution over strategy profiles that corresponds to a correlated equilibrium so that

$$\sum_{\mathbf{s}} p_{\mathbf{s}} u_i(\mathbf{s}) \geq \sum_{\mathbf{s}} p_{\mathbf{s}} u_i(s_i^*, \mathbf{s}_{-i}). \qquad (8.22)$$

Show that

$$\mathbb{E}_{\mathbf{s} \sim \mathbf{p}}[V(\mathbf{s})] \geq \frac{\lambda}{1 + \mu} V(\mathbf{s}^*).$$

8.13. Prove the analogue of Theorem 8.4.7, Part (3), for games in which we are interested in maximizing a global objective function such as social surplus: Suppose that $V(\cdot)$ is a global objective function that satisfies (8.20) and let G be a game that is (λ, μ)-smooth for $\mathbf{s}^*$, where $\mathbf{s}^*$ maximizes $V(\cdot)$. (See (8.21).) Suppose the game is played T times and each player uses a sublinear regret algorithm to determine his strategy in round t. Recall that if we let $\mathbf{s}^t = (s_1^t, s_2^t, \ldots, s_k^t)$ be the strategy profile employed by the k players in round t, then the guarantee of the algorithm is that, for any $\mathbf{s}_{-i}^t$, with $1 \leq t \leq T$,

$$\sum_{t=1}^{T} u_i(s_i^t, \mathbf{s}_{-i}^t) \geq \max_{s \in S_i} \sum_{t=1}^{T} u_i(s, \mathbf{s}_{-i}^t) - o(T). \qquad (8.23)$$

Show that

$$\frac{1}{T} \sum_{t=1}^{T} V(\mathbf{s}^t) \geq \left(\frac{\lambda}{1 + \mu} - o(1) \right) V(\mathbf{s}^*).$$

8.14. Suppose that there are n jobs, each job owned by a different player, and n machines. Each player chooses a machine to run its job on, and the cost that player incurs is the load on the machine, i.e., the number of jobs that selected that machine, since that determines the latency that player experiences. Suppose that it is desirable to minimize the maximum load (number of jobs) assigned to any machine. This is called the *makespan* of the allocation. Clearly it is a Nash equilibrium for each job to select a different machine, an allocation which achieves the optimal makespan of 1.
   - Show that it is a mixed-strategy Nash equilibrium for each player to select a random machine.
   - Show that the price of anarchy for this mixed Nash equilibrium (i.e., the expected makespan it achieves divided by the optimal makespan) is $\Theta(\log n / \log \log n)$.

8.15. Consider the following network formation game: There are n vertices each representing a player. The pure strategy of a player consists of choosing which other vertices to create a link to. A strategy profile induces a graph, where each edge is associated with the vertex that "created" it. Given a strategy profile $\mathbf{s}$. the cost incurred by player i is

$$\text{cost}_i(\mathbf{s}) := \alpha \cdot n_i(s_i) + \sum_{j \neq i} d_\mathbf{s}(i, j),$$

where $n_i(s_i)$ is the number of links i created (each link costs α to create) and $d_\mathbf{s}(i, j)$ is the distance from i to j in the graph resulting from strategy profile $\mathbf{s}$.

- Show that if $\alpha \geq 2$, then the graph which minimizes $\sum_i \text{cost}_i$ is a star, whereas if $\alpha < 2$, then it is a complete graph.
- Show that for $\alpha \leq 1$ or $\alpha > 2$, there is a Nash equilibrium with total cost equal to that of the optimum graph.
- Show that for $1 < \alpha < 2$, there is a Nash equilibrium with total cost at most $4/3$ times that of the optimum graph.

CHAPTER 9

Random-turn games

In Chapter 1 we considered combinatorial games, in which the right to move alternates between players; and in Chapter 2 and Chapter 4 we considered matrix-based games, in which both players (usually) declare their moves simultaneously and possible randomness decides what happens next. In this chapter, we consider some games which are combinatorial in nature, but the right to make the next move is determined by a random coin toss.

Let S be an n-element set, which will sometimes be called the *board*, and let f be a function from the 2^n subsets of S to $\mathbb{R}$. A *selection game* is played as follows: The first player selects an element of S, the second player selects one of the remaining $n-1$ elements, the first player selects one of the remaining $n-2$ elements, and so forth, until all elements have been chosen. Let S_1 and S_2 signify the sets chosen by the first and second players, respectively. Then player I receives a payoff of $f(S_1)$ and player II receives a payoff of $-f(S_1)$. Thus, selection games are zero-sum.

9.1. Examples

EXAMPLE 9.1.1 (**Random-turn Hex**). Let S be the set of hexagons on a rhombus-shaped $L \times L$ hexagonal grid. Define $f(S_1)$ to be 1 if S_1 contains a crossing connecting the two yellow sides, -1 otherwise. In this case, once S_1 contains a yellow crossing or S_2 contains blue crossing (which precludes the possibility of S_1 having a yellow crossing), the outcome is determined and there is no need to continue the game.

EXAMPLE 9.1.2 (**Team captains**). Two team captains are choosing baseball teams from a finite set S of n players for the purpose of playing a single game against each other. The payoff $f(S_1)$ for the first captain is the probability that the players in S_1 (together with the first captain) will win against the players in S_2 (together with the second captain). The payoff function may be very complicated (depending on which players know which positions, which players have played together before, which players get along well with which captain, etc.). Because we have not specified the payoff function, this game is as general as the class of selection games.

Portions of "Random-Turn Hex and Other Selection Games", Yuval Peres, Oded Schramm, Scott Sheffield, and David B. Wilson, The American Mathematical Monthly, Vol. 114, No. 5 (May 2007), pp. 373–387, have been used in this chapter, with permission.

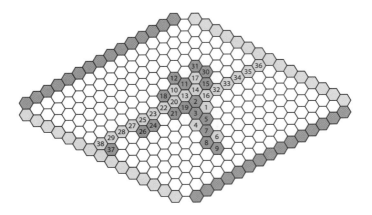

FIGURE 9.1. Random-turn Hex played on a 15×15 board.

EXAMPLE 9.1.3 (**Recursive Majority**). Suppose we are given a complete ternary tree of depth h. Let S be the set of leaves. In each step, a fair coin toss determines which player selects a leaf to label. Leaves selected by player I are marked with a $+$ and leaves selected by player II are marked with a $-$. A parent node in the tree acquires the same sign as the majority of its children. The player whose mark is assigned to the root wins. Example 1.2.15 discusses the alternating-turn version of this game.

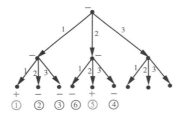

FIGURE 9.2. Here player II wins; the circled numbers give the order of the moves.

As we discussed in Chapter 1, determining optimal strategies in alternating move selection games, e.g., Hex, can be hard. Surprisingly, the situation is different in random-turn selection games.

9.2. Optimal strategy for random-turn selection games

A (pure) *strategy* for a given player in a random-turn selection game is a function M which maps each pair of disjoint subsets (T_1, T_2) of S to an element of $T_3 := S \backslash (T_1 \cup T_2)$, provided $T_3 \neq \emptyset$. Thus, $M(T_1, T_2)$ indicates the element that the player will pick if given a turn at a time in the game when player I has thus far picked the elements of T_1 and player II has picked the elements of T_2.

Denote by $E(T_1, T_2)$ the expected payoff for player I at this stage in the game, assuming that both players play optimally with the goal of maximizing expected payoff. As is true for all finite perfect-information, two-player games, E is well defined, and one can compute[1] E and the set of possible optimal strategies by induction on the size of T_3. First, if $T_3 = \emptyset$, then $E(T_1, T_2) = f(T_1)$. Next, suppose that we have computed $E(T_1, T_2)$ whenever $|T_3| \leq k$. Then if $|T_3| = k + 1$ and player I has the chance to move, player I will play optimally if and only if she chooses an s from T_3 for which $E(T_1 \cup \{s\}, T_2)$ is maximal. Similarly, player II plays optimally if and only if he minimizes $E(T_1, T_2 \cup \{t\})$ at each stage. Hence

$$E(T_1, T_2) = \frac{1}{2} \Big(\max_{s \in T_3} E(T_1 \cup \{s\}, T_2) + \min_{t \in T_3} E(T_1, T_2 \cup \{t\}) \Big). \qquad (9.1)$$

We will see that the maximizing and the minimizing moves are actually the same.

The foregoing analysis also demonstrates a well-known fundamental fact about finite, turn-based, perfect-information games: Both players have optimal pure strategies.[2] (This contrasts with the situation in which the players play "simultaneously" as they do in Rock-Paper-Scissors.)

THEOREM 9.2.1. *The value of a random-turn selection game is the expectation of $f(T)$ when a set T is selected randomly and uniformly among all subsets of S. Moreover, any optimal strategy for one of the players is also an optimal strategy for the other player.*

PROOF. For any player II strategy, player I can achieve the expected payoff $\mathbb{E}[f(T)]$ by playing exactly the same strategy (since, when both players play the same strategy, each element will belong to S_1 with probability $1/2$, independently). Thus, the value of the game is at least $\mathbb{E}[f(T)]$. However, a symmetric argument applied with the roles of the players interchanged implies that the value is no more than $\mathbb{E}[f(T)]$.

Since the remaining game in any intermediate configuration (T_1, T_2) is itself a random-turn selection game, it follows that for every T_1 and T_2

$$E(T_1, T_2) = \mathbb{E}(f(T_1 \cup T)),$$

where T is a uniform random subset of T_3. Thus, for every $s \in T_3$,

$$E(T_1, T_2) = \frac{1}{2} \Big(E(T_1 \cup \{s\}, T_2) + E(T_1, T_2 \cup \{s\}) \Big).$$

Therefore, if $s \in T_3$ is chosen to maximize $E(T_1 \cup \{s\}, T_2)$, then it also minimizes $E(T_1, T_2 \cup \{s\})$. We conclude that every optimal move for one of the players is an optimal move for the other. $\square$

If both players break ties the same way, then the final S_1 is equally likely to be any one of the 2^n subsets of S.

Theorem 9.2.1 is quite surprising. In the baseball team selection, for example, one has to think very hard in order to play the game optimally, knowing that at each stage the opponent can capitalize on any miscalculation. Yet, despite all of that mental effort by the team captains, the final teams look no different than they would look if at each step both captains chose players uniformly at random.

---

[1]This method is called dynamic programming.
[2]See also Theorem 6.1.7.

For example, suppose that there are only two players who know how to pitch and that a team without a pitcher always loses. In the alternating-turn game, a captain can always wait to select a pitcher until just after the other captain selects a pitcher. In the random-turn game, the captains must try to select the pitchers in the opening moves, and there is an even chance the pitchers will end up on the same team.

Theorem 9.2.1 generalizes to random-turn selection games in which the player to get the next turn is chosen using a biased coin. If player I gets each turn with probability p, independently, then the value of the game is $\mathbb{E}[f(T)]$, where T is a random subset of S for which each element of S is in T with probability p, independently. The proof is essentially the same.

9.3. Win-or-lose selection games

We say that a game is a *win-or-lose* game if $f(T)$ takes on precisely two values, which we assume to be -1 and 1. If $S_1 \subset S$ and $s \in S$, we say that s is *pivotal* for S_1 if $f(S_1 \cup \{s\}) \neq f(S_1 \setminus \{s\})$. A selection game is *monotone* if f is monotone; that is, $f(S_1) \geq f(S_2)$ whenever $S_1 \supset S_2$. Hex is an example of a monotone, win-or-lose game. For such games, the optimal moves have the following simple description.

LEMMA 9.3.1. *In a monotone, win-or-lose, random-turn selection game, a first move s is optimal if and only if s is an element of S that is most likely to be pivotal for a random-uniform subset T of S. When the position is (S_1, S_2), the move s in $S \setminus (S_1 \cup S_2)$ is optimal if and only if s is an element of $S \setminus (S_1 \cup S_2)$ that is most likely to be pivotal for $S_1 \cup T$, where T is a random-uniform subset of $S \setminus (S_1 \cup S_2)$.*

PROOF. This follows from monotonicity and the discussion of optimal strategies preceding (9.1).   □

For win-or-lose games, such as Hex, the players may stop making moves after the winner has been determined, and it is interesting to calculate how long a random-turn, win-or-lose, selection game will last when both players play optimally. Suppose that the game is a monotone game and that, when there is more than one optimal move, the players break ties in the same way. Then we may take the point of view that the playing of the game is a (possibly randomized) decision procedure for evaluating the payoff function f when the items in S are randomly allocated. Let $\vec{x}$ denote the allocation of the items, where $x_i = \pm 1$ according to whether the ith item goes to the first or second player. We may think of the x_i as input variables, and the playing of the game is one way to compute $f(\vec{x})$. The number of turns played is the number of variables of $\vec{x}$ examined before $f(\vec{x})$ is computed. We use some inequalities from the theory of Boolean functions to bound the average length of play.

DEFINITION 9.3.2. Let $f(\mathbf{x})$ be a Boolean function on n variables taking values in $\{-1, 1\}$. The **influence** $I_i(f)$ of the variable x_i on $f(\mathbf{x})$ is the probability that flipping x_i will change the value of $f(\mathbf{x})$, where $\mathbf{x}$ is uniformly distributed on $\{-1, 1\}^n$. Thus, recalling the notation $f(x_i, \mathbf{x}_{-i}) := f(\mathbf{x})$,

$$I_i(f) := \frac{1}{2^n} \sum_{\mathbf{x}_{-i}} \left| f(1, \mathbf{x}_{-i}) - f(-1, \mathbf{x}_{-i}) \right|.$$

For **monotone** functions, which have $f(1, \mathbf{x}_{-i}) \geq f(-1, \mathbf{x}_{-i})$ for all $\mathbf{x}_{-i}$, it follows that

$$I_i(f) := \frac{1}{2^n} \sum_{\mathbf{x}_{-i}} \Big(f(1, \mathbf{x}_{-i}) - f(-1, \mathbf{x}_{-i}) \Big) = \mathbb{E}\left[f(\vec{x})x_i\right]. \qquad (9.2)$$

DEFINITION 9.3.3. Given an unknown vector $\mathbf{x} = (x_1, \ldots, x_n)$, a *decision tree* for calculating $f(\mathbf{x})$ is a procedure for deciding, given the values of the variables already examined, which variable to examine next. The procedure ends when the value of $f(\mathbf{x})$ is determined. For example, if $n = 3$ and $f(\mathbf{x})$ is the majority function, then such a procedure might first examine x_1 and x_2. The third variable x_3 only needs to be examined if $x_1 \neq x_2$.

LEMMA 9.3.4. *Let* $f(\mathbf{x}) : \{-1, 1\}^n \to \{-1, 1\}$ *be a monotone function. Consider any decision tree for calculating* $f(\mathbf{x})$, *when* $\mathbf{x}$ *is selected uniformly at random from* $\{-1, 1\}^n$. *Then*

$$\mathbb{E}[\# \ variables \ examined] \geq \left[\sum_i I_i(f)\right]^2.$$

PROOF. Using (9.2) and Cauchy-Schwarz, we have

$$\sum_i I_i(f) = \mathbb{E}\left[\sum_i f(\vec{x})x_i\right] = \mathbb{E}\left[f(\vec{x}) \sum_i x_i 1_{x_i \text{ examined}}\right]$$

$$\leq \sqrt{\mathbb{E}[f(\vec{x})^2] \, \mathbb{E}\left[\left(\sum_{i: \ x_i \text{ examined}} x_i\right)^2\right]}$$

$$= \sqrt{\mathbb{E}\left[\left(\sum_{i: \ x_i \text{ examined}} x_i\right)^2\right]} = \sqrt{\mathbb{E}[\# \text{ bits examined}]}.$$

The last equality is justified by noting that $\mathbb{E}[x_i x_j 1_{x_i \text{ and } x_j \text{ both examined}}] = 0$ when $i \neq j$, which holds since conditioned on x_i being examined before x_j, conditioned on the value of x_i, and conditioned on x_j being examined, the expected value of x_j is zero. $\qquad \square$

Lemma 9.3.4 implies that in a win-or-lose random-turn selection game,

$$\mathbb{E}[\# \text{ turns}] \geq \left[\sum_i I_i(f)\right]^2.$$

9.3.1. Length of play for random-turn Recursive Majority. To apply Lemma 9.3.4 to Example 9.1.3, we need to compute the probability that flipping the sign of a given leaf changes the overall recursive majority. For any given node, the probability that flipping its sign will change the sign of its parent is just the probability that the signs of the other two siblings are distinct, which is $1/2$.

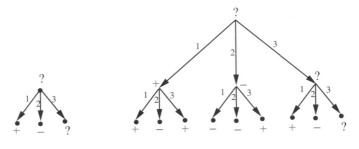

FIGURE 9.3

This holds all along the path to the root, so the probability that flipping the sign of leaf i will flip the sign of the root is just $I_i(f) = \left(\frac{1}{2}\right)^h$, where h is the height of the tree. Thus, since there are 3^h leaves,

$$\mathbb{E}[\# \text{ turns}] \geq \left[\sum_i I_i(f)\right]^2 = \left(\frac{3}{2}\right)^{2h}.$$

Notes

This chapter presents the work in [PSSW07]. Indeed, as we could not improve on the exposition there (mostly due to Oded Schramm, Scott Sheffield and David Wilson), we follow it almost verbatim. As noted in that paper, the game of Random-turn Hex was proposed by Wendelin Werner on a hike. Lemma 9.3.4 is from O'Donnell and Servedio [OS07]. Figure 9.1 and Figure 9.4 are due to David Wilson.

Figure 9.4 shows two optimally played games of Random-turn Hex. It is not known what is the expected length of such a game on an $L \times L$ board. [PSSW07] shows that it is at least $L^{3/2+o(1)}$, but the only upper bound known is the obvious one of L^2.

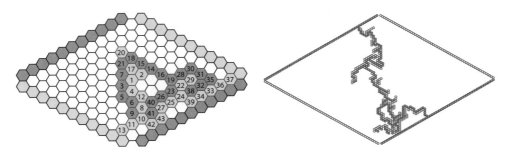

FIGURE 9.4. Random-turn Hex on boards of size 11×11 and 63×63 under (nearly) optimal play. (The caveat "nearly" is there because the probability that a hexagon is pivotal was estimated by Monte Carlo simulation.)

Exercises

9.1. Generalize the proof of Theorem 9.2.1 further so as to include the following two games:

    (a) Restaurant selection:
        Two people (with opposite food preferences) want to select a dinner location. They begin with a map containing 2^n distinct points in $\mathbb{R}^2$, indicating restaurant locations. At each step, the person who wins a coin toss may draw a straight line that divides the set of remaining restaurants exactly in half and eliminate all the restaurants on one side of that line. Play continues until one restaurant z remains, at which time player I receives payoff $f(z)$ and player II receives $-f(z)$.

    (b) Balanced team captains:
        Suppose that the captains wish to have the final teams equal in size (i.e., there are $2n$ players and we want a guarantee that each team will have exactly n players in the end). Then instead of tossing coins, the captains may shuffle a deck of $2n$ cards (say, with n red cards and n black cards). At each step, a card is turned over and the captain whose color is shown on the card gets to choose the next player.

9.2. Recursive Majority on b-ary trees: Let $b = 2r + 1$, $r \in \mathbb{N}$. Consider Recursive Majority on a b-ary tree of depth h. For each leaf, determine the probability that flipping the sign of that leaf would change the overall result (i.e., the influence of that leaf).

Part II: Designing games and mechanisms

CHAPTER 10

Stable matching and allocation

In 1962, David Gale and Lloyd Shapley published a seminal paper entitled
"College Admissions and the Stability of Marriage" [GS62]. This led to a rich
theory with numerous applications exploring the fundamental question of how to
find stable matchings, whether they are of men with women, students with schools,
or organ donors with recipients needing a transplant. In 2012, the Nobel Prize in
Economics was awarded to Alvin Roth and Lloyd Shapley for their research on "the
theory of stable allocations and the practice of market design'." In this chapter,
we describe and analyze stable allocation algorithms, including the Gale-Shapley
algorithm for stable matching.

10.1. Introduction

Suppose there are n men and m women. Every man has a preference order
over the m women, while every woman also has a preference order over the n men.
A **matching** is a one-to-one mapping from a subset of the men to a subset of the
women. A matching M is **unstable** if there exists a man and a woman who are
not matched to each other in M but prefer each other to their partners in M. We
assume every individual prefers being matched to being unmatched.[1] Otherwise,
the matching is called **stable**. Clearly, in any stable matching, the number of
matched pairs is $\min(n, m)$.

FIGURE 10.1. An unstable pair.

---

[1]Thus, there are four kinds of instability: (1) Alice and Bob are both matched in M but
prefer each other to their current matches, (2) Alice prefers Bob to her match in M and Bob is
unmatched in M, (3) similarly, with roles reversed, (4) both Alice and Bob are unmatched by M.

Consider the example shown in Figure 10.2 with three men x, y, and z, and three women a, b, and c. Their preference lists are:

$$x : a > b > c, \quad y : b > c > a, \quad z : a > c > b,$$

$$a : y > z > x, \quad b : y > z > x, \quad c : x > y > z.$$

Then, $x \leftrightarrow a$, $y \leftrightarrow b$, $z \leftrightarrow c$ is an unstable matching since z and a prefer each other to their partners.

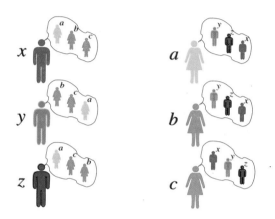

FIGURE 10.2. The figure shows three men and three women and their preference lists. For example, the green man y prefers b to c to a.

In the next section, we show that stable matchings exist for *any* preference profile and present an efficient algorithm for finding such a matching.

10.2. Algorithms for finding stable matchings

The following algorithm, called the **men-proposing algorithm**, was introduced by Gale and Shapley. At any point in time, there is some number of tentatively matched pairs.

(1) Initially all men and women are unmatched.
(2) Each man proposes to his most preferred woman who has not rejected him yet (or gives up if he's been rejected by all women).
(3) Each woman is tentatively matched to her favorite among her proposers and rejects the rest.
(4) Repeat steps (2) and (3) until a round in which there are no rejections. At that point the tentative matches become final.

REMARK 10.2.1. Note that if a man is tentatively matched to a woman in round k and the algorithm doesn't terminate, then he necessarily proposes to her again in round $k + 1$.

OBSERVATION 10.2.2. *From the first time a woman is proposed to, she remains tentatively matched (and is permanently matched at the end). Moreover, each tentative match is at least as good as the previous one from her perspective.*

THEOREM 10.2.3. *The men-proposing algorithm yields a stable matching.*

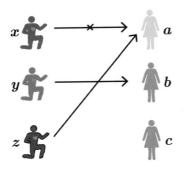

FIGURE 10.3. Arrows indicate proposals; cross indicates rejection.

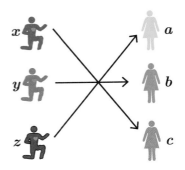

FIGURE 10.4. Stable matching is achieved in the second stage.

PROOF. The algorithm terminates because in every nonfinal round there is a rejection, and there are at most nm rejections possible. When it terminates, it clearly yields a matching, which we denote by M. To see that M is stable, consider a man, Bob, and a woman, Alice, not matched to each other, such that Bob prefers Alice to his match in M or Bob is single. This means that he was rejected by Alice at some point before the algorithm terminated. But then, by Observation 10.2.2, Alice is matched in M and prefers her match in M to Bob.   □

COROLLARY 10.2.4. *In the case $n = m$, the stable matching is* **perfect***; that is, all men and women are matched.*

REMARK 10.2.5. We could similarly define a women-proposing algorithm.

10.3. Properties of stable matchings

We say a woman j is **attainable** for a man i if there exists a stable matching M with $M(i) = j$.

THEOREM 10.3.1. *Let M be the stable matching produced by the men-proposing algorithm. Then*

  (a) *Every man is matched in M to his most preferred attainable woman.*
  (b) *Every woman is matched in M to her least preferred attainable man.*

PROOF. We prove (a) by contradiction. Suppose that M does not match each man with his most preferred attainable woman. Consider the first time during the execution of the algorithm that a man, say Bob, is rejected by his most preferred attainable woman, Alice, and suppose that Alice rejects Bob at that moment for David. Since this is the first time a man is rejected by his most preferred attainable woman,

David likes Alice at least as much as his most preferred attainable woman.

(10.1)

Also, since Alice is Bob's most preferred attainable women, there is another stable matching M' in which they are matched. In M', David is matched to someone other than Alice. But now we have derived a contradiction: By (10.1), David likes

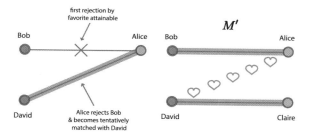

FIGURE 10.5. This figure shows the contradiction that results from assuming that some man, in this case, Bob, is the first to be rejected by his favorite attainable woman, Alice, when running the men-proposing algorithm. M' is the stable matching in which Bob and Alice are matched.

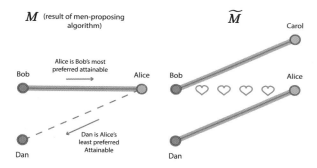

FIGURE 10.6. This figure shows the contradiction that results from assuming that in the men-proposing algorithm some woman, Alice, does not end up with her least preferred attainable, in this case Dan. $\widetilde{M}$ is the matching in which Alice is matched to Dan.

Alice more than his match in M', and Alice prefers David to Bob. Thus M' is unstable. (See Figure 10.5.)

We also prove part (b) by contradiction. Suppose that in M, Alice ends up matched to Bob, whom she prefers over her least preferred attainable man, Dan. Then, there is another stable matching $\widetilde{M}$ in which Dan and Alice are matched, and Bob is matched to a different woman, Carol. Then in $\widetilde{M}$, Alice and Bob are an unstable pair: By part (a), in M, Bob is matched to his most preferred attainable woman. Thus, Bob prefers Alice to Carol, and by assumption Alice prefers Bob to Dan. (See Figure 10.6.) □

COROLLARY 10.3.2. *If Alice is assigned to the same man in both the men-proposing and the women-proposing version of the algorithm, then this is the only attainable man for her.*

COROLLARY 10.3.3. *The set of women (and men) who get matched is the same in all stable matchings.*

PROOF. Consider the set of women matched by M, the matching resulting from the men-proposing algorithm. Suppose that one of these women, say Alice, is unmatched in some stable matching $\tilde{M}$. Then Bob, whom she was matched to in M, prefers her to whomever he is matched to in $\tilde{M}$, a contradiction. Since the number of matched women in both matchings is the same, namely $\min(n,m)$, this concludes the proof. $\qquad\square$

10.3.1. Preferences by compatibility. Suppose we seek stable matchings for n men and n women with preference order determined by a matrix $A = (a_{i,j})_{n\times n}$ where all entries in each row are distinct and all entries in each column are distinct. If in the ith row of the matrix we have

$$a_{i,j_1} > a_{i,j_2} > \cdots > a_{i,j_n},$$

then the preference order of man i is $j_1 > j_2 > \cdots > j_n$. Similarly, if in the jth column we have

$$a_{i_1 j} > a_{i_2 j} > \cdots > a_{i_n j},$$

then the preference order of woman j is $i_1 > i_2 > \cdots > i_n$. (Imagine that the number a_{ij} represents the compatibility of man i and woman j.)

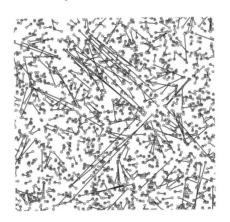 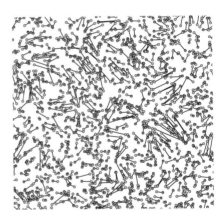

FIGURE 10.7. The left-hand figure shows a stable matching between red points $(x_i)_1^n$ and blue points $(y_j)_1^n$ randomly placed on a torus. Preferences are according to distance; the shorter the better. Thus, $a_{ij} = M - \text{dist}(x_i, y_j)$. The right-hand figure shows the minimum weight matching between the red points and the blue points.

LEMMA 10.3.4. *In this case, there exists a unique stable matching.*

PROOF. By Theorem 10.3.1, we know that the men-proposing algorithm produces a stable matching M in which each man obtains his most preferred attainable partner. In all other stable matchings, each man obtains at most the same value and at least one man obtains a lower value. Therefore, M is the unique maximizer of $\sum_i a_{i,M(i)}$ among all stable matchings. Similarly, the women-proposing algorithm produces a stable matching which maximizes $\sum_j a_{M^{-1}(j),j}$ among all stable matchings. Thus, the stable matchings produced by the two algorithms are the same. By Corollary 10.3.2, there exists a unique stable matching. $\qquad\square$

10.3.2. Truthfulness. Exercise 10.3 shows that if the men-proposing algorithm is implemented, a woman might benefit by misrepresenting her preferences. Our next goal is to show that in this setting, no man is incentivized to do so.

LEMMA 10.3.5. *Let μ be the men-optimal stable matching[2] and let ν be another matching. Denote by S the set of men who prefer their match in ν to their match in μ, i.e.,*

$$S := \{m \mid \nu(m) >_m \mu(m)\}. \tag{10.2}$$

Then there is a pair (m, w) which is unstable for ν, where $m \notin S$.

PROOF. We consider the execution of the men-proposing algorithm that generates μ.

Case 1: $\mu(S) \neq \nu(S)$ (i.e., the set of women matched to men in S is not the same in μ and ν): Let $w \in \nu(S) \setminus \mu(S)$ and let $m = \mu(w)$. Then (m, w) is unstable for ν: First, $m \notin S$, so m prefers w to $\nu(m)$. Second, w rejected $\nu(w)$ during the execution of the algorithm, so w prefers m to $\nu(w)$.

Case 2: $\mu(S) = \nu(S) = W_0$: Every woman in W_0 receives and rejects a proposal from her match in ν. Let w be the last woman in W_0 to receive a proposal from a man in S. Then w was tentatively matched to a man, say m, when she received this last proposal. Observe that $m \notin S$; otherwise, at some point after being rejected by w, he would have proposed to $\mu(m) \in W_0$, resulting in a later proposal in W_0. We claim (m, w) is unstable for ν: First, $w >_m \mu(m) \geq_m \nu(m)$. Second, $\nu(w)$ proposed to w and was rejected earlier than m, so w prefers m to $\nu(w)$. □

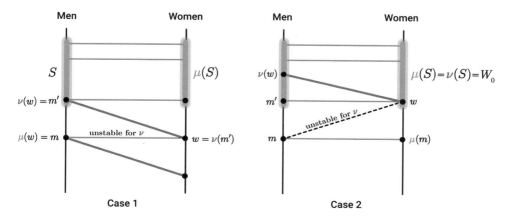

FIGURE 10.8. Red edges are in the male-optimal matching μ and blue edges are in the matching ν.

COROLLARY 10.3.6. *Let μ denote the men-optimal stable matching. Suppose that a set S_0 of men misrepresent their preferences. Then there is no stable matching for the resulting preference profile where all men in S_0 obtain strictly better matches than in μ (according to their original preference order).*

---

[2]that arises from the men-proposing algorithm

PROOF. Suppose that ν was such a matching. Then S as defined in (10.2) contains S_0. The pair (m, w) produced in Lemma 10.3.5 is unstable for both preference profiles because $m \notin S$. $\qquad\square$

10.4. Trading agents

The theory of stable matching concerns two-sided markets where decisions are made by both sides. Matching and allocation problems also arise in one-sided markets, for instance, workers trading shifts or teams trading players. For concreteness, we'll use the example of first-year graduate students being assigned offices.

Consider a set of n grad students, each initially assigned a distinct office when they arrive at graduate school. Each student has a total order over the offices. Two people who prefer each other's office would naturally swap. More generally, any permutation $\pi : [n] \to [n]$ defines an allocation where person i receives office $\pi(i)$ (i.e., the office originally assigned to person $\pi(i)$). Such an allocation is called **unstable** if there is a nonempty subset $A \subset [n]$ and a permutation $\sigma : A \to A$ (that is not identical to π on A) such that for each $i \in A$ where $\sigma(i) \neq \pi(i)$, person i prefers $\sigma(i)$ to $\pi(i)$. Otherwise, π is stable.

Is there always a stable allocation and, if so, how do we find it? The following **top trading cycle algorithm** finds such a stable allocation:

Define S_k inductively as follows:
Let $S_1 = [n]$. For each $k \geq 1$, as long as $S_k \neq \emptyset$:
- Let each person $i \in S_k$ point to her most preferred office in S_k, denoted $f_k(i)$.
- The resulting directed graph (with a vertex for each person and an edge from vertex i to vertex j if j's office is i's favorite in S_k) has one outgoing edge (it could be a self-loop) from each vertex, so it must contain directed cycles. Call their union C_k. Allocate according to these cycles; i.e., for each $i \in C_k$, set $\pi(i) = f_k(i)$.
- Set $S_{k+1} = S_k \setminus C_k$.

LEMMA 10.4.1.
(1) *The top trading cycle algorithm produces a stable allocation π.*
(2) *No person has an incentive to misreport her preferences.*

PROOF. (1): Fix a subset A of students and a permutation $\sigma : A \to A$ that defines an instability. Let $A_1 = \{i \in A : \sigma(i) \neq \pi(i)\}$, and let k be minimal such that there is a $j \in C_k \cap A_1$. Then $\sigma(j) \in S_k$, so j prefers $\pi(j)$ to $\sigma(j)$.

(2): Fix the reports of all people except person ℓ, and suppose that ℓ is in C_i. For any $j < i$ all people in C_j prefer their assigned office to office ℓ. Thus, person ℓ cannot obtain an office in $\cup_{j<i} C_j$ by misreporting her preferences. Since ℓ prefers her assigned office $\pi(\ell)$ to all the remaining offices S_i, this concludes the proof. $\quad\square$

REMARK 10.4.2. There is a unique stable allocation. See Exercise 10.13.

Notes

The stable matching problem was introduced and solved in a seminal paper by David Gale and Lloyd Shapley [GS62], though stable matching algorithms were developed and used as early as 1951 to match interns to hospitals [Sta53]. The shortest proof of the

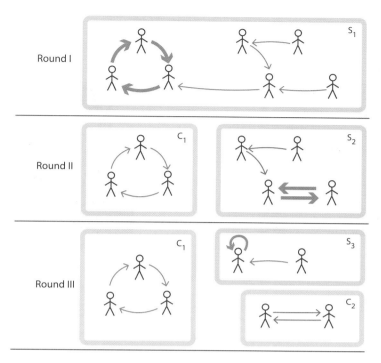

FIGURE 10.9. The figure shows the first few rounds of the top trading cycle algorithm.

existence of stable marriages is due to Marilda Sotomayor [Sot96]. See Exercise 10.6. The results on truthfulness in §10.3.2 were discovered by [DF81] and independently by [Rot82]. The proof we present is due to [GS85]. Roth [Rot82] gives an example showing that there is no mechanism to select a stable matching which incentivizes all participants to be truthful. Dubins and Freedman [DF81] present an example due to Gale where two men can falsify their preferences so that in the men-proposing algorithm, one will end up better off and the other's match will not change.

David Gale

Lloyd Shapley

See the books by Roth and Sotomayor [RS92], Gusfield and Irving [GI89], and Knuth [Knu97] for more information on the topic. For many examples of stable matching in the real world, see [Rot15].

The trading agents problem was introduced by Shapley and Scarf [SS74]. They attribute the top trading cycle algorithm to David Gale.

For a broader survey of these topics, see [NRTV07, Chapter 10] by Schummer and Vohra.

          Alvin Roth   Marilda Sotomayor

We learned about Exercise 10.9 from Alexander Holroyd. Exercise 10.11 is due to Al Roth [Rot86]. Exercise 10.12 is from [RRVV93]. The idea of using hearts in pictorial depictions of stable matching algorithms (as we have done in Figure 10.5 and Figure 10.6) is due to Stephen Rudich.

Figure 10.7 is from [HPPS09], where the distribution of distances for stable matchings on the torus is studied. An extension to stable allocation (see Figure 10.10) was analyzed in [HHP06].

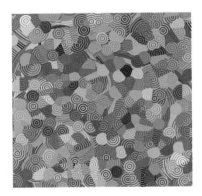

FIGURE 10.10. Given n random points on the torus, there is a unique stable allocation that assigns each point the same area where preferences are according to distance.

Exercises

10.1.   There are three men, called a, b, c, and three women, called x, y, z, with the following preference lists (most preferred on left):

for a :	$x > y > z,$	for x :	$c > b > a,$
for b :	$y > x > z,$	for y :	$a > b > c,$
for c :	$y > x > z,$	for z :	$c > a > b.$

Find the stable matchings that will be produced by the men-[roposing and by the women-proposing algorithm.

10.2.   Consider an instance of the stable matching problem, and suppose that M and M' are two distinct stable matchings. Show that the men who prefer their match in M to their match in M' are matched in M to women who prefer their match in M' to their match in M.

10.3.   Give an instance of the stable matching problem in which, by lying about her preferences during the execution of the men-proposing algorithm, a woman can end up with a man that she prefers over the man she would have ended up with had she told the truth.

10.4.   Consider a stable matching instance with n men and n women. Show that there is no matching (stable or not) that all men prefer to the male-optimal stable matching. Hint: In the men-proposing algorithm, consider the last woman to receive an offer.

10.5.   Consider the setting of §10.3.1, where the preferences are determined by a matrix. The Greedy Algorithm for finding a matching chooses the (i, j) for which a_{ij} is maximum, matches woman i to man j, removes row i and column j from the matrix, and repeats inductively on the resulting matrix. Show that the Greedy Algorithm finds the unique stable matching in this setting. Show also that the resulting stable matching is not necessarily a maximum weight matching.

10.6.   Consider an instance of the stable matching problem with n men and m women. Define a (partial) matching to be *simple* if the only unstable pairs involve an unmatched man. There exist simple matchings (e.g., the empty matching). Given a matching M, let r_j be the number of men that woman j prefers to her match in M (with $r(j) = n$ if she is unmatched). Let M^* be a simple matching with minimum $\sum_{j=1}^{m} r(j)$. Show that M^* is stable.

10.7.   Show that there is not necessarily a solution to the **stable roommates** problem: In this problem, there is a set of $2n$ people, each with a total preference order over all the remaining people. A matching of the people (each matched pair will become roommates) is stable if there is no pair of people that are not matched that prefer to be roommates with each other over their assigned roommate in the matching.

10.8.   Show that the Greedy Algorithm (defined in Exercise 10.5) gives a solution to the stable roommates problem when preferences are given by a matrix.

10.9.   Consider $2n$ points in the plane, n red and n blue. Alice and Bob play the
          following game, in which they alternate moves starting with Alice. Alice
          picks a point a_1 of either color, say red. Then Bob picks a point b_1 of the
          other color, in this case, blue. Then Alice picks a red point a_2, but this
          point must be closer to b_1 than a_1 is. They continue like this, alternating,
          with the requirement that the ith point that Alice picks is closer to b_{i-1}
          than a_{i-1} was, and the ith point b_i that Bob picks is closer to a_i than
          b_{i-1} was. The first person who can no longer pick a point that is closer to
          the other one's point loses. Show that the following strategy is a winning
          strategy for Bob: At each step pick the point b_i that is matched to a_i
          in the unique stable matching for the instance, where each point prefers
          points of the other color that are closer to it.

10.10.   Consider using stable matching in the National Resident Matching
          Program, for the problem of assigning medical students (as residents) to
          hospitals. In this setting, there are n hospitals and m students. Each
          hospital has a certain number of positions for residents, say k_i for hospital
          i. Each hospital has a ranking of all the students, and each student has a
          ranking of all the hospitals. Given an assignment of students to hospitals,
          a pair (H, s) is unstable if hospital H prefers student s to one of its
          assigned students (or has an unfilled slot), and s prefers hospital H to his
          current assignment. Describe an algorithm for finding a stable assignment
          (e.g., by reducing it to the stable matching problem).

10.11.   In the setting of the previous problem, show that if hospital H has at least
          one unfilled slot, then the set of students assigned to H is the same in all
          stable assignments.

10.12.   Consider the following integer programming[3] formulation of the stable
          matching problem. To describe the program, we use the following notation.
          Let m be a particular man and w a particular women. Then $j >_m w$ repre-
          sents the set of all women j that m prefers over w, and $i >_w m$ represents
          the set of all men i that w prefers over m. In the following program the
          variable x_{ij} will be selected to be 1 if man i and woman j are matched in
          the matching selected:

----

[3]Integer programming is linear programming in which the variables are required to take
integer values. See Appendix A for an introduction to linear programming.

$$\text{maximize} \sum_{i,j} x_{ij}$$

$$\text{subject to} \quad \sum_j x_{m,j} \leq 1 \text{ for all men } m, \tag{10.3}$$

$$\sum_i x_{i,w} \leq 1 \text{ for all women } w,$$

$$\sum_{j >_m w} x_{m,j} + \sum_{i >_w m} x_{i,w} + x_{m,w} \geq 1 \text{ for all pairs } (m,w),$$

$$x_{m,w} \in \{0,1\} \text{ for all pairs } (m,w).$$

- Prove that this integer program is a correct formulation of the stable matching problem.
- Consider the relaxation of the integer program that allows *fractional* stable matchings. It is identical to the above program, except that instead of each $x_{m,w}$ being either 0 or 1, $x_{m,w}$ is allowed to take any real value in $[0,1]$. Show that the following program is the dual program to the relaxation of (10.3).

$$\text{minimize} \sum_i \alpha_i + \sum_j \beta_j - \sum_{i,j} \gamma_{ij}$$

$$\text{subject to} \quad \alpha_m + \beta_w - \sum_{j <_m w} \gamma_{m,j} - \sum_{i <_w m} \gamma_{i,w} - \gamma_{m,w} \geq 1$$

$$\text{for all pairs } (m,w)$$

$$\alpha_i, \beta_j, \gamma_{i,j} \geq 0 \text{ for all } i \text{ and } j.$$

- Use complementary slackness to show that every feasible fractional solution to the relaxation of (10.3) is optimal and that setting

$$\alpha_m = \sum_j x_{m,j} \text{ for all } m,$$

$$\beta_w = \sum_i x_{i,w} \text{ for all } w,$$

and

$$\gamma_{ij} = x_{ij} \text{ for all } i,j$$

is optimal for the dual program.

10.13. Show that there is a unique stable allocation in the sense discussed in Section 10.4. Hint: Use a proof by contradiction, considering the first C_k in the top trading cycle algorithm where the allocations differ.

10.14. Consider a set of n teams, each with 10 players, where each team owner
        has a ranking of all $10n$ players. Define a notion of stable allocation in this
        setting (as in Section 10.4) and show how to adapt the top trading cycle
        algorithm to find a stable allocation. We assume that players' preferences
        play no role.

10.15. A weaker notion of instability than the one discussed in Section 10.4 re-
        quires that no set of graduate students can obtain better offices than they
        are assigned in π by reallocating among themselves the offices allocated to
        them in π. Show that this follows from stability as defined in Section 10.4.
        Note that the converse does not hold. For example, if there are two people
        who both prefer the same office, the only stable allocation is to give that
        office to its owner, but the alternative is also weakly stable.

CHAPTER 11

Fair division

A Jewish town had a shortage of men for wedding purposes, so they had to import men from other towns. One day a groom-to-be arrived on a train. As he disembarked, one lady proclaimed, "He's a perfect fit for my daughter!" Another lady disagreed, "No, he's a much better fit for my daughter!"

A rabbi was called to decide the matter. After hearing both ladies, he said, "Each of you has good arguments for why your daughter should be the one to marry this man. Let's cut him in two and give each of your daughters half of him." One of the ladies replied, "That sounds fair." The rabbi immediately declared, "That's the real mother-in-law!"

Suppose that several people need to divide an asset, such as a plot of land, between them. One person may assign a higher value to a portion of the asset than another. This is often illustrated with the example of dividing a cake.

11.1. Cake cutting

FIGURE 11.1. Two bears sharing a cake. One cuts; the other chooses.

The classical method for dividing a cake fairly between two people is to have one cut and the other choose. This method ensures that each player can get at least half the cake according to his preferences; e.g., a player who loves icing most will take care to divide the icing equally between the two pieces.

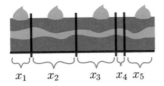

$$x_1 \quad x_2 \qquad x_3 \quad x_4\; x_5$$

FIGURE 11.2. This figure shows a possible way to cut a cake into five pieces. The ith piece is $B_i = [\sum_{k=1}^{i-1} x_k, \sum_{k=1}^{i} x_k)$. If the ith piece goes to player j (i.e., $A_j := B_i$), then his value for this piece is $\mu_j(B_i)$.

To divide a cake between more than two players, we first model the cake as the unit interval and assume that for each $i \in \{1, \ldots, n\}$, there is a distribution function $F_i(x)$ representing player i's value for the interval $[0, x]$. (See Figure 11.2 for a possible partition of the cake.) We assume these functions are continuous. Let $\mu_i(A)$ be the value player i assigns to the set $A \subset [0,1]$; in particular, $\mu_i([a, b]) = F_i(b) - F_i(a)$. We assume that μ_i is a probability measure.

DEFINITION 11.1.1. A partition $A_1, \ldots, A_n$ of the unit interval is called a **fair division**[1] if $\mu_i(A_i) \geq 1/n$. A crucial issue is which sets are allowed in the partition. For now, we assume that each A_i is an interval.

REMARK 11.1.2. The assumption that F_i is continuous is key since a discontinuity would represent an atom in the cake, and might preclude fair division.

Moving-knife Algorithm for fair division of a cake among n people

- Move a knife continuously over the cake from left to right until some player yells "Stop!"
- Give that player the piece of cake to the left of the knife.
- Iterate with the other $n - 1$ players and the remaining cake.

DEFINITION 11.1.3. The **safe strategy** for a player i is defined inductively as follows. If $n = 1$, take the whole cake. Otherwise, in the first round, i should yell "Stop" as soon as a $1/n$ portion of the cake is reached according to his measure. If someone else yells first, player i employs the safe strategy in the $(n - 1)$-person game on the remaining cake.

LEMMA 11.1.4. *Any player who plays the safe strategy is guaranteed to get a piece of cake that is worth at least $1/n$ of his value for the entire cake.*

PROOF. Any player i who plays the safe strategy either receives a piece of cake worth $1/n$ of his value in the first round or has value at least $(n - 1)/n$ for the remaining cake. In the latter case, by induction, i receives at least $1/(n - 1)$ of his value for the remaining cake and hence at least $1/n$ of his value for the whole cake. □

---

[1]This is also known as *proportional*.

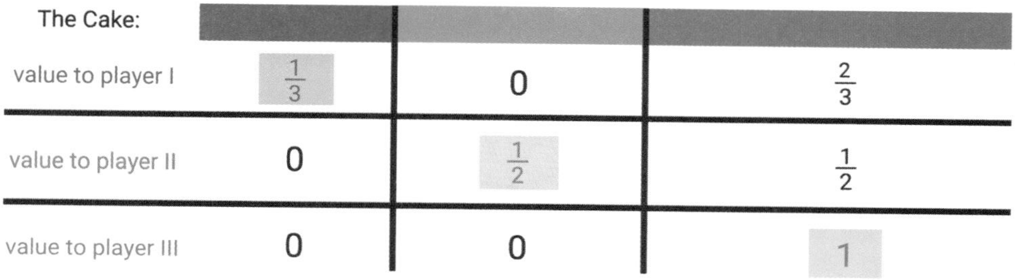

The Cake:			
value to player I	$\frac{1}{3}$	0	$\frac{2}{3}$
value to player II	0	$\frac{1}{2}$	$\frac{1}{2}$
value to player III	0	0	1

FIGURE 11.3. This figure shows an example of how the Moving-knife Algorithm might evolve with three players. The knife moves from left to right. Player I takes the first piece, then II, then III. In the end, player I is envious of player III.

While this cake-cutting algorithm guarantees a fair division if all participants play the safe strategy, it is not *envy-free*. It could be, when all is said and done, that some player would prefer the piece someone else received. See Figure 11.3 for an example.

11.1.1. Cake cutting via Sperner's Lemma. Let $\mu_1, \ldots, \mu_n$ and $F_1, \ldots, F_n$ be as above. In this section, we will show that there is a partition of the cake $[0,1]$ into n intervals that is envy-free, and hence fair, under the following assumption.

ASSUMPTION 11.1.5. Each of the n people prefers any piece of cake to no piece; i.e., $\mu_i(A) > 0$ for all i and any interval $A \neq \emptyset$.

We start by presenting an algorithm that constructs an ϵ-envy-free partition.

DEFINITION 11.1.6. A partition $A_1, \ldots, A_n$ is ϵ-**envy-free** if for all i, j we have $\mu_i(A_j) \leq \mu_i(A_i) + \epsilon$.

This means that player i, who was assigned interval A_i, does not prefer any other piece by more than ϵ.

11.1.1.1. *The construction.* Let e_i denote the ith standard basis vector and let $\Delta(e_1, e_2, \ldots, e_n)$ be the convex hull of $e_1, \ldots, e_n$. Each point $(x_1, \ldots, x_n)$ in the simplex $\Delta(e_1, e_2, \ldots, e_n)$ describes a partition of the cake (see Figure 11.2) where A_i is the piece of cake allocated to player i.

By Lemma 5.4.5, for any $\eta > 0$, there is a subdivision Γ of $\Delta(e_1, e_2, \ldots, e_n)$ for which all simplices in Γ have diameter less than η. By Corollary 5.4.6, there is a proper-coloring [2] with colors $\{c_1, \ldots, c_n\}$ of the vertices of Γ. If a vertex v has color c_i, we will say that player i *owns* that vertex. See Figure 11.4.

Next, construct a Sperner labeling $\ell(\cdot)$ of the vertices in the subdivision as follows: Given a vertex $\mathbf{x} = (x_1, \ldots, x_n)$ in Γ, define $B_i = B_i(\mathbf{x}) = [\sum_{k=1}^{i-1} x_k, \sum_{k=1}^{i} x_k]$. (Again, see Figure 11.2.) If $\mathbf{x}$ is owned by player j and $\mu_j(B_k)$ is maximal among $\mu_j(B_1), \ldots, \mu_j(B_n)$, then $\ell(\mathbf{x}) = k$. In other words, $\ell(\mathbf{x}) = k$ if B_k is player j's favorite piece among the pieces defined by $\mathbf{x}$. The fact that $\ell(\cdot)$ is a valid Sperner labeling follows from Assumption 11.1.5. See Figure 11.5.

---

[2] A coloring is proper if any two vertices in the same simplex $\Delta_1 \in \Gamma$ are assigned different colors.

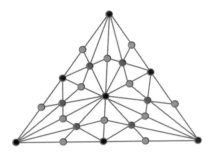

FIGURE 11.4. This picture shows the coloring of a subdivision Γ for three players. Each simplex in Γ has one black vertex (owned by player I), one purple vertex (owned by player II), and one green vertex (owned by player III).

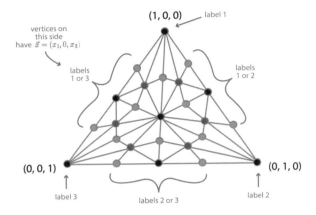

FIGURE 11.5. The coordinates $\mathbf{x} = (x_1, x_2, x_3)$ of a vertex represent a possible partition of the cake. The Sperner label of a vertex owned by a particular player is the index of the piece that player would choose given that partition of the cake. Notice that by Assumption 11.1.5, if, say, $x_i = 0$, then the label of vertex $\mathbf{x}$ is not i.

Finally, we apply Sperner's Lemma, from which we conclude that there is a fully labeled simplex in Γ.

THEOREM 11.1.7. *Let $\epsilon > 0$. There exists η such that if the maximal diameter of all simplices in Γ is less than η, then any vertex $\mathbf{x}$ of a fully labeled simplex in Γ determines an ϵ-envy-free partition.*

PROOF. Let $\Delta^* = \Delta(\mathbf{v}_1, \ldots, \mathbf{v}_n)$ be a fully labeled simplex in Γ, with $\mathbf{v}_i$ owned by player i. Let $\mathbf{x} := \mathbf{v}_1$ determine the partition $B_1, \ldots, B_n$. Write $\pi(k) := \ell(\mathbf{v}_k)$. The fact that Δ^* is fully labeled means that π is a permutation. For every k, assign player k the piece $A_k := B_{\pi(k)}$. This will be the ϵ-envy-free partition (provided η is sufficiently small).

Clearly A_i is the piece preferred by player i. Given another player j, use $\mathbf{v}_j$ to construct a partition $B'_1, \ldots, B'_n$. Observe that, for all k, the endpoints of B'_k are within $n\eta$ of the endpoints of B_k. By uniform continuity of F_r, there is $\delta > 0$ such

that
$$|t - t'| < \delta \Rightarrow |F_r(t) - F_r(t')| < \frac{\epsilon}{4}.$$
Thus, we will choose $\eta = \delta/n$, from which we can conclude
$$|\mu_j(B'_k) - \mu_j(B_k)| < \frac{\epsilon}{2}$$
for all k. Since
$$\mu_j(B'_{\pi(j)}) \geq \mu_j(B'_{\pi(k)}),$$
by the triangle inequality,
$$\mu_j(B_{\pi(j)}) \geq \mu_j(B_{\pi(k)}) - \epsilon.$$
The last part of the proof is illustrated in Figure 11.6.   □

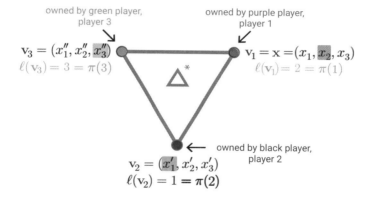

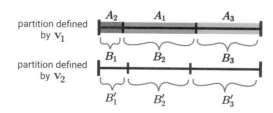

FIGURE 11.6. This figure illustrates the last part of the proof of Theorem 11.1.7. The big simplex at the top is a blown up version of Δ^*, the fully labeled simplex that is used to show that there is an ϵ-envy-free partition. In this figure, the distances between the vertices $\mathbf{v}_1, \mathbf{v}_2$ and $\mathbf{v}_3$ are at most η. The partition is determined by $\mathbf{v}_1 = (x_1, x_2, x_3)$.

COROLLARY 11.1.8. *If for each i the distribution function F_i defining player i's values is strictly increasing and continuous, then there exists an envy-free partition into intervals.*

PROOF. By the continuity of the F_i's, for every permutation $\boldsymbol{\pi}$ the set
$$\Lambda_{\boldsymbol{\pi}}(\epsilon) = \{\mathbf{x} \in \Delta \ : \ (B_{\pi(1)}(\mathbf{x}), \ldots, B_{\pi(n)}(\mathbf{x})) \text{ is } \epsilon\text{-envy-free}\}$$

is closed. The theorem shows that

$$\Lambda(\epsilon) = \bigcup_{\pi \in S_n} \Lambda_{\pi}(\epsilon)$$

is closed, nonempty, and monotone decreasing as $\epsilon \downarrow 0$. Thus,

$$\exists\, \mathbf{x}^* \in \bigcap_k \Lambda\left(1/k\right) \quad \text{so} \quad \forall k\; \exists\; \pi_k \text{ s.t. } \; \mathbf{x}^* \in \Lambda_{\pi_k}\left(1/k\right).$$

Finally, since some $\pi \in S_n$ repeats infinitely often in $\{\pi_k\}_{k \geq 1}$, the partition $A_i := B_{\pi(i)}(\mathbf{x}^*)$ is envy-free. $\square$

11.2. Bankruptcy

A debtor goes bankrupt. His total assets are less than his total debts. How should his assets be divided among his creditors?

DEFINITION 11.2.1. A **bankruptcy problem** is defined by the total available assets A and the claims $c_1, \ldots, c_n$ of the creditors, where c_i is the claim of the ith creditor, with $C := \sum_i c_i > A$. A **solution** to the bankruptcy problem is an allocation a_i to each creditor, where $a_i \leq c_i$ and $\sum_i a_i = A$.

One natural solution is **proportional division**, where each creditor receives the same fraction A/C of their claim; i.e., $a_i = c_i A/C$.

However, consider a problem of partitioning a garment[3] between two people, one claiming half the garment and the other claiming the entire garment. Since only half of the garment is contested, it's reasonable to partition that half between the two claimants, and assign the uncontested half to the second claimant. For $A = 1$, $c_1 = 0.5$, and $c_2 = 1$, this yields $a_1 = 0.25$ and $a_2 = 0.75$. This solution was proposed in the Talmud, an ancient Jewish text.

FIGURE 11.7. The question of how to split a talit is discussed in Tractate Baba Metzia, Chapter 1, Mishnah 1.

---

[3]or a plot of land.

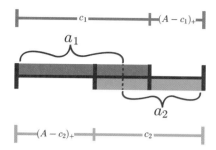

FIGURE 11.8. This picture shows the partitioning proposed by the Talmud.

FIGURE 11.9. Three widows in dispute over how the estate of their husband should be divided among them.

Formally, this is the principle of **equal division of contested amounts**, which we will refer to as the **garment rule** for $n = 2$: Since $(A - c_1)_+$ is not contested by 1 and $(A - c_2)_+$ is not contested by 2, each claimant receives his "uncontested portion" and half of the contested portion. See Figure 11.8.

The Talmud also presents solutions, without explanation, to the 3-claimant scenario shown in Table 1.

An explanation for the numbers in this table remained a conundrum for over 1,500 years. To address this, let's explore other fairness principles.

(1) **Constrained equal allocations (CEA):** Allocate the same amount to each creditor up to his claim; i.e., $a_i = a \wedge c_i$, where a is chosen so that $\sum_i a_i = A$. (Recall that $a \wedge b = \min(a, b)$.)

(2) **Constrained equal losses (CEL):** Assign to each creditor the same loss $\ell_i := c_i - a_i$ up to his claim; i.e., $\ell_i = \ell \wedge c_i$ where $\sum_i \ell_i = C - A$.

Neither of these principles yields the Talmud allocations (see Table 2), but they both share a consistency property, which will be the key to solving the puzzle.

TABLE 1. This table presents the solution proposed in the Talmud for partitioning the estate of a man who has died among his three wives.

Creditors' claims: Estate Size:	100	200	300
100	$33\frac{1}{3}$	$33\frac{1}{3}$	$33\frac{1}{3}$
200	50	75	75
300	50	100	150

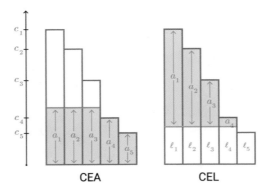

FIGURE 11.10. An example illustrating constrained equal allocations (CEA) and constrained equal losses (CEL). The shaded area is the total assets A.

DEFINITION 11.2.2. An **allocation rule**[4] is a function F mapping bankruptcy problems $(c_1, \ldots, c_n; A)$, for arbitrary n, to solutions $(a_1, \ldots, a_n)$. Such a rule is called **pairwise consistent** if

$$F(c_1, \ldots, c_n; A) = (a_1, \ldots, a_n) \text{ implies that } \forall i \neq j, \ F(c_i, c_j; a_i + a_j) = (a_i, a_j).$$

More generally, an allocation rule is **consistent** if for any subset $S \subset [1, n]$, the total allocation $\sum_{i \in S} a_i$ is split by F exactly to $(a_i)_{i \in S}$.

EXERCISE 11.a. Verify that proportional division, constrained equal allocations, and constrained equal losses are all consistent. Also, show that these allocation rules are **monotone**: For a fixed set of claims $(c_1, \ldots, c_n)$, the allocation a_i to each claimant is monotone in the available assets A.

THEOREM 11.2.3. *There is a unique pairwise consistent rule* $T(c_1, \ldots, c_n; A)$ *(the **Talmud rule**) which reduces to the garment rule for two creditors. This rule is:*

- *If* $A \leq C/2$, *then* $a_i = a \wedge \frac{c_i}{2}$ *with a chosen so that* $\sum_i a_i = A$. *I.e.,*

$$T(c_1, \ldots, c_n; A) := CEA\left(\frac{c_1}{2}, \ldots, \frac{c_n}{2}; A\right).$$

---

[4]This is for bankruptcy problems.

TABLE 2. The allocations of Proportional Division, Constrained Equal Allocations (CEA), and Constrained Equal Losses (CEL) for the scenario shown in Table 1.

Creditors' claims:		100	200	300
Estate Size:				
100	Proportional	$16\frac{2}{3}$	$33\frac{1}{3}$	50
	CEA	$33\frac{1}{3}$	$33\frac{1}{3}$	$33\frac{1}{3}$
	CEL	0	0	100
200	Proportional	$33\frac{1}{3}$	$66\frac{2}{3}$	100
	CEA	$66\frac{2}{3}$	$66\frac{2}{3}$	$66\frac{2}{3}$
	CEL	0	50	150
300	Proportional	50	100	150
	CEA	100	100	100
	CEL	0	100	200

- If $A > C/2$, let $\ell_i = \ell \wedge \frac{c_i}{2}$, with ℓ chosen so that $\sum_i \ell_i = C - A$, and set $a_i = c_i - \ell_i$. I.e.,

$$T(c_1, \ldots, c_n; A) := CEL\left(\frac{c_1}{2}, \ldots, \frac{c_n}{2}; A\right).$$

Moreover, the Talmud rule is consistent.

PROOF OF THEOREM 11.2.3. It follows from Exercise 11.1 that the Talmud rule is the garment rule for $n = 2$.

Consistency follows from the fact that if $A \leq C/2$, then $\sum_{i \in S} a_i \leq \sum_{i \in S} \frac{c_i}{2}$ for every S, so the Talmud rule applied to S is CEA with claims $(\frac{c_i}{2})_{i \in S}$. Consistency of the CEA rule (Exercise 11.a) completes the argument. A similar argument works for the case of $A > C/2$ using consistency of CEL.

For uniqueness, suppose there are two different pairwise consistent rules that reduce to the garment rule for $n = 2$. Then on some bankruptcy problem $(c_1, \ldots, c_n; A)$ they produce different allocations, say $(a_1, \ldots, a_n)$ and $(b_1, \ldots, b_n)$. Since $\sum_i a_i = \sum_i b_i$, there is a pair i, j with $a_i < b_i$ and $a_j > b_j$. Without loss of generality suppose that $a_i + a_j \geq b_i + b_j$. Then the fact that $a_i < b_i$ is a contradiction to the monotonicity in assets of the garment rule.   □

REMARK 11.2.4. The proof of Theorem 11.2.3 shows that any monotone rule is uniquely determined by its restriction to pairs.

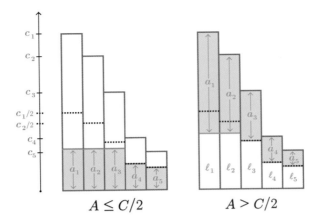

$$A \leq C/2 \qquad\qquad A > C/2$$

FIGURE 11.11. A depiction of the Talmud rule in the two cases.

REMARK 11.2.5. In this section, we have assumed that the claims are verifiable and not subject to manipulation by participants.

Notes

The divide and choose algorithm for cake cutting goes back to the Old Testament (Genesis 13):

> So Abraham said to Lot, "Let us not have any quarreling between you and me, or between your herders and mine, for we are brothers. Is not the whole land before you? Let us part company. If you go to the left, I will go to the right; if you go to the right, I will go to the left."
>
> Lot looked and saw that the whole plain of the Jordan toward Zoar was well watered, like the garden of the Lord.... So Lot chose for himself the whole plain of the Jordan and set out toward the east.

The Moving-knife Algorithm for cake cutting is due to Dubins and Spanier [DS61]. A discrete version of the Moving-knife Algorithm was discovered earlier by Banach and Knaster [Ste48]: The first player cuts a slice, and each of the other players, in turn, is given the opportunity to diminish it. The last diminisher gets the slice, and then the procedure is applied to the remaining $n - 1$ players.

The use of Sperner's Lemma to solve the envy-free cake cutting problem is due to Su [Su99].

In the setting of n nonatomic measures[5], Lyapunov showed [Lya40] that there always is a partition of the cake into n measurable slices $A_1, A_2, \ldots, A_n$, such that $\mu_i(A_j) = 1/n$ for all i and j. In particular, there is an envy-free partition of the cake. Note though that even if the cake is $[0, 1]$, the resulting slices can be complicated measurable sets, and no algorithm is given to find them. An elegant proof of Lyapunov's theorem was given by Lindenstrauss [Lin66].

Alon [Alo87] proved a theorem about "splitting necklaces", which implies that if the cake is $[0, 1]$, then a perfect partition as in Lyapunov's Theorem can be obtained by cutting the cake into $(n - 1)n + 1$ intervals and assigning each participant a suitable subset of these intervals.

---

[5] A measure $\mu(\cdot)$ is nonatomic if $\mu(A) > 0$ implies that there is $B \subset A$ with $0 < \mu(B) < \mu(A)$.

Lyapunov, Alon and Su's theorems are non-constructive. Selfridge and Conway (see e.g., [BT96]) presented a constructive procedure for finding an envy-free partition when $n = 3$. In 1995, Brams and Taylor [BT95] described a procedure that produces an envy-free partition for any n, but the number of steps it takes is finite but unbounded. Only in 2016, Aziz and Mackenzie [AM16] discovered a procedure whose complexity is bounded as a function of n.

The resolution of the conundrum regarding bankruptcy is due to Aumann and Maschler [AM85]. Another rule, proposed by O'Neill [O'N82], is the *random arrival rule*: Consider an arrival order for the claimants, and allocate to each one the minimum of his claim and what is left of the estate. To make this fair, the final allocation is the average of these allocations over all $n!$ orderings. For an extensive discussion of fair division, see the books by Brams and Taylor [BT96] and Robertson and Webb [RW98] and the survey by Procaccia [Pro13].

Exercise 11.2 is from [BT96].

Exercises

S   11.1.   Show that the Talmud rule is monotone in A for all n and coincides with the garment rule for $n = 2$.

11.2.   Consider the following procedure to partition a round cake among three players: Alice, Barbara and Carol. We assume that each player has a continuous measure on the cake that allocates zero measure to every line segment (radius) that starts at the center of the cake. (Alternatively, these could be continuous measures on a circle.)

  • Alice positions three knives on the cake (like the hands of a clock). She then rotates them clockwise, with the requirement that if some knife reaches the initial position of the next knife, then the same must hold for the other two knives. (At all times the tips of the knives meet at the center of the cake).

  • Alice continues rotating the knives until either Barbara yells "stop", or each knife has reached the initial position of the next knife.

  • When Alice stops rotating the knives, Carol chooses one of the three slices determined by the knives, and then Barbara selects one of the two remaining slices, leaving the last slice to Alice.

Show that each player has a strategy ensuring that (no matter how the others play), the slice she obtains is at least as large as any other slice according to her measure. Thus if all three players adhere to these strategies, an envy-free partition will result.

Hint: Alice should rotate the knives so that, at all times, the three slices they determine are of equal size according to her measure. Barbara should yell stop when the two largest slices (according to her measure) are tied.

CHAPTER 12

Cooperative games

In this chapter, we consider multiplayer games where players can form coalitions. These come in two flavors: *transferable utility* (TU) games where side payments are allowed, and *nontransferable utility* games (NTU). The latter includes settings where payments are not allowed (e.g., between voters and candidates in an election) or where players have different utility for money. In this chapter, we mostly focus on the former.

12.1. Transferable utility games

We review the example discussed in the preface. Suppose that three people are selling their wares in a market. One of them is selling a single, left-handed glove, while the other two are each selling a right-handed glove. A wealthy tourist arrives at the market in dire need of a pair of gloves, willing to pay one Bitcoin[1] for a pair of gloves. She refuses to deal with the glove-bearers individually, and thus, these sellers have to come to some agreement as to how to make a sale of a left- and right-handed glove to her and how to then split the one Bitcoin among themselves. Clearly, the first player has an advantage because his commodity is in scarcer supply. This means that he should be able to obtain a higher fraction of the payment than either of the other players. However, if he holds out for too high a fraction of the earnings, the other players may act as a coalition and require him to share more of the revenue.

The question then is, in their negotiations prior to the purchase, how much can each player realistically demand out of the total payment made by the customer?

To resolve this question, we introduce a *characteristic function* v, defined on subsets of the player set. In the glove market, $v(S)$, where S is a subset of the three players, is 1 if, just among themselves, the players in S have both a left glove and a right glove. Thus,

$$v(123) = v(12) = v(13) = 1,$$

and the value is 0 on every other subset of $\{1, 2, 3\}$. (We abuse notation in this chapter and sometimes write $v(12)$ instead of $v(\{1, 2\})$, etc.)

More generally, a **cooperative game** with **transferable utilities** is defined by a **characteristic function** v on subsets of the n players, where $v : 2^S \to \mathbb{R}$ is the value, or payoff, that subset S of players can achieve on their own regardless of what the remaining players do. This value can then be split among the players in any way that they agree on. The characteristic function satisfies the following properties:

- $v(\varnothing) = 0$.
- *Monotonicity:* If $S \subseteq T$, then $v(S) \leq v(T)$.

---

[1] A Bitcoin is a unit of digital currency that was worth \$100 at the time of the transaction.

FIGURE 12.1

Given a characteristic function v, each possible outcome of the game is an allocation vector $\boldsymbol{\psi}(v) \in \mathbb{R}^n$, where $\psi_i(v)$ is the share of the payoff allocated to player i.

What is a plausible outcome of the game? We will see several different solution concepts.

12.2. The core

An allocation vector $\boldsymbol{\psi} = \boldsymbol{\psi}(v)$ is in the **core** if it satisfies the following two properties:

- **Efficiency:** $\sum_{i=1}^{n} \psi_i = v(\{1, \ldots, n\})$. This means, by monotonicity, that, between them, the players extract the maximum possible total value.
- **Stability:** Each coalition is allocated at least the payoff it can obtain on its own; i.e., for every set S,

$$\sum_{i \in S} \psi_i \geq v(S).$$

For the glove market, an allocation vector in the core must satisfy

$$\psi_1 + \psi_2 \geq 1,$$
$$\psi_1 + \psi_3 \geq 1,$$
$$\psi_1 + \psi_2 + \psi_3 = 1.$$

This system has only one solution: $\psi_1 = 1$ and $\psi_2 = \psi_3 = 0$. g

EXAMPLE 12.2.1 (**Miners and Gold:**). Consider a set of miners who have discovered large bars of gold. The value of the loot to the group is the number of bars that they can carry home. It takes two miners to carry one bar, and thus the value of the loot to any subset of k miners is $\lfloor k/2 \rfloor$.

If the total number of miners is even, then the vector $\boldsymbol{\psi} = (1/2, \ldots, 1/2)$ is in the core. On the other hand, if n is odd, say 3, then the core conditions require that

$$\psi_1 + \psi_2 \geq 1,$$
$$\psi_1 + \psi_3 \geq 1,$$
$$\psi_2 + \psi_3 \geq 1,$$
$$\psi_1 + \psi_2 + \psi_3 = 1.$$

This system has no solution.

EXAMPLE 12.2.2 (**Splitting a dollar:**). A parent offers his two children $100 if they can agree on how to split it. If they can't agree, they will each get $10. In this case $v(12) = 100$, whereas $v(1) = v(2) = 10$. The core conditions require that

$$\psi_1 \geq 10 \qquad \psi_2 \geq 10 \quad \text{and} \quad \psi_1 + \psi_2 = 100,$$

which clearly has multiple solutions.

The drawback of the core, as we saw in these examples, is that it may be empty or it might contain multiple allocation vectors. This motivates us to consider alternative solution concepts.

12.3. The Shapley value

Another way to choose the allocation $\psi(\mathbf{v})$ is to adopt an axiomatic approach, wherein a set of desirable properties for the solution is enumerated.

12.3.1. Shapley's axioms.

(1) **Symmetry**: If $v(S \cup \{i\}) = v(S \cup \{j\})$ for all S with $i, j \notin S$, then $\psi_i(v) = \psi_j(v)$.
(2) **Dummy**: A player that doesn't add value gets nothing; i.e., if $v(S \cup \{i\}) = v(S)$ for all S, then $\psi_i(v) = 0$.
(3) **Efficiency**: $\sum_{i=1}^{n} \psi_i(v) = v(\{1, \ldots, n\})$.
(4) **Additivity**: $\psi_i(v + u) = \psi_i(v) + \psi_i(u)$.

The first three axioms are self-explanatory. To motivate the additivity axiom, imagine the same players engage in two consecutive games, with characteristic functions v and u, respectively. This axiom states that the outcome in one game should not affect the other, and thus, in the combined game, the allocation to a player is the sum of his allocations in the component games.

We shall see that there is a *unique* choice for the allocation vector, given these axioms. This unique choice for each $\psi_i(v)$ is called the **Shapley value** of player i in the game defined by characteristic function v.

EXAMPLE 12.3.1 (**The S-veto game**). Consider a coalitional game with n players, in which a fixed subset S of the players holds all the power. We will denote the characteristic function here by w_S, defined as follows: $w_S(T)$ is 1 if T contains S and it is 0 otherwise. Suppose $\psi(v)$ satisfies Shapley's axioms. By the dummy axiom,

$$\psi_i(w_S) = 0 \qquad \text{if } i \notin S.$$

Then, for $i, j \in S$, the symmetry axiom gives $\psi_i(w_S) = \psi_j(w_S)$. Finally, the efficiency axiom implies that

$$\psi_i(w_S) = \frac{1}{|S|} \qquad \text{if } i \in S.$$

Similarly, we can derive that $\psi_i(c\, w_S) = c\, \psi_i(w_S)$ for any $c \in [0, \infty)$. Note that to derive this, we did not use the additivity axiom.

Glove Market, again: We can now use our understanding of the S-veto game to solve for the Shapley values $\boldsymbol{\psi}(v)$, where v is the characteristic function of the Glove Market game.

Observe that for bits and for $\{0, 1\}$-valued functions

$$u \vee w = \max(u, w) = u + w - u \cdot w.$$

With w_{12}, etc, defined as in Example 12.3.1, we have that for every S,

$$v(S) = w_{12}(S) \vee w_{13}(S) = w_{12}(S) + w_{13}(S) - w_{123}(S).$$

Thus, the additivity axiom gives

$$\psi_i(v) = \psi_i(w_{12}) + \psi_i(w_{13}) - \psi_i(w_{123}).$$

We conclude from this that $\psi_1(v) = 1/2 + 1/2 - 1/3 = 2/3$, whereas $\psi_2(v) = \psi_3(v) = 0 + 1/2 - 1/3 = 1/6$. Thus, under Shapley's axioms, player 1 obtains a two-thirds share of the payoff, while players 2 and 3 equally share one-third between them.

The calculation of $\psi_i(v)$ we just did for the glove game relied on the representation of v as a linear combination of S-veto functions w_S. Such a representation always exists.

LEMMA 12.3.2. *For any characteristic function* $v : 2^{[n]} \to \mathbb{R}$, *there is a unique choice of coefficients* c_S *such that*

$$v = \sum_{S \neq \varnothing} c_S w_S.$$

PROOF. The system of $2^n - 1$ equations in the $2^n - 1$ unknowns c_S, that is, for all nonempty $T \subseteq [n]$

$$v(T) = \sum_{\varnothing \neq S \subseteq [n]} c_S w_S(T), \tag{12.1}$$

has a unique solution. To see this, observe that if the subsets of $[n]$ are ordered in increasing cardinality, then the matrix $w_S(T)$ is upper triangular, with 1's along the diagonal. For example, with $n = 2$, rows indexed by S and columns indexed by T, the matrix $w_S(T)$ is

$$
\begin{array}{c c}
 & \begin{array}{ccc} \{1\} & \{2\} & \{12\} \end{array} \\
\begin{array}{c} \{1\} \\ \{2\} \\ \{12\} \end{array} &
\left[\begin{array}{ccc}
1 & 0 & 1 \\
0 & 1 & 1 \\
0 & 0 & 1
\end{array} \right]
\end{array} .
$$

$\qquad\qquad\qquad\qquad\qquad\qquad\qquad\qquad\qquad\qquad\qquad\qquad\qquad\qquad\qquad\quad\square$

EXAMPLE 12.3.3 (**Four Stockholders**). Four people own stock in ACME. Player i holds i units of stock, for each $i \in \{1, 2, 3, 4\}$. Six shares are needed to pass a resolution at the board meeting. Here $v(S)$ is 1 if subset S of players have enough shares of stock among them to pass a resolution. Thus,

$$1 = v(1234) = v(24) = v(34),$$

while $v = 1$ on any 3-tuple and $v = 0$ in each other case. In this setting, the Shapley value $\psi_i(v)$ for player i represents the power of player i and is known as the **Shapley-Shubik power index.** By Lemma 12.3.2, we know that the system of equations

$$v = \sum_{S \neq \varnothing} c_S w_S$$

has a solution. Solving this system, we find that

$$v = w_{24} + w_{34} + w_{123} - w_{234} - w_{1234},$$

from which

$$\psi_1(v) = 1/3 - 1/4 = 1/12$$

and

$$\psi_2(v) = 1/2 + 1/3 - 1/3 - 1/4 = 1/4,$$

while $\psi_3(v) = 1/4$, by symmetry with player 2. Finally, $\psi_4(v) = 5/12$. It is interesting to note that the person with two shares and the person with three shares have equal power.

EXERCISE 12.a. Show that Four Stockholders has no solution in the core.

12.3.2. Shapley's Theorem. Consider a fixed ordering of the players, defined by a permutation π of $[n] = \{1, \ldots, n\}$. Imagine the players arriving one by one according to this permutation π, and define $\phi_i(v, \pi)$ to be the marginal contribution of player i at the time of his arrival assuming players arrive in this order. That is,

$$\phi_i(v, \pi) = v\big(\pi\{1, \ldots, k\}\big) - v\big(\pi\{1, \ldots, k-1\}\big) \quad \text{where } \pi(k) = i. \qquad (12.2)$$

Notice that if we were to set $\psi_i(v) = \phi_i(v, \pi)$ for any fixed π, the dummy, efficiency, and additivity axioms would be satisfied.

To satisfy the symmetry axiom as well, we will instead imagine that the players arrive in a random order and define $\psi_i(v)$ to be the expected value of $\phi_i(v, \pi)$ when π is chosen uniformly at random.

REMARK 12.3.4. If $v(\cdot)$ is $\{0, 1\}$-valued, then $\psi_i(v)$ is the probability that player i's arrival converts a losing coalition to a winning coalition.

THEOREM 12.3.5. *Shapley's four axioms uniquely determine the functions ψ_i. They are given by the random arrival formula:*

$$\psi_i(v) = \frac{1}{n!} \sum_{\pi \in S_n} \phi_i(v, \pi). \qquad (12.3)$$

REMARK 12.3.6. $\phi_i(v, \pi)$ depends on π only via the set $S = \{j \ : \ \pi^{-1}(j) < \pi^{-1}(i)\}$ of players that precede i in S. Therefore

$$\psi_i(v) = \sum_{S \subseteq N \setminus \{i\}} \frac{|S|!(n - |S| - 1)!}{n!} (v(S \cup \{i\}) - v(S)).$$

PROOF. First, we prove that the functions $\psi_i(v)$ are uniquely determined by v and the four axioms. By Lemma 12.3.2, we know that any characteristic function v can be uniquely represented as a linear combination of S-veto characteristic functions w_S.

Recalling that $\psi_i(w_S) = 1/|S|$ if $i \in S$ and $\psi_i(w_S) = 0$ otherwise, we apply the additivity axiom and conclude that $\psi_i(v)$ is uniquely determined:

$$\psi_i(v) = \psi_i\left(\sum_{\varnothing \neq S \subseteq [n]} c_S w_S\right) = \sum_{\varnothing \neq S \subseteq [n]} \psi_i\left(c_S w_S\right) = \sum_{S \subseteq [n], i \in S} \frac{c_S}{|S|}.$$

We complete the proof by showing that the specific values given in the statement of the theorem satisfy all of the axioms. Recall the definition of $\phi_i(v, \pi)$ from (12.2). By averaging over all permutations π and then defining $\psi_i(v)$ as in (12.3), we claim that all four axioms are satisfied. Since averaging preserves the dummy, efficiency, and additivity axioms, we only need to prove the intuitive fact that by averaging over all permutations, we obtain symmetry.

To this end, suppose that i and j are such that

$$v\left(S \cup \{i\}\right) = v\left(S \cup \{j\}\right)$$

for all $S \subseteq [n]$ with $S \cap \{i, j\} = \varnothing$. For every permutation π, define π^* to be the same as π except that the positions of i and j are switched. Then

$$\phi_i(v, \pi) = \phi_j(v, \pi^*).$$

Using the fact that the map $\pi \mapsto \pi^*$ is a one-to-one map from S_n to itself for which $\pi^{**} = \pi$, we obtain

$$\psi_i(v) = \frac{1}{n!} \sum_{\pi \in S_n} \phi_i(v, \pi) = \frac{1}{n!} \sum_{\pi \in S_n} \phi_j(v, \pi^*)$$

$$= \frac{1}{n!} \sum_{\pi^* \in S_n} \phi_j(v, \pi^*) = \psi_j(v).$$

Therefore, $\boldsymbol{\psi}(v) = (\psi_1(v), \ldots, \psi_n(v))$ are indeed the unique Shapley values. $\square$

12.3.3. Additional examples.

EXAMPLE 12.3.7 (**A fish with little intrinsic value**). A seller s has a fish having little intrinsic value to him; i.e., he values it at \$2. A buyer b values the fish at \$10. Thus, $v(s) = 2$ and $v(b) = 0$. Denote by x the sale price. Then $v(s, b) = x + (10 - x) = 10$ for $x \geq 2$.

In this game, any allocation of the form $\psi_s(v) = x$ and $\psi_b(v) = 10 - x$ (for $2 \leq x \leq 10$) is in the core. On the other hand, the Shapley values are $\psi_s(v) = 6$ and $\psi_b(v) = 4$.

Note, however, that the value of the fish to b and s is private information, and if the price is determined by the formula above, they would have an incentive to misreport their values.

EXAMPLE 12.3.8 (**Many right gloves**). Consider the following variant of the glove game. There are $n = r + 2$ players. Players 1 and 2 have left gloves. The remaining players each have a right glove. Thus, the characteristic function $v(S)$ is the maximum number of proper and disjoint pairs of gloves owned by players in S.

We compute the Shapley value. Note that $\psi_1(v) = \psi_2(v)$ and that $\psi_r(v) = \psi_3(v)$ for each $r \geq 3$. By the efficiency axiom, we have

$$2\psi_1(v) + r\psi_3(v) = 2$$

provided that $r \geq 2$. To determine the Shapley value of the third player, we consider all permutations π with the property that the third player adds value to the group of players that precede him in π. These are the following orders:

$$13, 23, \{1, 2\}3, \{1, 2, j\}3,$$

where j is any value in $\{4, \ldots, n\}$ and the curly brackets mean that each permutation of the elements in curly brackets is included. The number of permutations corresponding to each of these possibilities is $r!$, $r!$, $2(r-1)!$, and $6(r-1) \cdot (r-2)!$. Thus,

$$\psi_3(v) = \frac{2r! + 8(r-1)!}{(r+2)!} = \frac{2r + 8}{(r+2)(r+1)r}.$$

12.4. Nash bargaining

EXAMPLE 12.4.1. The owner of a house and a potential buyer are negotiating over the price. The house is worth one (million) dollars to the seller, but it is worth $1 + s$ (million) dollars to the buyer. Thus, any price p they could agree on must be in $[1, 1 + s]$. However, the seller already has an offer of $1 + d_1$, and the buyer has an alternative house that she could buy (also worth $1 + s$ to her) for $1 + s - d_2$. (Assume that $d_1 + d_2 < s$.) If they come to agreement on a price p, then the utility to the buyer will be $p - 1$ and the utility to the seller will be $1 + s - p$. If the negotiation breaks down, they can each accept their alternative offers, resulting in a utility of d_1 for the seller and d_2 for the buyer. At what price might we expect their bargaining to terminate?

DEFINITION 12.4.2. A two-person **bargaining problem** is defined by a closed, bounded convex set $S \subset \mathbb{R}^2$ and a point $\mathbf{d} = (d_1, d_2)$. The set S represents the possible payoffs that the two players could potentially come to an agreement on, and $\mathbf{d}$ represents the payoffs that will result if they are unable to come to an agreement, sometimes called the *disagreement point*. We assume there is a point $(x_1, x_2) \in S$ such that $x_1 > d_1$ and $x_2 > d_2$ and that $\mathbf{x} \geq \mathbf{d}$ for all $\mathbf{x} \in S$ (since no player will accept an outcome below his disagreement value).

In Example 12.4.1, $S = \{(x_1, x_2) \mid x_1 + x_2 \leq s, \ x_1 \geq d_1, \ x_2 \geq d_2\}$ and $\mathbf{d} = (d_1, d_2)$.

DEFINITION 12.4.3. **A solution to a bargaining problem** is a mapping F that takes each instance $(S, \mathbf{d})$ and outputs an agreement point

$$F(S, \mathbf{d}) = \mathbf{a} \in S.$$

For $\mathbf{a} = (a_1, a_2)$, the final payoff for player I is a_1 and the final payoff for player II is a_2.

What constitutes a fair/reasonable solution? The approach taken by John Nash was to formulate a set of axioms that a fair solution should satisfy. These are known as the **Nash bargaining axioms**:

- *Affine covariance:* Let

$$\Psi(x_1, x_2) = (\alpha_1 x_1 + \beta_1, \alpha_2 x_2 + \beta_2). \tag{12.4}$$

We say that $F(\cdot)$ is affine covariant if for any real $\alpha_1, \alpha_2, \beta_1, \beta_2$ with $\alpha_1, \alpha_2 > 0$ and bargaining problem $(S, \mathbf{d})$,

$$F(\Psi(S), \Psi(\mathbf{d})) = \Psi(F(S, \mathbf{d})).$$

- *Pareto optimality:* If $F(S, \mathbf{d}) = \mathbf{a}$, and if $\mathbf{a}' = (a'_1, a'_2) \in S$ satisfies $a'_1 \geq a_1$ and $a'_2 \geq a_2$, then $\mathbf{a} = \mathbf{a}'$.
- *Symmetry:* For any bargaining problem $(S, \mathbf{d})$ such that $d_1 = d_2$ and $(x, y) \in S \to (y, x) \in S$, we have $F(S, \mathbf{d}) = (a, a)$ for some $(a, a) \in S$.
- *Independence of Irrelevant Alternatives (IIA):* If $(S, \mathbf{d})$ and $(S', \mathbf{d})$ are two bargaining problems such that $S \subseteq S'$ and if $F(S', \mathbf{d}) \in S$, then $F(S, \mathbf{d}) = F(S', \mathbf{d})$.

DEFINITION 12.4.4. **The Nash bargaining solution** $F^N(S, \mathbf{d}) = \mathbf{a} = (a_1, a_2)$ is the solution to the following maximization problem:

$$\text{maximize} \qquad \prod_{i=1}^{2}(x_i - d_i)$$

$$\text{subject to} \qquad x_1 \geq d_1, \quad x_2 \geq d_2,$$

$$(x_1, x_2) \in S. \qquad (12.5)$$

REMARK 12.4.5. The Nash bargaining solution $F^N(\cdot)$ always exists since S is closed and bounded and contains a point with $x_1 > d_1$ and $x_2 > d_2$. Moreover, the solution is unique. To see this, without loss of generality, assume that $d_1 = d_2 = 0$, and suppose there are two optimal solutions (x, y) and (w, z) to (12.5) with

$$x \cdot y = w \cdot z = \alpha. \qquad (12.6)$$

Observe that the function $f(t) = \frac{\alpha}{t}$ is strictly convex and therefore

$$f\left(\frac{x+w}{2}\right) < \frac{f(x) + f(w)}{2}.$$

Using (12.6), this is equivalent to

$$\frac{\alpha}{\frac{x+w}{2}} < \frac{y+z}{2}.$$

This contradicts the assumption that (x, y) and (w, z) are optimal solutions to the maximization problem (12.5) because $\left(\frac{x+w}{2}, \frac{y+z}{2}\right)$ (also feasible due to the convexity of S) yields a larger product.

EXERCISE 12.b. Check that in Example 12.4.1, where

$$S = \{(x_1, x_2) \mid x_1 + x_2 \leq s, \ x_1 \geq d_1, \ x_2 \geq d_2\} \quad \text{and} \quad \mathbf{d} = (d_1, d_2),$$

the Nash bargaining solution and the Shapley values are both $\left(\frac{s+d_1-d_2}{2}, \frac{s+d_2-d_1}{2}\right)$.

THEOREM 12.4.6. *The solution $F^N(\cdot)$ is the unique function satisfying the Nash bargaining axioms.*

PROOF. We first observe that $F^N(\cdot)$ satisfies the axioms.

- Affine covariance follows from the identity

$$\prod_i [(\alpha_i x_i + \beta_i) - (\alpha_i d_i + \beta_i)] = \prod_i \alpha_i \prod_i (x_i - d_i).$$

- To check Pareto optimality, observe that[2] if $\mathbf{y} \geq \mathbf{x} \geq \mathbf{d}$ and $\mathbf{y} \neq \mathbf{x}$, then $\prod_i (y_i - d_i) > \prod_i (x_i - d_i)$.
- Symmetry: Let $\mathbf{a} = (a_1, a_2) = F^N(S, \mathbf{d})$ be the solution to (12.5), where $(S, \mathbf{d})$ is symmetric. Then (a_2, a_1) is also an optimal solution to (12.5), so, by the uniqueness of the solution, we must have $a_1 = a_2$.
- IIA: Consider any $S \subseteq S'$. If $F^N(S', \mathbf{d}) \in S$, then it must be a solution to (12.5) in S. By uniqueness, it must coincide with $F^N(S, \mathbf{d})$.

Next we show that any $F(\cdot)$ that satisfies the axioms is equal to the Nash bargaining solution. We first prove this assuming that the bargaining problem that we are considering has disagreement point $\mathbf{d} = (0,0)$ and $F^N(S, \mathbf{d}) = (1,1)$. We will argue that this assumption, together with the symmetry axiom and IIA, imply that $F(S, \mathbf{d}) = (1,1)$.

To this end, let S' be the convex hull of $S \cup \{\mathbf{x}^T | \mathbf{x} \in S\}$. For every $\mathbf{x} \in S$, convexity implies that $(1-\lambda)(1,1) + \lambda \mathbf{x} \in S$, so $\varphi(\lambda) := (1-\lambda+\lambda x_1)(1-\lambda+\lambda x_2) \leq 1$. Since $\varphi(0) = 1$, we infer that $0 \geq \varphi'(0) = x_1 + x_2 - 2$. (See also Figure 12.2 for an alternative argument.)

Thus, $S \cup \{\mathbf{x}^T | \mathbf{x} \in S\} \subset \{\mathbf{x} \geq 0 \mid x_1 + x_2 \leq 2\}$; the set on the right is convex, so it must contain S' as well. Therefore $(1,1)$ is Pareto optimal in S' so the symmetry axiom yields $F(S', (0,0)) = (1,1)$. Since $(1,1) \in S$, the IIA axiom gives $F(S, (0,0)) = (1,1)$.

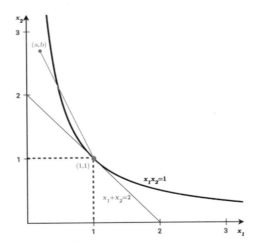

FIGURE 12.2. The line $x_1 + x_2 = 2$ is tangent to the hyperbola $x_1 \cdot x_2 = 1$ at $(1,1)$. Therefore, any line segment from a point (a, b) with $a + b > 2$ to the point $(1,1)$ must intersect the region $x_1 \cdot x_2 > 1$.

Finally, we argue that $F(S, \mathbf{d}) = F^N(S, \mathbf{d})$ for an arbitrary bargaining problem $(S, \mathbf{d})$ with $F^N(S, \mathbf{d}) = \mathbf{a}$. To this end, find an affine function Ψ as in (12.4) such that $\Psi(\mathbf{d}) = \mathbf{0}$ and $\Psi(\mathbf{a}) = (1,1)$.

By the affine covariance axiom,

$$\Psi(F(S, \mathbf{d})) = F(\Psi(S), \mathbf{0}) = F^N(\Psi(S), \mathbf{0}) = \Psi(F^N(S, \mathbf{d})),$$

---

[2]The vector notation $\mathbf{y} \geq \mathbf{x}$ means that $y_i \geq x_i$ for all i.

which means that

$$F(S, \mathbf{d}) = F^N(S, \mathbf{d}).$$

$\square$

Suppose that player i has strictly increasing utility $U_i(x_i)$ for an allocation x_i. Often it is assumed that these utility functions are concave[3], since the same gain is often worth less to a player when he is rich than when he is poor.

When some players have non-linear utility functions, it is not reasonable to require the affine covariance axiom for monetary allocations. Rather we require affine covariance of the vector of utilities obtained by the players. The same considerations apply to the symmetry axiom. The other axioms (Pareto and IIA) are not affected by applying a monotone bijection to each allocation.

Thus, we will seek the Nash bargaining solution in utility space. That is, we will apply Nash bargaining to the set $S_U = \{(U_1(x_1), U_2(x_2)) \mid (x_1, x_2) \in S\}$ with disagreement point $(U_1(d_1), U_2(d_2))$.

EXAMPLE 12.4.7. Consider two players that need to split a dollar between them. Suppose that player I is risk-neutral ($U_1(x_1) = x_1$) and the other has a strictly increasing, concave utility function; his utility for a payoff of x_2 is $U_2(x_2)$, where $U_2(0) = 0$, $U_2(1) = 1$.

The Nash bargaining solution is the maximum of $U_1(x_1)U_2(x_2) = x_1 U_2(x_2)$ over $\{\mathbf{x} \geq 0 : x_1 + x_2 \leq 1\}$. Since $U := U_2$ is increasing, this reduces to maximizing $f(x) := xU(1-x)$ over $x \in [0, 1]$. Observe that for all $x \leq 1/2$,

$$f'(x) = U(1-x) - xU'(1-x) > 0,$$

and therefore $f(x)$ is maximized at $x > 1/2$. In other words, at the Nash bargaining solution, the risk-neutral player gets *more* than half of the dollar, and the risk-averse player loses out.

For example, with $U_2(x_2) = \sqrt{x_2}$, the Nash bargaining solution is obtained by maximizing $f(x) = x\sqrt{1-x}$ over $[0, 1]$. The optimal choice is $x = 2/3$, which yields a Nash bargaining solution of $(2/3, 1/3)$.

Notes

The notion of a cooperative game is due to von Neumann and Morgenstern [vNM53]. Many different approaches to defining allocation rules, i.e., the shares $\boldsymbol{\psi}(v)$, have been proposed, based on either stability (subgroups should not have an incentive to abandon the grand coalition) or fairness. The notion of core is due to Gilles [Gil59]. The definition and axiomatic characterization of the Shapley value is due to Shapley [Sha53b]. See also [Rot88]. Another index of the power of a player in a cooperative game where $v(\cdot) \in \{0, 1\}$ is due to Banzhaf [BI64]. Recalling Definition 9.3.2, the Banzhaf index of player i is defined as $I_i(2v - 1)/\sum_j I_j(2v - 1)$. See Owen [Owe95] for more details.

An important solution concept omitted from this chapter is the *nucleolus*, due to Schmeidler [Sch69]. For a possible allocation vector $\boldsymbol{\psi}(v) = (\psi_1, \dots, \psi_n)$ with $\sum_i \psi_i = v([n])$, define the *excess of coalition S with respect to $\boldsymbol{\psi}$* to be $e(S, \boldsymbol{\psi}) := v(S) - \sum_{i \in S} \psi_i$, i.e., how unhappy coalition S is with the allocation $\boldsymbol{\psi}$. Among all allocation vectors, consider the ones that minimize the largest excess. Of these, consider those that minimize the second largest excess, and so on. The resulting allocation is called the nucleolus. (Note that when core allocations exist, the nucleolus and core allocations coincide.)

---

[3]A player with a concave utility function is called *risk-averse* because he prefers an allocation of $(x + y)/2$ to a lottery where he would receive x or y with probability $1/2$ each.

Recall the setup of bankruptcy problems from §11.2. There is an associated cooperative game defined as follows: Given any set of creditors S, let $v(S) := \max(A - \sum_{i \notin S} c_i, 0)$. The corresponding Shapley values coincide with O'Neill's solution of the bankruptcy problem [O'N82], and the nucleolus coincides with the Talmud rule [AM85]. For more on this topic and on cooperative games in general, see [MSZ13].

Lloyd Shapley

John Nash

The Nash bargaining solution is from [NJ50]. The IIA axiom is controversial. For instance in Figure 12.3, player II might reasonably feel he is entitled to a higher allocation than player I in S' than in S.

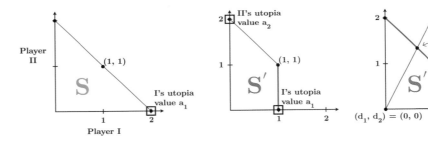

FIGURE 12.3. In the leftmost figure, each player gets half his utopia value, the largest value he could obtain at any point in S. By IIA, in the middle figure, the Nash bargaining solution gives player I her utopia value but gives player II only half his utopia value. The rightmost figure shows the Kalai-Smorodinsky solution for this example.

Kalai and Smorodinsky [KS75] addressed this by proposing another solution. If $a_i := \max_{x \in S} x_i$ is the *utopia value* for player i, then the Kalai-Smorodinsky (KS) bargaining solution $F^{KS}(S, \mathbf{d})$ is the point in S closest to $\mathbf{a} = (a_1, a_2)$ on the line connecting $\mathbf{d}$ with $\mathbf{a}$. The KS solution satisfies the affine covariance, Pareto optimality, and symmetry axioms. While it doesn't satisfy IIA, it does satisfy a monotonicity assumption: If $S \subset T$ and $\mathbf{a}(S, \mathbf{d}) = \mathbf{a}(T, \mathbf{d})$, then $F^{KS}(S, \mathbf{d}) \leq F^{KS}(T, \mathbf{d})$. See Figure 12.3.

Another criticism of the axiomatic approach to bargaining is that it doesn't describe how the players will arrive at the solution. This was addressed by Rubinstein [Rub82], who described the process of bargaining as an extensive form game in which players take turns making offers and delays are costly. In the first round, player I makes an offer in S. If it is accepted, the game ends. If not, then all outcomes (both S and $\mathbf{d}$) are scaled by a factor of $1 - \delta$, and it is player II's turn to make an offer. This continues until an offer is accepted. See Exercise 12.2.

In his book [Leo10], R. Leonard recounts the following story[4] about Shapley: John von Neumann, the preeminent mathematician of his day, was standing at the blackboard in the Rand Corporation offices explaining a complicated proof that a certain game had no solution. "No! No!" interjected a young voice from the back of the room, "That can be done much more simply!"

You could have heard a pin drop. Even years later, Hans Speier remembered the moment:

> Now my heart stood still, because I wasn't used to this sort of thing. Johnny von Neumann said, "Come up here, young man. Show me." He goes up, takes the piece of chalk, and writes down another derivation, and Johnny von Neumann interrupts and says, "Not so fast, young man. I can't follow."
>
> Now he was right; the young man was right. Johnny von Neumann, after this meeting, went to John Williams and said, "Who is this boy?"

Exercises

12.1. **The glove market revisited**. A proper pair of gloves consists of a left glove and a right glove. There are n players. Player 1 has two left gloves, while each of the other $n - 1$ players has one right glove. The payoff $v(S)$ for a coalition S is the number of proper pairs that can be formed from the gloves owned by the members of S.

(a) For $n = 3$, determine $v(S)$ for each of the seven nonempty sets $S \subset \{1, 2, 3\}$. Then find the Shapley value $\psi_i(v)$ for each of the players $i = 1, 2, 3$.

(b) For a general n, find the Shapley value $\psi_i(v)$ for each of the n players $i = 1, 2, \ldots, n$.

12.2. **Rubinstein bargaining:** Consider two players deciding how to split a cake of size 1. The players alternate making offers for how to split the cake until one of the players accepts the offer made by the other. Suppose also that there is a cost to delaying: If a player rejects the current offer by the other, the value of the cake goes down by a factor of $1 - \delta$. Consider a strategy profile where each player offers a fraction x of the cake (in his turn) and accepts no offer in which he receives a fraction less than x. Determine which values of x yield an equilibrium and which of these are subgame perfect.

---

[4]This story is adapted from page 294 in Leonard's book.

CHAPTER 13

Social choice and voting

Suppose that the individuals in a society are presented with a list of alternatives (e.g., which movie to watch, or who to elect as president) and have to choose one of them. Can a selection be made so as to truly reflect the preferences of the individuals? What does it mean for a social choice to be fair?

When there are only two options to choose from, **majority rule** can be applied to yield an outcome that more than half of the individuals find satisfactory. When the number of options is three or more, pairwise contests may fail to yield a consistent ordering. This *paradox*, shown in Figure 13.1, was first discovered by the Marquis de Condorcet in the late eighteenth century.

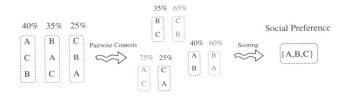

FIGURE 13.1. In pairwise contests A defeats C and C defeats B, yet B defeats A.

13.1. Voting and ranking mechanisms

We begin with two examples of voting systems.

EXAMPLE 13.1.1 (**Plurality voting**). In plurality voting, each voter chooses his or her favorite candidate, and the candidate who receives the most votes wins (with some tie-breaking rule). The winner need not obtain a majority of the votes. In the U.S., congressional elections are conducted using plurality voting.

To compare this to other voting methods, it's useful to consider extended plurality voting, where each voter submits a rank-ordering of the candidates and the candidate with the most first-place votes wins the election (with some tie-breaking rule).

This voting system is attractively simple but, as shown in Figure 13.2, it has the disturbing property that the candidate that is elected can be the least favorite for a majority of the population. This occurred in the 1998 election for governor of Minnesota when former professional wrestler Jesse Ventura won the election with 37% of the vote. Exit polls showed that he would have lost (with a wide margin) one-on-one contests with each of the other two candidates.

Comparing Figure 13.2 and Figure 13.3 indicates that plurality is strategically vulnerable. The voters who prefer C to B to A can change the outcome from A to B without changing the relative ordering of A and B in the rankings they submit.

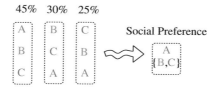

FIGURE 13.2. Option A is preferred by 45% of the population, option B by 30%, and option C by 25%, and thus A wins a plurality vote. However, A is the least favorite for 55% of the population.

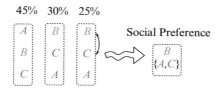

FIGURE 13.3. When 25% strategically switch their votes from C to B, the relative ranking of A and B in the outcome changes.

EXAMPLE 13.1.2 (**Runoff elections**). A modification of plurality that avoids the Minnesota scenario mentioned above is runoff elections. If in the first round no candidate has a majority, then the two leading candidates compete in a second round. This system is used in many countries including India, Brazil, and France.

FIGURE 13.4. In the first round C is eliminated. When votes are redistributed, B gets the majority.

This method is also strategically vulnerable. If voters in the second group from Figure 13.4 knew the distribution of preferences, they could ensure a victory for B by having some of them conceal their true preference and move C to the top of their rankings, as shown in Figure 13.5.

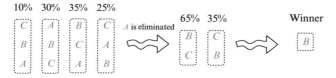

FIGURE 13.5. Some of the voters from the second group in Figure 13.4 misrepresent their true preferences, ensuring that A is eliminated. As a result, B wins the election.

13.2. Definitions

We consider settings in which there is a set of candidates Γ, a set of n voters, and a **rule** that describes how the voters' preferences are used to determine an outcome. We consider two different kinds of rules. A **voting rule** produces a single winner, and a **ranking rule** produces a **social ranking** over the candidates. Voting rules are obviously used for elections, or, more generally, when a group needs to select one of several alternatives. A ranking rule might be used when a university department is ranking faculty candidates based on the preferences of current faculty members.

In both cases, we assume that the ranking of each voter is represented by a preference relation $\succ$ on the set of candidates Γ which is **complete** ($\forall A, B$, either $A \succ B$ or $B \succ A$) and **transitive** ($A \succ B$ and $B \succ C$ implies $A \succ C$). This definition does not allow for ties; we discuss rankings with ties in the notes.

We use $\succ_i$ to denote voter i's preference relation: $A \succ_i B$ if voter i strictly prefers candidate A to candidate B. A preference profile $(\succ_1, \ldots, \succ_n)$ describes the preference relations of all n voters.

DEFINITION 13.2.1. A **voting rule** f maps each **preference profile** $\pi = (\succ_1, \ldots, \succ_n)$ to an element of Γ, the winner of the election.

DEFINITION 13.2.2. A **ranking rule** R associates to each preference profile, $\pi = (\succ_1, \ldots, \succ_n)$, a **social ranking**, another complete and transitive preference relation $\rhd = R(\pi)$. ($A \rhd B$ means that A is strictly preferred to B in the social ranking.)

REMARK 13.2.3. An obvious way to obtain a voting rule from a ranking rule is to output the top ranked candidate. (For another way, see Exercise 13.3.) Conversely, a voting rule yields an **induced ranking rule** as follows. Apply the voting rule to select the top candidate. Then apply the voting rule to the remaining candidates to select the next candidate and so on. However, not all ranking rules can be obtained this way; see Exercise 13.2.

An obvious property that we would like a ranking rule R to have is **unanimity**: If for every voter i we have $A \succ_i B$, then $A \rhd B$. In words, if every voter strictly prefers candidate A to B, then A should be strictly preferred to B in the social ranking.

Kenneth Arrow introduced another property called **Independence of Irrelevant Alternatives**[1] **(IIA)**: For any two candidates A and B, the preference between A and B in the social ranking depends only on the voters' preferences between A and B. Formally, if $\pi = \{\succ_i\}$ and $\pi' = \{\succ_i'\}$ are two profiles for which each voter has the same preference between A and B, i.e., $\{i \mid A \succ_i B\} = \{i \mid A \succ_i' B\}$, then $A \rhd B$ implies $A \rhd' B$.

IIA seems appealing at first glance, but as we shall see later, it is problematic. Indeed, almost all ranking rules violate IIA. The next lemma shows that if IIA fails, then there exist profiles in which some voter is better off submitting a ranking that differs from his ideal ranking.

DEFINITION 13.2.4. A ranking rule R is **strategically vulnerable** at the profile $\pi = (\succ_1, \ldots, \succ_n)$ if there is a voter i and alternatives A and B so that $A \succ_i B$ and $B \rhd A$ in $R(\pi)$, yet replacing $\succ_i$ by $\succ_i^*$ yields a profile π^* such that $A \rhd^* B$ in $R(\pi^*)$.

---

[1]This is similar to the notion by the same name from §12.4.

LEMMA 13.2.5. *If a ranking rule R violates IIA, then it is strategically vulnerable.*

PROOF. Let $\boldsymbol{\pi} = \{\succ_i\}$ and $\boldsymbol{\pi}' = \{\succ_i'\}$ be two profiles that are identical with respect to preferences between candidates A and B but differ on the social ranking of A relative to B. That is, $\{j \mid A \succ_j B\} = \{j \mid A \succ_j' B\}$, but $A \rhd B$ in $R(\boldsymbol{\pi})$, whereas $B \rhd' A$ in $R(\boldsymbol{\pi}')$. Let $\boldsymbol{\sigma}_i = (\succ_1', \ldots, \succ_i', \succ_{i+1}, \ldots, \succ_n)$, so that $\boldsymbol{\sigma}_0 = \boldsymbol{\pi}$ and $\boldsymbol{\sigma}_n = \boldsymbol{\pi}'$. Let $i \in [1, n]$ be such that $A \rhd B$ in $R(\boldsymbol{\sigma}_{i-1})$, but $B \rhd A$ in $R(\boldsymbol{\sigma}_i)$. If $B \succ_i A$, then R is strategically vulnerable at $\boldsymbol{\sigma}_{i-1}$ since voter i can switch from $\succ_i$ to $\succ_i'$. Similarly, if $A \succ_i B$, then R is vunerable at $\boldsymbol{\sigma}_i$ since voter i can switch from $\succ_i'$ to $\succ_i$. $\qquad\square$

For plurality voting, as we saw in the example of Figures 13.2 and 13.3, the induced ranking rule violates IIA. Similarly, Figure 13.8 shows that the ranking rule induced by runoff elections also violates IIA, since it allows for the relative ranking of A and B to be switched without changing any of the individual A-B preferences.

There is one ranking rule that obviously does satisfy IIA:

EXAMPLE 13.2.6 (**Dictatorship**). A ranking rule is a *dictatorship* if there is a voter v whose preferences are reproduced in the outcome. In other words, for every pair of candidates A and B, we have $A \succ_v B$ if and only if $A \rhd B$.

13.3. Arrow's Impossibility Theorem

THEOREM 13.3.1. *Any ranking rule that satisfies unanimity and independence of irrelevant alternatives is a dictatorship.*

What does the theorem mean? If we want to avoid dictatorship, then it need not be optimal for voters to submit their ideal ranking; the same applies to voting by Theorem 13.4.2. Thus, strategizing is an inevitable part of ranking and voting. See §13.7 for a proof of Arrow's theorem.

REMARK 13.3.2. Not only is IIA impossible to achieve under any reasonable voting scheme; it is doubtful if it is desirable because it ignores key information in the rankings, namely the strengths of preferences. See Figure 13.6.

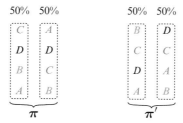

FIGURE 13.6. Given the profile $\boldsymbol{\pi}$, it seems that society should rank A above B since, for the second group, A is their top-ranked candidate. In profile $\boldsymbol{\pi}'$, the situation is reversed, yet IIA dictates that the relative social ranking of A and B is the same in both profiles.

13.4. The Gibbard-Satterthwaite Theorem

Arrow's Impossibility Theorem applies to the setting where a social ranking is produced. A similar phenomenon arises even when the goal is to select a single candidate. Consider n voters in a society, each with a complete ranking of a set of m candidates Γ and a voting rule f mapping each profile $\boldsymbol{\pi} = (\succ_1, \ldots, \succ_n)$ of n rankings of Γ to a candidate $f(\boldsymbol{\pi}) \in \Gamma$. A voting rule for which no voter can benefit by misreporting his ranking is called **strategy-proof**.

DEFINITION 13.4.1. A voting rule f from profiles to Γ is **strategy-proof** if for all profiles $\boldsymbol{\pi}$, candidates A and B, and voters i, the following holds: If $A \succ_i B$ and $f(\boldsymbol{\pi}) = B$, then all $\boldsymbol{\pi}'$ that differ from $\boldsymbol{\pi}$ only in voter i's ranking satisfy $f(\boldsymbol{\pi}') \neq A$.

THEOREM 13.4.2. *Let f be a strategy-proof voting rule onto Γ, where $|\Gamma| \geq 3$. Then f is a **dictatorship**. That is, there is a voter i such that for every profile $\boldsymbol{\pi}$ voter i's highest ranked candidate is equal to $f(\boldsymbol{\pi})$.*

The proof of the theorem is in §13.8.

13.5. Desirable properties for voting and ranking

Arrow's theorem is often misconstrued to imply that all voting systems are flawed and hence it doesn't matter which voting system is used. In fact, there are many dimensions on which to evaluate voting systems and some systems are better than others.

The following are desirable properties of voting systems beyond unanimity and IIA:

(1) **Anonymity (i.e., symmetry):** The identities of the voters should not affect the results. I.e., if the preference orderings of voters are permuted, the society ranking should not change. This is satisfied by most reasonable voting systems, but not by the US electoral college or other regional based systems. Indeed, switching profiles between very few voters in California and Florida would have changed the results of the 2000 election between Bush and Gore.

(2) **Monotonicity:** If a voter moves candidate A higher in his ranking without changing the order of other candidates, this should not move A down in the society ranking.

(3) **Condorcet winner criterion:** If a candidate beats all other candidates in pairwise contests, then he should be the winner of the election. A related, and seemingly[2] weaker, property is the **Condorcet loser criterion**: The system should never select a candidate that loses to all others in pairwise contests.

(4) **IIA with preference strengths:** If two profiles have the same preference strengths for A versus B in all voter rankings, then they should yield the same preference order between A and B in the social ranking. (The *preference strength* of A versus B in a ranking is the number of places where A is ranked above B, which can be negative.)

---

[2]See Exercise 13.7.

(5) **Cancellation of ranking cycles:** If there is a subset of N candidates, and N voters whose rankings are the N cyclic shifts of one another (e.g. three voters each with a different ranking from Figure 13.1), then removing these N voters shouldn't change the outcome.

13.6. Analysis of specific voting rules

Next we examine the extent to which the properties just described are satisfied by specific voting and ranking rules.

In **instant runoff voting (IRV),** also called *plurality with elimination*, each voter submits a ranking, and the winner in an election with N candidates is determined as follows. If $m = 2$, majority vote is used. If $m > 2$, the candidate with the fewest first-place votes is eliminated and removed from all the rankings. An instant runoff election is then run on the remaining $m - 1$ candidates. When there are three candidates, this method is equivalent to runoff elections. See Figure 13.7.

FIGURE 13.7. In the first round C is eliminated. When votes are redistributed, B gets the majority.

IRV satisfies anonymity but fails monotonicity, as shown in Figure 13.8.

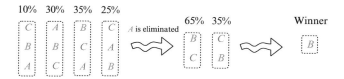

FIGURE 13.8. When some of the voters from the second group in Figure 13.7 switch their preferences by moving B below C, B switches from being a loser to a winner.

IRV does not satisfy the Condorcet winner criterion, as shown in Figure 13.9, but satisfies the Condorcet loser criterion since the Cordorcet loser would lose the runoff if he gets there. IRV fails IIA with preference strengths and cancellation of ranking cycles. See Exercise 13.4.

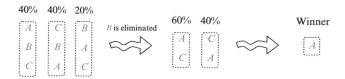

FIGURE 13.9. B is a Condorcet winner but loses the election.

The Burlington, Vermont, mayoral election of 2009 used IRV. The IRV winner (Bob Kiss) was neither the same as the plurality winner (Kurt Wright) nor the Condorcet winner (Andy Montroll, who was also the Borda count winner; see definition below). As a consequence, the IRV method was repealed in Burlington by a vote of 52% to 48% in 2010.

Borda count is a ranking rule in which each voter's ranking is used to assign points to the candidates. If there are m candidates, then m points are assigned to each voter's top-ranked candidate, $m-1$ points to the second-ranked candidate, and so on. The candidates are then ranked in decreasing order of their point totals (with ties broken arbitrarily). Borda count is equivalent to giving each candidate the sum of the votes he would get in pairwise contests with all other candidates.

The Borda count satisfies anonymity, IIA with preference strengths, monotonicity, the Condorcet loser criterion, and cancellation of ranking cycles. It does not satisfy the Condorcet winner criterion, e.g., if 60% of the population has preferences $A \succ B \succ C$ and the remaining 40% have preferences $B \succ C \succ A$. This example illustrates a weakness of the Condorcet winner criterion: It ignores the strength of preferences.

By Arrow's Theorem, the Borda count violates IIA and is strategically vulnerable. In the example shown in Figure 13.10, A has an unambiguous majority of votes and is also the winner.

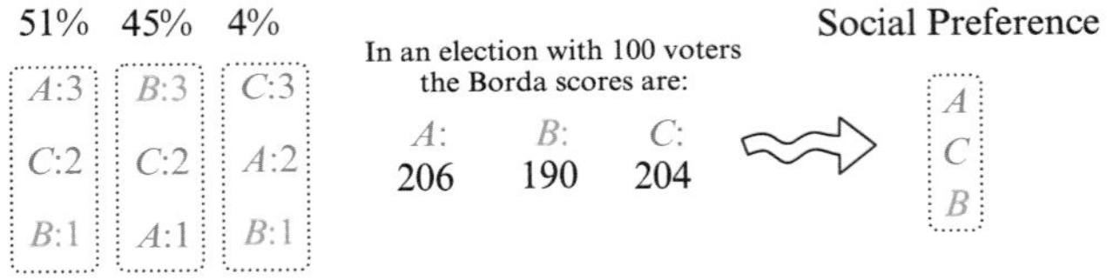

51% 45% 4%
In an election with 100 voters the Borda scores are: Social Preference

A:3	B:3	C:3
C:2	C:2	A:2
B:1	A:1	B:1

A: B: C:
206 190 204

A
C
B

FIGURE 13.10. Alternative A has the overall majority and is the winner under Borda count.

However, if supporters of C (the third group) were to strategically rank B above A, they could ensure a victory for C. This is also a violation of IIA since none of the individual A-C preferences had been changed.

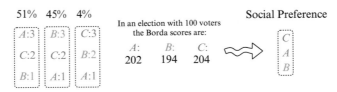

51% 45% 4%
In an election with 100 voters the Borda scores are: Social Preference

A:3	B:3	C:3
C:2	C:2	B:2
B:1	A:1	A:1

A: B: C:
202 194 204

C
A
B

FIGURE 13.11. Supporters of C can ensure his win by moving B up in their rankings.

A **positional voting method** is determined by a fixed vector $\mathbf{a} = a_1 \geq a_2 \geq \cdots \geq a_N$ as follows: Each voter assigns a_1 points to his top candidate, a_2 to the second, etc; the social ranking is determined by the point totals. Plurality and Borda count are positional voting methods. Every positional voting method satisfies anonymity, monotonicity, and cancellation of ranking cycles. See Figure 13.12 for a relevant example. No positional method satisfies the Condorcet winner criterion (see Figure 13.12), and the only one that satisfies IIA with preference strengths is the Borda count.

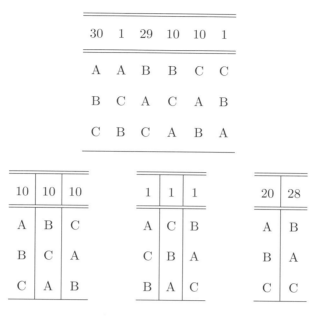

30	1	29	10	10	1
A	A	B	B	C	C
B	C	A	C	A	B
C	B	C	A	B	A

10	10	10
A	B	C
B	C	A
C	A	B

1	1	1
A	C	B
C	B	A
B	A	C

20	28
A	B
B	A
C	C

FIGURE 13.12. The table on top shows the rankings for a set of voters. (The top line gives the number of voters with each ranking.) One can readily verify that A wins all pairwise contests. However, for this voter profile, the Borda count winner is B. The three tables below provide a different rationale for B winning the election by dividing the voters into groups. The left and middle groups are ranking cycles. After cancellation of ranking cycles, the only voters remaining are in the bottom rightmost table. In this group, B is the clear winner. It follows that, in the original ranking, B is the winner for any positional method.

Approval voting is a voting scheme used by various professional societies, such as the American Mathematical Society and the American Statistical Association. In this procedure, the voters can approve as many candidates as they wish. Candidates are then ranked in the order of the number of approvals they received.

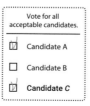

FIGURE 13.13. Candidates A and C will each receive one vote.

Given the ranking of the candidates by a voter, we may assume the voter selects some k and approves his top k choices. Approval voting satisfies anonymity, monotonicity, and a version of IIA: If voter i approved candidate A and disapproved of B, yet B was ranked above A by society, then there is no modification of voter i's input that can reverse this. There is no contradiction to Arrow's theorem because voters are not submitting a ranking. However, given a voter's preference ranking, he must choose how many candidates to approve, and the optimal choice depends on other voters' approvals. So strategizing is important here as well.

In close elections, approval voting often reduces to plurality, as each voter only approves his top choice.

13.7. Proof of Arrow's Impossibility Theorem*

Recall Theorem 13.3.1 which says that any ranking rule that satisfies unanimity and independence of irrelevant alternatives is a dictatorship.

Fix a ranking rule R that satisfies unanimity and IIA. The proof we present requires that we consider extremal candidates, those that are either most preferred or least preferred by each voter. The proof is written so that it applies verbatim to rankings with ties, as discussed in the notes; therefore, we occasionally refer to "strict" preferences.

LEMMA 13.7.1 (**Extremal Lemma**). *Consider an arbitrary candidate B. For any profile π in which B has an extremal rank for each voter (i.e., B is strictly preferred to all other candidates or all other candidates are strictly preferred to B), B has an extremal rank in the social ranking $R(\pi)$.*

PROOF. Suppose not. Then for such a profile π, with $\rhd = R(\pi)$, there are two candidates A and C such that $A \rhd B$ and $B \rhd C$. Consider a new profile $\pi' = (\succ'_1, \ldots, \succ'_n)$ obtained from π by having every voter move C just above A in his ranking (if C is not already above A). See Figure 13.14. None of the AB or BC preferences change since B started out and stays in the same extremal rank. Hence, by IIA, in the outcome $\rhd' = R(\pi')$, we have $A \rhd' B$ and $B \rhd' C$, and hence $A \rhd' C$. But this violates unanimity since for all voters i in π', we have $C \succ'_i A$.   □

DEFINITION 13.7.2. Let B be a candidate. Voter i is said to be B-**pivotal** for a ranking rule $R(\cdot)$ if there exist profiles π_1 and π_2 such that

- B is extremal for all voters in both profiles;
- the only difference between π_1 and π_2 is that B is strictly lowest ranked by i in π_1 and B is strictly highest ranked by i in π_2;
- B is ranked strictly lowest in $R(\pi_1)$ and strictly highest in $R(\pi_2)$.

Such a voter has the "power" to move candidate B from the very bottom of the social ranking to the very top.

LEMMA 13.7.3. *For every candidate B, there is a B-pivotal voter $v(B)$.*

PROOF. Consider an arbitrary profile in which candidate B is ranked strictly lowest by every voter. By unanimity, all other candidates are strictly preferred to B in the social ranking. Now consider a sequence of profiles obtained by letting the voters, one at a time, move B from the bottom to the top of their rankings. By the extremal lemma, for each one of these profiles, B is either at the top or at

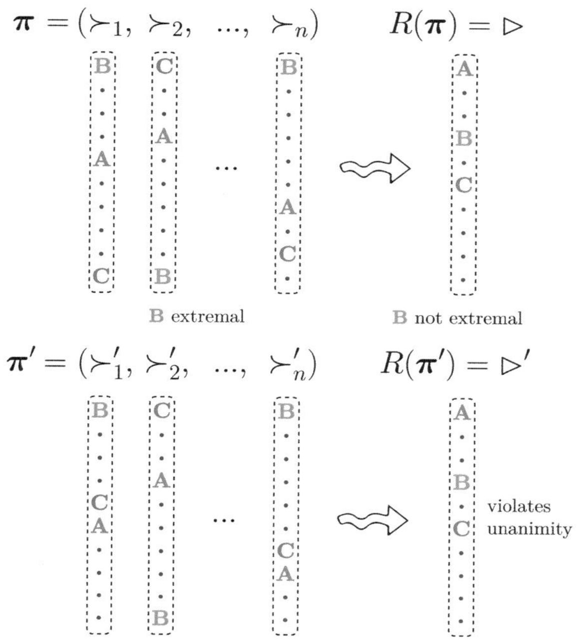

$$\boldsymbol{\pi} = (\succ_1, \succ_2, \ldots, \succ_n) \qquad R(\boldsymbol{\pi}) = \rhd$$

B extremal   **B** not extremal

$$\boldsymbol{\pi}' = (\succ'_1, \succ'_2, \ldots, \succ'_n) \qquad R(\boldsymbol{\pi}') = \rhd'$$

violates unanimity

FIGURE 13.14. Illustration of the proof of Lemma 13.7.1. The bottom figure shows what happens when every voter whose preference in $\boldsymbol{\pi}$ has C below A moves C just above A.

the bottom of the social ranking. Also, by unanimity, as soon as all the voters put B at the top of their rankings, so must the social ranking. Hence, there is a first voter v whose change in preference precipitates the change in the social ranking of candidate B. This change is illustrated in Figure 13.15, where $\boldsymbol{\pi}_1$ is the profile just before v has switched B to the top with $\rhd_1 = R(\boldsymbol{\pi}_1)$ and $\boldsymbol{\pi}_2$ the profile immediately after the switch with $\rhd_2 = R(\boldsymbol{\pi}_2)$. This voter v is B-pivotal. $\square$

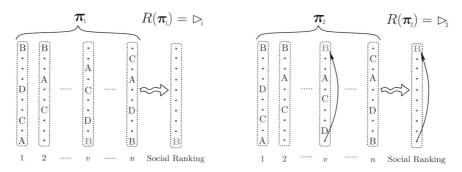

FIGURE 13.15. When, in $\boldsymbol{\pi}_1$, voter v moves B to the top of his ranking, resulting in $\boldsymbol{\pi}_2$, B moves to the top of the social ranking.

LEMMA 13.7.4. *If voter v is B-pivotal, then v is a dictator on $\Gamma \setminus \{B\}$; i.e., for any profile $\boldsymbol{\pi}$, if A, B, and C are distinct and $A \succ_v C$ in $\boldsymbol{\pi}$, then $A \rhd C$ in $R(\boldsymbol{\pi})$.*

PROOF. Since v is B-pivotal, there are profiles $\boldsymbol{\pi}_1$ and $\boldsymbol{\pi}_2$ such that B is extremal for all voters, B is ranked lowest by v in $\boldsymbol{\pi}_1$ and in $R(\boldsymbol{\pi}_1)$, and B is ranked highest by v in $\boldsymbol{\pi}_2$ and in $R(\boldsymbol{\pi}_2)$.

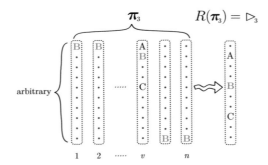

FIGURE 13.16. In $\boldsymbol{\pi}_3$, voter B is in the same extremal position as $\boldsymbol{\pi}_2$, except that A is just above B for voter v. Otherwise, the preferences in $\boldsymbol{\pi}_3$ are arbitrary.

Suppose that A, B, and C are distinct and let $\boldsymbol{\pi}$ be a profile in which $A \succ_v C$. Construct profile $\boldsymbol{\pi}_3$ from $\boldsymbol{\pi}$ as follows:

- For each voter $i \neq v$, move B to the extremal position he has in $\boldsymbol{\pi}_2$.
- Let v rank A first and B second.

Let $\rhd_3 = R(\boldsymbol{\pi}_3)$. Then the preferences between A and B in $\boldsymbol{\pi}_3$ are the same as in $\boldsymbol{\pi}_1$ and thus, by IIA, we have $A \rhd_3 B$. Also, the preferences between B and C in $\boldsymbol{\pi}_3$ are the same as in $\boldsymbol{\pi}_2$ and thus, by IIA, we have $B \rhd_3 C$. Hence, by transitivity, we have $A \rhd_3 C$. See Figure 13.16. Since the A-C preferences of all voters in $\boldsymbol{\pi}$ are the same as in $\boldsymbol{\pi}_3$, we must have $A \rhd C$. $\square$

PROOF OF THEOREM 13.3.1. By Lemmas 13.7.3 and 13.7.4, there is a B-pivotal voter $v = v(B)$ that is a dictator on $\Gamma \setminus \{B\}$. Let $\boldsymbol{\pi}_1$ and $\boldsymbol{\pi}_2$ be the profiles from the definition of v being B-pivotal. We claim that for any other candidate, say C, the C-pivotal voter $v' = v(C)$ is actually the same voter; i.e., $v = v'$.

To see this, consider A distinct from both B and C. We know that in $\rhd_1$, we have $A \rhd B$, and in $\rhd_2$, we have $B \rhd A$. Moreover, by Lemma 13.7.4, v' dictates the strict preference between A and B in both of these outcomes. But in both profiles, the strict preference between A and B is the same for all voters other than v. Hence $v' = v$, and thus v is a dictator (over all of Γ). $\square$

13.8. Proof of the Gibbard-Satterthwaite Theorem*

Recall Theorem 13.4.2 which says that if f is a strategy-proof voting rule onto Γ, where $|\Gamma| \geq 3$, then f is a **dictatorship**. That is, there is a voter i such that for every profile $\boldsymbol{\pi}$, voter i's highest ranked candidate is equal to $f(\boldsymbol{\pi})$.

We deduce this theorem from Arrow's theorem, by showing that if f is strategy-proof and is not a dictatorship, then it can be extended to a ranking rule that satisfies unanimity, IIA, and that is not a dictatorship, a contradiction.

The following notation will also be useful.

DEFINITION 13.8.1. Let $\pi = (\succ_1, \ldots, \succ_n)$ and $\pi' = (\succ'_1, \ldots, \succ'_n)$ be two profiles and let $r_i(\pi, \pi')$ denote the profile $(\succ'_1, \ldots, \succ'_i, \succ_{i+1}, \ldots, \succ_n)$. Thus $r_0(\pi, \pi') = \pi$ and $r_n(\pi, \pi') = \pi'$.

We will repeatedly use the following lemma.

LEMMA 13.8.2. *Suppose that f is strategy-proof. Consider two profiles $\pi = (\succ_1, \ldots, \succ_n)$ and $\pi' = (\succ'_1, \ldots, \succ'_n)$ and two candidates X and Y such that*

- *all preferences between X and Y in π and π' are the same (i.e., $X \succ_i Y$ iff $X \succ'_i Y$ for all i);*
- *in π' all voters prefer X to all candidates other than Y (i.e., $X \succ'_i Z$ for all $Z \notin \{X, Y\}$);*
- *$f(\pi) = X$.*

Then $f(\pi') = X$.

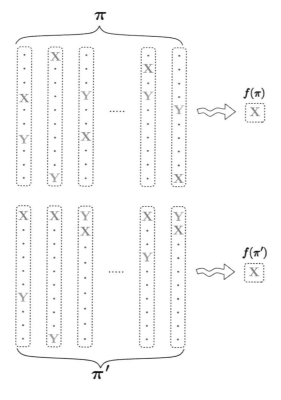

FIGURE 13.17. An illustration of the statement of Lemma 13.8.2.

PROOF. Let $r_i := r_i(\pi, \pi')$. We have $f(r_0) = X$ by assumption. We prove by induction on i that $f(r_i) = X$ or else f is not strategy-proof. To this end, suppose that $f(r_{i-1}) = X$. Observe that r_{i-1} and r_i differ only in voter i's ranking: In r_{i-1} it is $\succ_i$ and in r_i it is $\succ'_i$.

There are two cases to rule out: If $f(r_i) = Z \notin \{X, Y\}$, then in profile r_i, voter i has an incentive to lie and report $\succ_i$ instead of $\succ'_i$.

On the other hand, suppose $f(r_i) = Y$. If $X \succ_i Y$, then in profile r_i, voter i has an incentive to lie and report $\succ_i$ instead of $\succ'_i$. On the other hand, if $Y \succ_i X$, then in profile r_{i-1}, voter i has an incentive to lie and report $\succ'_i$ instead of $\succ_i$. $\square$

We also need the following definition.

DEFINITION 13.8.3. Let S be a subset of the candidates Γ, and let π be a ranking of the candidates Γ. Define a new ranking π^S by moving all candidates in S to the top of the ranking, maintaining the same relative ranking between them, as well as the same relative ranking between all candidates not in S.

CLAIM 13.8.4. *Let f be strategy-proof and onto Γ. Then for any profile π and every subset S of the candidates Γ, it must be that $f(\pi^S) \in S$.*

PROOF. Take any $A \in S$. Since f is onto, there is a profile $\tilde{\pi}$ such that $f(\tilde{\pi}) = A$. Consider the sequence of profiles $r_i = r_i(\tilde{\pi}, \pi^S)$, with $0 \leq i \leq n$. We claim that $f(r_{i-1}) \in S$ implies that $f(r_i) \in S$. Otherwise, on profile r_i, voter i has an incentive to lie and report $\tilde{\succ}_i$ instead of $\succ_i^S$. Thus, since $f(r_0) = f(\tilde{\pi}) \in S$, we conclude that $f(r_n) = f(\pi^S) \in S$ as well. $\square$

PROOF OF THEOREM 13.4.2. Let f be strategy-proof, onto, and a nondictatorship. Define a ranking rule $R(\pi)$ as follows. For each pair of candidates A and B, let $A \rhd B$ if $f(\pi^{\{A,B\}}) = A$ and $B \rhd A$ if $f(\pi^{\{A,B\}}) = B$. (Claim 13.8.4 guarantees that these are the only two possibilities.)

To see that this is a bona fide ranking rule, we observe that these pairwise rankings are transitive. If not, there is a triple of candidates such that $A \rhd B$, $B \rhd C$, and $C \rhd A$. Let $S = \{A, B, C\}$. We know that $f(\pi^S) \in S$; without loss of generality $f(\pi^S) = A$. Applying Lemma 13.8.2, with $\pi = \pi^S$ and $\pi' = \pi^{\{A,C\}}$, $X = A$ and $Y = C$, we conclude that $f(\pi^{\{A,C\}}) = A$ and $A \rhd C$, a contradiction.

Next, we verify that the ranking rule R satisfies unanimity, IIA, and is not a dictatorship.

Unanimity follows from the fact that if in π all voters have $A \succ_i B$, then $(\pi^{\{A,B\}})^A = \pi^{\{A,B\}}$, and thus by Claim 13.8.4, $f(\pi^{\{A,B\}}) = A$.

To see that IIA holds, let π_1 and π_2 be two profiles that agree on all of their AB preferences. Then by Lemma 13.8.2, with $\pi = \pi_1^{\{A,B\}}$ and $\pi' = \pi_2^{\{A,B\}}$ and Claim 13.8.4, we conclude that $f(\pi_1^{\{A,B\}}) = f(\pi_2^{\{A,B\}})$, and hence IIA holds.

Finally, the ranking rule R is not a dictatorship because f is not a dictatorship: For every voter v, there is a candidate A and a profile π in which A is v's highest ranked candidate, but $f(\pi) = B \neq A$. Then, applying Lemma 13.8.2 to the pair of profiles π and $\pi^{\{A,B\}}$, with $X = B$ and $Y = A$, we conclude that $f(\pi^{\{A,B\}}) = B$, and thus $B \rhd A$ in the outcome of the election. Hence voter v is not a dictator relative to the ranking rule R. $\square$

Notes

The history of social choice and voting is fascinating [ASS02, ASS11, AH94, Pro11]; we mention only a few highlights here: Chevalier de Borda proposed the Borda count in 1770 when he discovered that the Plurality method then used by the French Academy of Sciences was vulnerable to strategic manipulation. The Borda count was then used by the

Academy for the next two decades. The method of pairwise contests referred to in the beginning of this chapter was proposed by the Marquis de Condorcet after he demonstrated that the Borda count was also vulnerable to strategic manipulation. Apparently, Borda's response was that his scheme is "intended only for honest men." Condorcet proceeded to show a vulnerability in his own method—a tie in the presence of a preference cycle [dC85].

Besides the properties of voting systems discussed in the text, other important properties include:

- **Consistency:** If separate elections are run with two groups of voters and yield the same social ranking, then combining the groups should yield the same ranking. See Exercise 13.5.
- **Participation:** The addition of a voter who strictly prefers candidate A to B should not change the winner from candidate A to candidate B. See Exercise 13.6.

   Moulin showed that every method that elects the Condorcet winner, when there is one, must violate the participation property, assuming four or more candidates [Mou88b].
- **Reversal symmetry:** If all the rankings are reversed, then the social ranking should also be reversed.
- **Invariance to candidates dropping out:** If a candidate who loses an election on a given profile drops out, the winner of the election should not change.

   This property sounds reasonable, but none of the voting methods we discuss satisfy it. (It is satisfied by approval voting only if the candidate dropping out does not affect the other approvals.) Moreover, like IIA, it is not clear that this property is desirable: When a candidate drops out, some information is lost on the strength of preferences between other candidates.

The example in Figure 13.12 is from [dC85]; the analysis is due to [Saa95].

Since there need not be a Condorcet winner, various methods known as **Condorcet completions** have been proposed. For instance, a voting system known as Black's method elects the Condorcet winner if there is one and otherwise applies Borda count. A method known as Copeland's rule elects the candidate who wins the most pairwise comparisons. Charles Dodgson (also known as Lewis Carroll) proposed the following Condorcet completion. For each candidate, count how many adjacent swaps in voters' rankings are needed to make him a Condorcet winner. The Dodgson winner is the candidate that minimizes this count. However, it is NP-hard to determine the Dodgson winner [BTT89].

Kenneth Arrow

Donald Saari

Among known voting methods, Borda count and approval voting stand out for satisfying key desirable properties. Borda count has the advantage of allowing voters to submit a full ranking. Donald Saari [Saa95] showed that when the number of strategic voters is small, the Borda count is the least susceptible to strategic manipulation among positional methods; i.e., the number of profiles susceptible to manipulation is smallest.

Approval voting does not allow voters to submit a full ranking. At the other extreme, in some voting methods, voters provide information beyond a ranking. For example, in **cumulative voting,** each voter is given ℓ votes, which he can distribute among the candidates at will. Candidates are then ranked by vote totals.

Cumulative voting satisfies anonymity and monotonicity. It also satisfies IIA with preference strengths, if preference strengths are interpreted as the difference in the number of votes.

Finally, cumulative voting enables **minority representation:** Given k, a coordinated minority comprising a proportion p of the voters is able to determine the selection of $\lfloor kp \rfloor$ among the top k in the society ranking. See Exercise 13.8.

In 1949, Kenneth Arrow proved his famous Impossibility Theorem [Arr51]. The proof presented here is adapted from Geanakoplos [Gea04]. The Gibbard-Satterthwaite Theorem is from [Gib73] and [Sat75].

In 1972, Arrow was awarded the Nobel Prize in Economics (jointly with John Hicks), for "their pioneering contributions to general economic equilibrium theory and welfare theory."

For more on voting and social choice, see [Pac15, Saa01b, Saa01a, ASS02, ASS11, AH94]. Recent work in the computer science community has focused on the use of approximation to bypass some impossibility results and on connections of social choice with computational complexity, noise sensitivity, and sharp thresholds. See, e.g., [KM15, Kal10, Pro11, MBC⁺16].

For more on the 2009 Burlington, Vermont, mayoral election, see [GHS09].

Arrow's Impossibility Theorem

We have presented here a simplified proof of Arrow's theorem that is due to Geanakoplos [Gea04]. The version in the text assumes that each voter has a complete ranking of all the candidates. However, in many cases voters are indifferent between certain subsets of candidates. To accommodate this possibility, one can generalize the setting as follows.

Assume that the preferences of each voter are described by a relation $\succeq$ on the set of candidates Γ which is **reflexive** ($\forall A, A \succeq A$), **complete** ($\forall A, B, A \succeq B$ or $B \succeq A$ or both), and **transitive** ($A \succeq B$ and $B \succeq C$ implies $A \succeq C$).

As in the chapter, we use $\succeq_i$ to denote the preference relation of voter i: $A \succeq_i B$ if voter i weakly prefers candidate A to candidate B. However, we can now distinguish between strict preferences and indifference. As before, we use the notation $A \succ_i B$ to denote a strict preference; i.e., $A \succeq_i B$ but $B \nsucceq_i A$. (If $A \succeq_i B$ and $B \succeq_i A$, then voter i is indifferent between the two candidates.)

A reflexive, complete, and transitive relation $\succeq$ can be described in two other equivalent ways:

- It is a set of equivalence classes (each equivalence class is a set of candidates that the voter is indifferent between), with a total order on the equivalence classes. In other words, it is a ranking that allows for ties.
- It is the ranking induced by a function $g : \Gamma \to \mathbb{R}$ from the candidates to the reals, such that $A \succeq B$ if and only if $g(A) \geq g(B)$. Obviously, many functions induce the same preference relation.

A **ranking rule** R associates to each **preference profile**, $\boldsymbol{\pi} = (\succeq_1, \ldots, \succeq_n)$, another reflexive, complete, and transitive preference $\trianglerighteq = R(\boldsymbol{\pi})$.

In this more general setting, the definitions of unanimity and IIA are essentially unchanged. (Formally, IIA states that if $\boldsymbol{\pi} = \{\succeq_i\}$ and $\boldsymbol{\pi}' = \{\succeq_i'\}$ are two profiles such that $\{i \mid A \succeq_i B\} = \{i \mid A \succeq_i' B\}$ and $\{i \mid B \succeq_i A\} = \{i \mid B \succeq_i' A\}$, then $A \unrhd B$ implies $A \unrhd' B$.)

Arrow's Impossibility Theorem in this setting is virtually identical to the version given in the text: Any ranking rule that satisfies unanimity and IIA is a dictatorship. The only difference is that, in the presence of ties, voters other than the dictator can influence the outcome with respect to candidates that the dictator is indifferent between. Formally, in this more general setting, a dictator is a voter v all of whose *strict* preferences are reproduced in the outcome.

It is straightforward to check that the proof presented in Section 13.3 goes through unchanged.

Exercises

13.1.   Give an example where one of the losing candidates in a runoff election would have a greater support than the winner in a one-on-one contest.

13.2.   Describe a ranking rule that is not the induced ranking rule of any voting rule.

13.3.   Another way to go from a ranking rule to a voting rule is the following: Use the ranking rule. Eliminate the lowest ranked candidate. Repeat until one candidate remains. Apply this procedure and the one in the text to vote counting. What voting rule do you get in the two cases?

S 13.4.   Show that Instant Runoff violates the Condorcet winner criterion, IIA with preference strengths and cancellation of ranking cycles.

13.5.   The consistency property of a voting rule says: If separate elections are run with two groups of voters and yield the same social ranking, then combining the groups should yield the same ranking. Show that plurality, Borda count, and approval voting satisfy consistency but IRV does not.

13.6.   The participation criterion requires that the addition of a voter who strictly prefers candidate A to B should not change the winner from candidate A to candidate B. Show that plurality, approval voting, and Borda count satisfy the participation criterion, whereas instant runoff voting doesn't.

13.7.   Consider the following voting method: If there is a Condorcet winner, he is selected. Otherwise, plurality is used. Clearly this satisfies the Condorcet winner condition. Give an example with four candidates where it doesn't satisfy the Condorcet loser criterion.

13.8.   Show that cumulative voting enables minority representation: Given k, a coordinated minority comprising a proportion p of the voters can determine the selection of $\lfloor kp \rfloor$ among the top k in the society ranking.

13.9.   Determine which of the voting methods in the text satisfy reversal symmetry; that is, if all the rankings are reversed, then the social ranking should also be reversed.

13.10.   Show that the assumption in Theorem 13.4.2 that f is onto Γ, where $|\Gamma| \geq 3$, can be weakened to the assumption that the image of f has size at least 3. Show that 3 cannot be replaced by 2.

CHAPTER 14

Auctions

Auctions are an ancient mechanism for buying and selling goods. Indeed, the entire Roman Empire was sold by auction in 193 A.D. The winner was Didius Julianus. He became the next emperor but was in power only two months before he was overthrown and executed by Septimius Severus.

In modern times, many economic transactions are conducted through auctions: The US government runs auctions to sell treasury bills, spectrum licenses, and timber and oil leases, among other things. Christie's and Sotheby's run auctions to sell art. In the age of the Internet, we can buy and sell goods and services via auction, using the services of companies like eBay. The advertisement auctions that companies like Google and Microsoft run in order to sell advertisement slots on their webpages bring in significant revenue.

Why might a seller use an auction as opposed to simply fixing a price? Primarily because sellers often don't know how much buyers value their goods and don't want to risk setting prices that are either too low, thereby leaving money on the table, or so high that nobody will want to buy the item. An auction is a technique for dynamically setting prices. Auctions are particularly important these days because of their prevalence in Internet settings where the participants in the auction are computer programs or individuals who do not know each other.

14.1. Single item auctions

We are all familiar with the famous **English or ascending, auction** for selling a single item: The auctioneer starts by calling out a low price p. As long as there are at least two people willing to pay the price p, he increases p by a small amount. This continues until there is only one player left willing to pay the current price, at which point that player "wins" the auction, i.e., receives the item at that price.

When multiple rounds of communication are inconvenient, the English auction is sometimes replaced by other formats. For example, in a **sealed-bid first-price auction**, the participants submit sealed bids to the auctioneer. The auctioneer allocates the item to the highest bidder who pays the amount she bid.

We'll begin by examining auctions from two perspectives: What are equilibrium bidding strategies and what is the resulting revenue of the auctioneer?

To answer these questions, we need to know what value each bidder places on the item and what bidders know about each other. For example, in an art auction, the value a bidder places on a painting is likely to depend on other people's values for that painting, whereas in an auction for fish among restaurant owners, each bidder's value is known to him before the auction and is roughly independent of other bidder's valuations.

FIGURE 14.1. The Tsukiji Fish Market in Tokyo is one of the largest and busiest fish markets in the world. Each day before 5 a.m. a tuna auction is conducted there. The right-hand figure depicts the 2012 auction at Sotheby's for "The Scream" by Edvard Munch. The painting sold for 120 million dollars, about 40 million dollars more than the preauction estimates.

14.1.1. Bidder model. For most of this chapter, we will assume that each player has a private **value** v for the item being auctioned. This means that he would not want to pay more than v for the item: If he gets the item at a price p, his **utility**[1] is $v - p$, and if he doesn't get the item (and pays nothing), his utility is 0. Given the rules of the auction and any knowledge he has about other players' bids, he will bid so as to maximize his utility.

In the ascending auction, it is a **dominant strategy** for a bidder to increase his bid as long as the current price is below his value; i.e., doing this maximizes his utility *no matter what the other bidders do*. But how should a player bid in a sealed-bid first-price auction? Clearly, bidding one's value makes no sense since even upon winning, this would result in a utility of 0. So a bidder will want to bid lower than his true value. But how much lower? Low bidding has the potential to increase a player's gain, but at the same time it increases the risk of losing the auction. In fact, the optimal bid in such an auction depends on how the other players are bidding, which, in general, a bidder will not know.

DEFINITION 14.1.1. A **(direct) single-item auction** $\mathcal{A}$ with n bidders is a mapping that assigns to any vector of bids $\mathbf{b} = (b_1, \ldots, b_n)$ a winner and a set of prices. The allocation rule[2] of auction $\mathcal{A}$ is denoted by $\boldsymbol{\alpha}^{\mathcal{A}}[\mathbf{b}] = (\alpha_1[\mathbf{b}], \ldots, \alpha_n[\mathbf{b}])$, where $\alpha_i[\mathbf{b}]$ is 1 if the item is allocated to bidder i and 0 otherwise.[3] The payment rule of $\mathcal{A}$ is denoted by $\mathscr{P}^{\mathcal{A}}[\mathbf{b}] = (\mathscr{P}_1[\mathbf{b}], \ldots, \mathscr{P}_n[\mathbf{b}])$ where $\mathscr{P}_i[\mathbf{b}]$ is the payment of bidder i when the bid vector is $\mathbf{b}$.

A **bidding strategy** for agent[4] i is a mapping $\beta_i : [0, \infty) \to [0, \infty)$ which specifies agent i's bid $\beta_i(v_i)$ for each possible value v_i she may have.

DEFINITION 14.1.2 (**Private Values**). Suppose that n bidders are competing in a (direct) single-item auction. The bidders' values $V_1, V_2, \ldots, V_n$ are independent

---

[1]This is called a *quasilinear utility model*.

[2]Later we will also consider randomized allocation rules.

[3]When the auction is clear from the context, we will drop the superscripts and write $\boldsymbol{\alpha}[\cdot]$ and $\mathscr{P}[\cdot]$ for the allocation rule and payment rule.

[4]We use the terms "agent", "bidder", and "player" interchangeably.

and for each i the distribution F_i of V_i is common knowledge.[5] Each bidder i also knows the realization v_i of his own value V_i. Fix a bidding strategy $\beta_i : [0, \infty) \to [0, \infty)$ for each agent i. Note that we may restrict β_i to the support of V_i.[6]

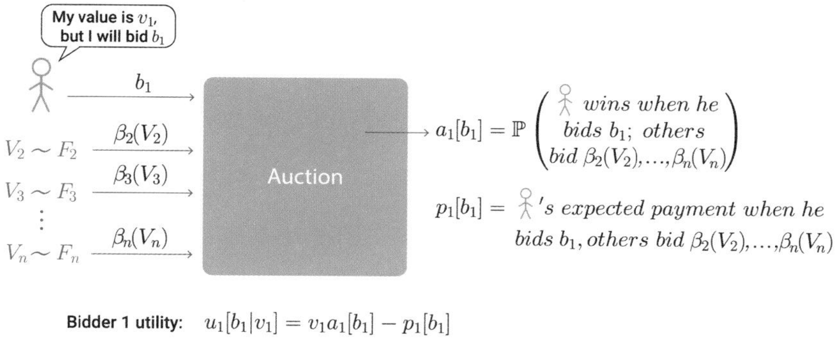

Bidder 1 utility: $u_1[b_1|v_1] = v_1 a_1[b_1] - p_1[b_1]$

FIGURE 14.2. Illustration of basic definitions from the perspective of bidder 1. In this figure, bidder 1 knows that each bidder i, for $2 \le i \le n$, has a value drawn independently from F_i and is bidding according to the bidding strategy $\beta_i(\cdot)$. The allocation probability, expected payment, and expected utility of bidder 1 are expressed in terms of bidder 1's *bid*.

The **allocation probabilities** are

$$a_i[b] := \mathbb{P}\,[\text{bidder } i \text{ wins bidding } b \text{ when other bids are } \beta_j(V_j),\ \forall j \ne i]$$
$$= \mathbb{E}\,[\alpha_i[b_i, \boldsymbol{\beta}_{-i}(V_{-i})]].$$

The **expected payments** are:

$$p_i[b] := \mathbb{E}\,[\text{payment of bidder } i \text{ bidding } b \text{ when other bids are } \beta_j(V_j),\ \forall j \ne i]$$
$$= \mathbb{E}\,[\mathscr{P}_i[b_i, \boldsymbol{\beta}_{-i}(V_{-i})]].$$

The **expected utility** of bidder i with value v_i bidding b is

$$u_i[b|v_i] = v_i a_i[b] - p_i[b].$$

The bidding strategy profile $(\beta_1, \ldots, \beta_n)$ is in **Bayes-Nash equilibrium** if for all i,

$$u_i[\beta_i(v_i)|v_i] \ge u_i[b|v_i] \quad \text{for all } v_i \text{ and } b.$$

In words, *for each bidder i, the bidding strategy β_i maximizes i's expected utility, given that for all $j \ne i$, bidder j is bidding $\beta_j(V_j)$.*

---

[5]I.e., all participants know the bidder distributions and know that each other knows and so on.

[6]The support of a random variable V with distribution function F is "the set of values it takes"; formally it is defined as $\text{supp}(V) = \text{supp}(F) := \cap_{\epsilon>0}\{x | F(x + \epsilon) - F(x - \epsilon) > 0\}$.

14.2. Independent private values

Consider a first-price auction in which each player's value V_i is drawn independently from a distribution F_i. If each other bidder j bids $\beta_j(V_j)$ and bidder i bids b, his expected utility is

$$u_i[b|v_i] = (v_i - b) \cdot a_i[b] = (v_i - b) \cdot \mathbb{P}\left[b > \max_{j \neq i} \beta_j(V_j)\right]. \tag{14.1}$$

EXAMPLE 14.2.1 (**Two-bidder first-price auction with uniformly distributed values**). Consider a two-bidder first-price auction where the V_i are independent and uniform on $[0, 1]$. Suppose that $\beta_1 = \beta_2 = \beta$ is an equilibrium, with $\beta : [0, 1] \to [0, \beta(1)]$ differentiable and strictly increasing. Bidder 1 with value v_1, knowing that bidder 2 is bidding $\beta(V_2)$, compares the utility of alternative bids b to $\beta(v_1)$. (We may assume that $b \in [0, \beta(1)]$ since higher bids are dominated by bidding $\beta(1)$.) With this bid, the expected utility for bidder 1 is

$$u_1[b|v_1] = (v_1 - b) \cdot \mathbb{P}[b > \beta(V_2)]. \tag{14.2}$$

We can write $b = \beta(w)$ for some $w \neq v_1$ and then $\mathbb{P}[b > \beta(V_2)] = w$. Using the notation[7]

$$u_1(w|v_1) := u_1[\beta(w)|v_1], \tag{14.3}$$

equation (14.2) becomes

$$u_1(w|v_1) = (v_1 - \beta(w)) \cdot w. \tag{14.4}$$

For β to be an equilibrium, the utility $u_1(w|v_1)$ must be maximized when $w = v_1$, and consequently,

$$\forall v_1 \qquad \frac{\partial u_1(w|v_1)}{\partial w}\bigg|_{w=v_1} = v_1 - \beta'(w)w - \beta(w)\bigg|_{w=v_1} = 0.$$

Thus, for *all* v_1,

$$v_1 = \beta'(v_1)v_1 + \beta(v_1) = (v_1\beta(v_1))'.$$

Integrating both sides, we obtain

$$\frac{v_1^2}{2} = v_1\beta(v_1) \quad \text{and so} \quad \beta(v_1) = \frac{v_1}{2}.$$

(We have dropped the constant term, as we assume that $\beta(0) = 0$.)

We now verify that $\beta_1 = \beta_2 = \beta$ is an equilibrium with $\beta(v) = v/2$. Bidder 1's utility when her value is v_1, she bids b, and bidder 2 bids $\beta(V_2) = V_2/2$ is

$$u_1[b|v_1] = \mathbb{P}\left[\frac{V_2}{2} \leq b\right](v_1 - b) = 2b(v_1 - b).$$

The choice of b that maximizes this utility is $b = v_1/2$. Since the bidders are symmetric, $\beta(v) = v/2$ is indeed an equilibrium.

Thus, in this Bayes-Nash equilibrium, the auctioneer's expected revenue is

$$\mathbb{E}\left[\max\left(\frac{V_1}{2}, \frac{V_2}{2}\right)\right]. \tag{14.5}$$

---

[7]It is worth emphasizing this notational convention: We use square brackets to denote functions of bids and regular parentheses to denote functions of alternative valuations.

In the example above of an equilibrium for the first-price auction, bidders must bid below their values, taking the distribution of competitors' values into account. This contrasts with the English auction, where no such strategizing is needed. Is strategic bidding (that is, considering competitors' values and potential bids) a necessary consequence of the convenience of sealed-bid auctions? No. In 1960, William Vickrey discovered that one can combine the low communication cost[8] of sealed-bid auctions with the simplicity of the optimal bidding rule in ascending auctions. We can get a hint on how to construct this combination by determining the revenue of the auctioneer in the ascending auction when all bidders act rationally: The item is sold to the highest bidder when the current price exceeds what other bidders are willing to offer; this threshold price is approximately the value of the item to the second highest bidder.

DEFINITION 14.2.2. In a **(sealed-bid) second-price auction** (also known as a **Vickrey auction**), the highest bidder wins the auction at a price equal to the second highest bid.

THEOREM 14.2.3. *The second-price auction is* **truthful**.[9] *In other words, for each bidder i and for any fixed set of bids of all other bidders, bidder i's utility is maximized by bidding her true value v_i.*

PROOF. Suppose the maximum of the bids submitted by bidders other than i is m. Then bidder i's utility in the auction is at most $\max(v_i - m, 0)$, where the first term is the utility for winning and the second term is the utility for losing. For each possible value v_i, this maximum is achieved by bidding truthfully. $\square$

REMARK 14.2.4. We emphasize that the theorem statement is not merely saying that truthful bidding is a Nash equilibrium, but rather the much stronger statement that bidding truthfully is a **dominant strategy**; i.e., it maximizes each bidder's utility **no matter what the other bids are**.

In Chapter 15 and Chapter 16, we show that a variant of this auction applies much more broadly. For example, when an auctioneer has k identical items to sell and each bidder wants only one, the following auction is also truthful.

DEFINITION 14.2.5. In a **(sealed-bid) k-unit Vickrey auction** the top k bidders win the auction at a price equal to the $(k+1)^{\text{st}}$ highest bid.

EXERCISE 14.a. Prove that the k-unit Vickrey auction is truthful.

14.3. Revenue in single-item auctions

From the perspective of the bidders in an auction, a second-price auction is appealing. They don't need to perform any complex strategic calculations. The appeal is less clear, however, from the perspective of the auctioneer. Wouldn't the auctioneer make more money running a first-price auction?

EXAMPLE 14.3.1. We return to our earlier example of two bidders, each with a value drawn independently from a U[0,1] distribution. From that analysis, we know

---

[8]Each bidder submits just one bid to the auctioneer.

[9]An alternative term often used is *incentive-compatible*.

that if the auctioneer runs a first-price auction, then in equilibrium his expected revenue will be

$$\mathbb{E}\left[\max\left(\frac{V_1}{2}, \frac{V_2}{2}\right)\right] = \frac{1}{3}.$$

On the other hand, suppose that in the exact same setting, the auctioneer runs a second-price auction. Since we can assume that the bidders will bid truthfully, the auctioneer's revenue will be the expected value of the second-highest bid, which is

$$\mathbb{E}\left[\min(V_1, V_2)\right] = \frac{1}{3},$$

exactly the same as in the 1ˢᵗ price auctions!
     In fact, in *both* cases, bidder i with value v_i has probability v_i of winning the auction, and the conditional expectation of his payment given winning is $v_i/2$: in the case of the first-price auction, this is because he bids $v_i/2$ and in the case of the second-price auction, this is because the expected bid of the other player is $v_i/2$. Thus, overall, in both cases, his expected payment is $v_i^2/2$.

     Coincidence? No. As we shall see next, the amazing *Revenue Equivalence Theorem* shows that any two auction formats that have the same allocation rule in equilibrium yield the same auctioneer revenue! (This applies even to funky auctions like the **all-pay auction**; see below.)

14.4. Toward revenue equivalence

     To test whether a strategy profile $\boldsymbol{\beta} = (\beta_1, \beta_2, \ldots, \beta_n)$ is an equilibrium, it will be important to determine the utility for bidder i when he bids *as if* his value is $w \neq v_i$; we need to show that he does not benefit from this deviation (in expectation). We adapt the notation[10] of Definition 14.1.2 as follows:

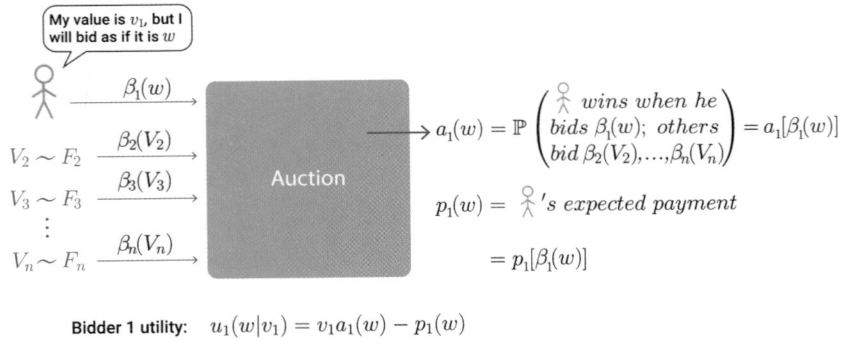

Bidder 1 utility: $u_1(w|v_1) = v_1 a_1(w) - p_1(w)$

FIGURE 14.3. Illustration of Definition 14.4.1 from the perspective of bidder 1. Here, in contrast to Figure 14.2, the allocation probability, expected payment, and expected utility of bidder 1 are expressed in terms of the value bidder 1 is "pretending" to have, as opposed to being expressed in terms of his bid.

---

[10] As in Example 14.2.1, we use square brackets to denote functions of bids and regular parentheses to denote functions of alternative valuations.

DEFINITION 14.4.1. Let $(\beta_i)_{i=1}^{n}$ be a strategy profile for n bidders with independent values $V_1, V_2, \ldots, V_n$. We assume that the distribution of each V_i, $i = 1, \ldots, n$, is common knowledge among all the bidders. Suppose bidder i, knowing his own value v_i and knowing that the other bidders $j \neq i$ are bidding $\beta_j(V_j)$, bids $\beta_i(w)$. Recalling Definition 14.1.1, we define the following:

- The **allocation probability** to bidder i is $a_i(w) := a_i[\beta_i(w)]$.
- His **expected payment**[11] is $p_i(w) := p_i[\beta_i(w)]$.
- His **expected utility** is $u_i(w|v_i) := u_i[\beta_i(w)|v_i] = v_i a_i(w) - p_i(w)$.

In this notation, $(\beta_i)_{i=1}^{n}$ is in **Bayes-Nash equilibrium (BNE)** only if [12]

$$u_i(v_i|v_i) \geq u_i(w|v_i) \quad \text{for all } i, v_i, \text{ and } w. \tag{14.6}$$

14.4.1. I.I.D. bidders. Consider the setting of n bidders, with i.i.d. values drawn from a distribution with strictly increasing distribution function F. Since the bidders are all symmetric, it's natural to look for *symmetric equilibria*; i.e., $\beta_i = \beta$ for all i. As above (dropping the subscript in $a_i(\cdot)$ due to symmetry), let

$$a(v) = \mathbb{P}\left[\text{item allocated to } i \text{ with bid } \beta(v) \text{ when other bidders bid } \beta(V_j)\right].$$

Consider any auction in which the item goes to the highest bidder (as in a first-price or second-price auction). If $\beta(\cdot)$ is strictly increasing, then

$$a(w) = \mathbb{P}\left[\beta(w) > \max_{j \neq i} \beta(V_j)\right] = \mathbb{P}\left[w > \max_{j \neq i} V_j\right] = F^{n-1}(w).$$

If bidder i bids $\beta(w)$, his expected utility is

$$u(w|v_i) = v_i a(w) - p(w). \tag{14.7}$$

Assume $p(w)$ and $a(w)$ are differentiable. For β to be an equilibrium, it must be that for all v_i, the derivative $v_i a'(w) - p'(w)$ vanishes at $w = v_i$, so

$$p'(v_i) = v_i a'(v_i) \quad \text{for all } v_i.$$

Hence, if $p(0) = 0$, we get

$$p(v_i) = \int_0^{v_i} v a'(v) dv,$$

which yields, via integration by parts, that

$$p(v_i) = v_i a(v_i) - \int_0^{v_i} a(w) dw. \tag{14.8}$$

In other words, *the expected payment of a bidder with value v is the same in any auction that allocates to the highest bidder. Hence, all such auctions yield the same expected revenue to the auctioneer.*

---

[11]We will usually assume that $p_i(0) = 0$, as this holds in most auctions.

[12]Conversely, if (14.6) holds and $u_i(v_i|v_i) \geq u_i(b|v_i)$ for all i, $b \notin \text{Image}(\beta_i)$, and v_i, then $(\beta_i)_{i=1}^{n}$ is a BNE.

14.4.2. Payment and revenue equivalence. The following theorem formalizes the discussion in the previous section.

THEOREM 14.4.2 (**Revenue Equivalence**). *Suppose that each agent's value V_i is drawn independently from the same strictly increasing distribution $F \in [0, h]$. Consider any n-bidder single-item auction in which the item is allocated to the highest bidder, and $u_i(0) = 0$ for all i. Assume that the bidders employ a symmetric strategy profile $\beta_i := \beta$ for all i, where β is strictly increasing in $[0, h]$.*

(i) *If $(\beta, \dots, \beta)$ is a Bayes-Nash equilibrium, then for a bidder with value v,*

$$a(v) = F(v)^{n-1} \quad and \quad p(v) = va(v) - \int_0^v a(w)dw. \qquad (14.9)$$

(ii) *If (14.9) holds for the strategy profile $(\beta, \dots, \beta)$, then for any bidder i with utility $u(\cdot|\cdot)$ and all $v, w \in [0, h]$,*

$$u(v|v) \geq u(w|v). \qquad (14.10)$$

REMARK 14.4.3. Part (ii) of the theorem implies that if (14.9) holds for a symmetric strategy profile, then these bidding strategies are an equilibrium relative to alternatives in the image of $\beta(\cdot)$. In fact, showing that (14.9) holds can be an efficient way to prove that $(\beta, \dots, \beta)$ is a Bayes-Nash equilibrium since strategies that are outside the range of β can often be ruled out directly. We will see this in the examples below.

PROOF OF THEOREM 14.4.2. In the previous section, we proved (i) under the assumption that $a(\cdot)$ and $p(\cdot)$ are differentiable; a proof of (i) without these assumptions is given in §14.6. For (ii), it follows from (14.9) that

$$u(v|v) = va(v) - p(v) = \int_0^v a(z)dz,$$

whereas

$$u(w|v) = va(w) - p(w) = (v - w)a(w) + wa(w) - p(w)$$
$$= (v - w)a(w) + \int_0^w a(z)dz.$$

If $v > w$, then

$$u(v|v) - u(w|v) = \int_w^v \big[a(z) - a(w)\big]dz > 0$$

since $a(v) = F(v)^{n-1}$ is an increasing function. The case $v < w$ is similar. Thus, for all $v, w \in [0, h]$, $u(v|v) \geq u(w|v)$. □

COROLLARY 14.4.4. *Under the assumptions of Theorem 14.4.2,*

$$p(v) = F(v)^{n-1}\mathbb{E}\left[\max_{i \leq n-1} V_i \;\Big|\; \max_{i \leq n-1} V_i \leq v\right]. \qquad (14.11)$$

PROOF. Since the truthful second-price auction allocates to the highest bidder, we can use it to calculate $p(v)$: The probability that a bidder wins is $a(v) = F(v)^{n-1}$, and if he wins, his payment has the distribution of $\max_{i \leq n-1} V_i$ given that this maximum is at most v. □

14.4.3. Applications. We now use this corollary to **derive** equilibrium strategies in a number of auctions.

First-price auction: (a) Suppose that β is strictly increasing on $[0, h]$ and defines a symmetric equilibrium. Then $a(v)$ and $p(v)$ are given by (14.9). Since the expected payment $p(v)$ in a first-price auction is $F(v)^{n-1}\beta(v)$, it follows that

$$\beta(v) = \mathbb{E}\left[\max_{i \leq n-1} V_i \mid \max_{i \leq n-1} V_i \leq v\right] = \int_0^v 1 - \left(\frac{F(w)}{F(v)}\right)^{n-1} dw. \qquad (14.12)$$

The rightmost expression in (14.12) follows from the general formula

$$\mathbb{E}[Z] = \int_0^\infty \mathbb{P}[Z \geq w] dw$$

for nonnegative random variables, applied to $Z = \max_i V_i$ conditioned on $Z \leq v$.

(b) Suppose that β is defined by the preceding equation. We verify that this formula actually defines an equilibrium. Since F is strictly increasing, by the preceding equation, β is also strictly increasing. Therefore $a(v) = F(v)^{n-1}$ and (14.9) holds. Hence (14.10) holds. Finally, bidding more than $\beta(h)$ is dominated by bidding $\beta(h)$. Hence this bidding strategy is in fact an equilibrium.

Examples:

With n bidders, each with a value that is $U[0, 1]$, we obtain that $\beta(v) = \frac{n-1}{n}v$.

With 2 bidders, each with a value that is exponential with parameter 1 (that is, $F(v) = 1 - e^{-v}$), we obtain

$$\beta(v) = \int_0^v 1 - \left(\frac{1 - e^{-w}}{1 - e^{-v}}\right) dw = 1 - \frac{ve^{-v}}{1 - e^{-v}}.$$

This function is always below 1. Thus, even a bidder with a value of 100 will bid below 1 in equilibrium!

All-pay auction: This auction allocates to the player that bids the highest but charges *every* player their bid. For example, architects competing for a construction project submit design proposals. While only one architect wins the contest, all competitors expend the effort to prepare their proposals. Thus, participants need to make the strategic decision as to how much effort to put in.

In an all-pay auction, $\beta(v) = p(v)$, and therefore by Corollary 14.4.4, we find that the only symmetric increasing equilibrium is given by

$$\beta(v) = F(v)^{n-1}\mathbb{E}\left[\max_{i \leq n-1} V_i \mid \max_{i \leq n-1} V_i \leq v\right].$$

For example, if F is uniform on $[0, 1]$, then $\beta(v) = \frac{n-1}{n}v^n$.

War of attrition auction: This auction allocates to the highest bidder, charges him the second-highest bid, and charges all other bidders their bid. For example, animals fighting over territory expend energy. A winner emerges when the fighting ends, and each animal has expended energy up to the point at which he dropped out or, in the case of the winner, until he was the last one left. See the third remark after Example 4.2.3 for another example of a war of attrition and Exercise 14.5 for the analysis of this auction.

FIGURE 14.4. A war of attrition auction.

EXERCISE 14.b. In the India Premier League (IPL), cricket franchises can acquire a player by participating in the annual auction. The rules of the auction are as follows. An English auction is run until either only one bidder remains or the price reaches m (for example m could be \$750,000). In the latter case, a sealed-bid first-price auction is run with the remaining bidders. (Each of these bidders knows how many other bidders remain).

Use the Revenue Equivalence Theorem to determine equilibrium bidding strategies in an IPL cricket auction for a player with n competing franchises. Assume that the value each franchise has for this player is uniform from 0 to 1 million.

14.5. Auctions with a reserve price

We have seen that in equilibrium, with players whose values for the item being sold are drawn independently from the same distribution, the expected seller revenue is the same for any auction that always allocates to the highest bidder. How should the seller choose which auction to run? As we have discussed, an appealing feature of the second-price auction is that it induces truthful bidding. On the other hand, the auctioneer's revenue might be lower than his own value for the item. A notorious example was the 1990 New Zealand sale of spectrum licenses in which a second-price auction was used; the winning bidder bid \$100,000 but paid only \$6! A natural remedy for situations like this is for the auctioneer to impose a **reserve price**.

DEFINITION 14.5.1. The **Vickrey auction (or second-price auction) with a reserve price** r is a sealed-bid auction in which the item is not allocated if all bids are below r. Otherwise, the item is allocated to the highest bidder, who pays the maximum of the second-highest bid and r.

A virtually identical argument to that of Theorem 14.2.3 shows that the Vickrey auction with a reserve price is truthful. Alternatively, the truthfulness follows by imagining that there is an extra bidder whose value/bid is the reserve price.

Perhaps surprisingly, an auctioneer may want to impose a reserve price even if his own value for the item is zero. For example, we have seen that for two bidders with values independent and drawn from $U[0, 1]$, all auctions that allocate to the highest bidder have an expected auctioneer revenue of $1/3$.

Now consider the expected revenue if, instead, the auctioneer uses the Vickrey auction with a reserve of r. Relative to the case of no reserve price, the auctioneer loses an expected revenue of $r/3$ if both bidders have values below r, for a total expected loss of $r^3/3$. On the other hand, he gains if one bidder is above r and one below. This occurs with probability $2r(1 - r)$, and the gain is r minus the expected value of the bidder below r; i.e., $r - r/2$. Altogether, the expected revenue is

$$\frac{1}{3} - \frac{r^3}{3} + 2r(1 - r)\frac{r}{2} = \frac{1}{3} + r^2 - \frac{4}{3}r^3.$$

Differentiating shows that this is maximized at $r = 1/2$, yielding an expected auctioneer revenue of $5/12$. (This is not a violation of the Revenue Equivalence Theorem because imposition of a reserve price changes the allocation rule.)

Remarkably, this simple auction optimizes the auctioneer's expected revenue over *all* possible auctions. It is a special case of *Myerson's optimal auction*, a broadly applicable technique for maximizing auctioneer revenue when agents' values are drawn from known prior distributions. We develop the theory of optimal auctions in §14.9.

14.5.1. Revenue equivalence with reserve prices. Theorem 14.4.2 generalizes readily to the case of single-item auctions in which the item is allocated to the highest bidder, as long as the bid is above the reserve price r. (See also Theorem 14.6.1). The only change in the theorem statement is that the allocation probability becomes

$$a(v) = F(v)^{n-1}\mathbb{1}_{\{v \geq r\}}.$$

As in §14.4.3, this enables us to solve for equilibrium bidding strategies in auctions with a reserve price. See Exercise 14.3.

14.5.2. Entry fee versus reserve price. Consider a second-price auction with an entry fee of δ. That is, to enter the auction, a bidder must pay δ, and then a standard second-price auction (no reserve) is run. Fix a bidder i and suppose that all bidders $j \neq i$ employ the following threshold strategy: Bidder j with value v_j enters the auction if and only if $v_j \geq r := r(\delta)$ and then bids truthfully.

Then it is a best response for bidder i to employ the same strategy if

$$rF(r)^{n-1} = \delta.$$

To see this, let $u_E(v)$ be the overall utility to bidder i with value v if he pays the entry fee and bids truthfully[13]. If $v > r$, then $u_E(v) \geq -\delta + vF(r)^{n-1} \geq 0$, since the probability that none of the other bidders enter the auction is $F(r)^{n-1}$. On the other hand, if $v < r$, then $u_E(v) = -\delta + vF(r)^{n-1} < 0$.

---

[13]Given that he enters, he is participating in a second-price auction and it is a dominant strategy to bid truthfully.

Now, let us compare the second-price auction with entry fee of $\delta = rF(r)^{n-1}$ to a second-price auction with a reserve price of r. Clearly, in both cases,

$$a(v) = F(v)^{n-1} \mathbb{1}_{\{v \geq r\}}.$$

Moreover, the expected payment for a bidder in both cases is the same: If $v < r$, the expected payment is 0 for both. For $v \geq r$ and the auction with entry fee,

$$p(v) = \delta + \left(F(v)^{n-1} - F(r)^{n-1} \right) \mathbb{E} \left[\max_{i \leq n-1} V_i \,\middle|\, r \leq \max_{i \leq n-1} V_i \leq v \right],$$

whereas with the reserve price,

$$p(v) = rF(r)^{n-1} + \left(F(v)^{n-1} - F(r)^{n-1} \right) \mathbb{E} \left[\max_{i \leq n-1} V_i \,\middle|\, r \leq \max_{i \leq n-1} V_i \leq v \right].$$

This means that $u(w|v) = va(w) - p(w)$ is the same in both auctions. (This gives another proof that the threshold strategy is a Bayes-Nash equilibrium in the entry-fee auction.)

Notice, however, that the payment of a bidder with value v can differ in the two auctions. For example, if bidder 1 has value $v > r$ and all other bidders have value less than r, then in the entry-fee auction, bidder 1's payment is $\delta = rF(r)^{n-1}$, whereas in the reserve-price auction it is r. Moreover, if bidder 1 has value $v > r$, but there is another bidder with a higher value, then in the entry-fee auction, bidder 1 loses the auction but still pays δ, whereas in the reserve-price auction he pays nothing. Thus, when the entry-fee auction is over, a bidder may regret having participated. This means that this auction is *ex-interim individually rational*, but not *ex-post individually rational*. See the definitions in §14.5.4.

14.5.3. Evaluation fee.

EXAMPLE 14.5.2. The queen is running a second-price auction to sell her crown jewels. However, she plans to charge an evaluation fee: A bidder must pay ϕ in order to examine the jewels and determine how much he values them prior to bidding in the auction.

Assume that bidder i's value V_i for the jewels is a random variable and that the V_i's are i.i.d. Prior to the evaluation, he only knows the distribution of V_i. He will learn the realization of V_i only if he pays the evaluation fee. In this situation, as long as the fee is below the bidder's expected utility in the second-price auction, i.e.,

$$\phi < \mathbb{E}\left[u(V_i|V_i)\right],$$

the bidder has an incentive to pay the evaluation fee. Thus, the seller can charge an evaluation fee equal to the bidder's expected utility minus some $\epsilon > 0$. The expected auctioneer revenue from bidder i's evaluation fee and his payment in the ensuing second-price auction is

$$\mathbb{E}\left[u(V_i|V_i)\right] - \epsilon + \mathbb{E}\left[p(V_i)\right] = \mathbb{E}\left[a_i(V_i) \cdot V_i\right] - \epsilon.$$

Since, in a second-price auction, the allocation is to the bidder with the highest value, the seller's expected revenue is

$$\mathbb{E}\left[\max_i (V_1, \ldots, V_n)\right] - \epsilon n,$$

which is essentially best possible.

FIGURE 14.5. The queen has fallen on hard times and must sell her crown jewels.

14.5.4. Ex-ante versus ex-interim versus ex-post. An auction, with an associated equilibrium $\boldsymbol{\beta}$, is called **individually rational** (IR) if each bidder's expected utility is nonnegative. The examples given above illustrate three different notions of individual rationality: *ex-ante*, when bidders know only the value distributions; *ex-interim*, when each bidder also knows his own value; and *ex-post*, after the auction concludes.

To formalize this, recall that $u(w_i|v_i)$ denotes the expectation (over $\mathbf{V}_{-i}$) of the utility of bidder i when he bids $\beta_i(w_i)$, his value is v_i, and each other bidder j bids $\beta_j(V_j)$.

- An auction, with an associated equilibrium $\boldsymbol{\beta}$, is **ex-ante** individually rational if, knowing only the distribution of his value and of the other bidder's values, each bidder i's expected utility is nonnegative. I.e.,

$$\mathbb{E}\left[u(V_i|V_i)\right] \geq 0. \tag{14.13}$$

The outside expectation here is over V_i. The evaluation fee auction of §14.5.3 is ex-ante IR.

- An auction, with an associated equilibrium, is **ex-interim** individually rational if for each bidder i,

$$u(v_i|v_i) \geq 0.$$

The entry-fee auction of §14.5.2 is ex-interim IR.

- An auction, with an associated equilibrium $\boldsymbol{\beta}$, is **ex-post** individually rational if for each bidder i,

$$u[\beta_i(v_i), \mathbf{b}_{-i}|v_i] \geq 0,$$

where $u[b_i, \mathbf{b}_{-i}|v_i]$ is the utility of player i when his value is v_i, he bids b_i and the other players bid $\mathbf{b}_{-i}$. The standard first-price and second-price auctions are ex-post IR since in these auctions a bidder never pays more than his bid, and in the equilibrium bidding strategy, a bidder never bids above his value.

14.6. Characterization of Bayes-Nash equilibrium

In §14.4.2, we saw that, with i.i.d. bidders, all auctions that allocate to the highest bidder result in the same auctioneer revenue in equilibrium. Revenue equivalence holds for other allocation rules, e.g., with reserve prices and randomization[14] if bidders are assumed to be i.i.d.

The next theorem generalizes Theorem 14.4.2 and will allow us to design revenue-maximizing auctions in §14.9.

THEOREM 14.6.1. *Let $\mathcal{A}$ be a (possibly randomized) auction for selling a single item, where bidder i's value V_i is drawn independently from F_i. Suppose that F_i is strictly increasing and continuous on $[0, h_i]$, with $F(0) = 0$ and $F(h_i) = 1$. (h_i can be ∞.)*

(a) *If $(\beta_1, \ldots, \beta_n)$ is a Bayes-Nash equilibrium, then for each agent i:*

(1) *The probability of allocation $a_i(v)$ is weakly increasing in v.*

(2) *The utility $u_i(v)$ is a convex function of v, with*

$$u_i(v) = \int_0^v a_i(z)dz + u_i(0).$$

(3) *The expected payment is determined by the allocation probabilities up to a constant $p_i(0)$:*

$$p_i(v) = v a_i(v) - \int_0^v a_i(z)dz - p_i(0) \tag{14.14}$$

(b) *Conversely, if $(\beta_1, \ldots, \beta_n)$ is a set of bidder strategies for which (1) and (14.14) hold, or (1) and (2), then for all bidders i and values v and w*

$$u_i(v|v) \geq u_i(w|v). \tag{14.15}$$

REMARK 14.6.2. See Figure 14.6 for an illustration of the theorem statement and a "proof by picture".

REMARK 14.6.3. This theorem implies that two auctions with the *same* allocation rule yield the same expected payments and auctioneer revenue in equilibrium (assuming $p_i(0) = 0$ for all i), without making any smoothness assumptions. However, if bidders valuations are not i.i.d., the equilibrium will not be symmetric (i.e. $\beta_i(\cdot) \neq \beta_j(\cdot)$ for $i \neq j$), so the first-price auction need not allocate to the highest bidder. Thus, the first and second-price auction are not revenue equivalent for asymmetric bidders.

---

[14]A *randomized* auction is defined as in Definition 14.1.1, but $\alpha_i[\mathbf{b}]$ represents the allocation probability to bidder i and takes values in $[0, 1]$. In a randomized auction, $\mathscr{P}_i[\mathbf{b}]$ is the expected payment of bidder i on bid vector $\mathbf{b}$. Definition 14.1.2 and Definition 14.4.1 remain unchanged.

PROOF. **(a):** Suppose that $(\beta_1, \ldots, \beta_n)$ is a Bayes-Nash equilibrium. In what follows, all quantities refer to bidder i, so for notational simplicity we usually drop the subscript i. Also, we use the shorthand notation $u(v) := u(v|v)$.

Consider two possible values that bidder i might have. If he has value v, then he has higher utility bidding $\beta_i(v)$ than $\beta_i(w)$; i.e.,

$$u(w|v) = va(w) - p(w) \leq u(v) = va(v) - p(v). \tag{14.16}$$

Similarly, if he has value w,

$$wa(w) - p(w) = u(w) \geq wa(v) - p(v) = u(v|w). \tag{14.17}$$

Subtracting (14.17) from (14.16), we get

$$(v - w)a(w) \leq u(v) - u(w) \leq (v - w)a(v).$$

Comparing the right and left sides of this inequality shows that $a(\cdot)$ is (weakly) increasing. Moreover, the left-hand inequality means that $a(w)$ is a subgradient of $u(\cdot)$ at w, as defined in (8) of Appendix C. It then follows from (9)–(11) of Appendix C that $u(\cdot)$ is convex and satisfies

$$u(v) = \int_0^v a(z)dz + u(0).$$

Finally, since $u(v) = va(v) - p(v)$, (14.14) follows.

(b): For the converse, from condition (14.14) (or (2)) it follows that

$$u(v) = \int_0^v a(z)dz,$$

whereas

$$u(w|v) = va(w) - p(w) = (v - w)a(w) + \int_0^w a(z)dz,$$

whence, by condition (1)

$$u(v) \geq u(w|v). \qquad \square$$

REMARK 14.6.4. Another way to see that $u(\cdot)$ is convex is to observe that

$$u(v) := u(v|v) = \sup_w u(w|v) = \sup_w \{va(w) - p(w)\},$$

and thus $u(v)$ is the supremum of affine functions. (See Appendix C for more on convex functions.)

EXERCISE 14.c. Extend Theorem 14.6.1 to the case of an auction for selling k identical items, where each bidder wants only one item. (Other than the fact that k items are being sold, as opposed to one, the theorem statement is unchanged.)

EXAMPLE 14.6.5. **Uniform price auction:** The industry which sells tickets to concerts and athletic events does not necessarily operate efficiently or, in some cases, profitably due to complex and unknown demand. This can lead to tickets being sold on secondary markets at exorbitant prices and/or seats being unfilled. An auction format, known as a uniform price auction, has been proposed in order to mitigate these problems. The auction works as follows: Prices start out high and drop until the tickets sell out; however, all buyers pay the price at which the final ticket is sold.

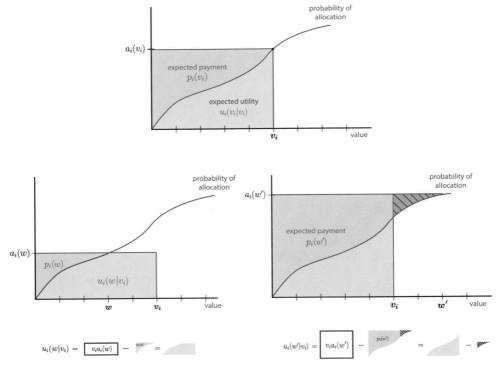

FIGURE 14.6. The top figure illustrates how the expected payment $p_i(\cdot)$ and utility are determined by the allocation function $a_i(\cdot)$ via equation (14.14). The bottom figures shows how a monotone allocation function and payments determined by (14.14) ensures that no bidder has an incentive to bid as if he had a value other than his true value. Consider a bidder with value v_i. On the left side, we see what happens to the bidder's utility if he bids as if his value is $w < v_i$. In this case, $u_i(w|v_i) = v_i a_i(w) - p_i(w) \le u_i(v_i|v_i)$. On the right side we see what happens to the bidder's expected utility if he bids as if his value is $w' > v_i$. In this case, $u_i(w'|v_i) = v_i a_i(w') - p_i(w')$ is less than the expected utility $u_i(v_i|v_i)$ he would have obtained by bidding $\beta_i(v)$ by an amount equal to the small blue striped area.

*   EXERCISE 14.d. Use the result of Exercise 14.c, the Revenue Equivalence Theorem (Corollary 14.4.2) and the equilibrium of the k-unit Vickrey auction to show that in a uniform price auction with n bidders and k tickets, where each bidder wants one ticket, the following bidding strategy $\beta(v)$ is in Bayes-Nash equilibrium:

$$\beta(v) = \mathbb{E}\left[\max_{i \le n-k} V_i \ \Big| \ \max_{i \le n-k} V_i \le v \right].$$

14.7. Price of anarchy in auctions

Consider an auctioneer selling a single item via a **first-price auction**. The **social surplus** $V(\mathbf{b})$ of the auction[15], when the submitted bid vector is $\mathbf{b}$, is the sum of the utilities of the bidders and the auctioneer utility (i.e., revenue). Since only the winning bidder has nonzero utility and the auctioneer revenue equals the winning bid, we have

$$V(\mathbf{b}) := \text{value of winning bidder.}$$

For bidders with i.i.d. values, we derived in (14.12) symmetric equilibrium strategies. For such equilibria, the winning bidder is always the bidder with the highest value; i.e., social surplus is maximized.

It is more difficult to solve for equilibria when bidders' values are not i.i.d.: In general, this is an open problem. Moreover, equilibria are no longer symmetric. When there are two bidders, say bidder 1 with $V_1 \sim U[0,1]$ and bidder 2 with $V_2 \sim U[0,2]$, the bidder with the higher value may not win in equilibrium. The intuition is that from the perspective of bidder 1, the weaker bidder, the competition is more fierce than it was when he faced a bidder whose value was drawn from the same distribution as his. Thus, bidder 1 will have to bid a bit more aggressively. On the other hand, bidder 2 faces weaker competition than he would in an i.i.d. environment and so can afford to bid less aggressively. This suggests that there will be valuations $v_1 < v_2$ for which $\beta_1(v_1) > \beta_2(v_2)$, and in such a scenario the bidder with the higher value will lose the auction. See Exercise 14.10.

In this section, without actually deriving equilibrium strategies, we will show that the expected social surplus in any Bayes-Nash equilibrium is still within a constant factor of optimal.

THEOREM 14.7.1. *Let $\mathcal{A}$ be a first-price auction (with an arbitrary tie-breaking rule) for selling a single item. Suppose the bidder values $(V_1, \ldots, V_n)$ are drawn from the joint distribution F. (The values could be correlated.) Let $(\beta_1(\cdot), \ldots, \beta_n(\cdot))$ be a Bayes-Nash equilibrium and let V^* be the value of the winning bidder. Then*

$$\mathbb{E}\left[V^*\right] \geq \frac{\mathbb{E}\left[\max_i V_i\right]}{2}.$$

That is, the price of anarchy[16] (with respect to social surplus) in BNE is at most 2.

PROOF. For any bid vector $\mathbf{b}$, let $u_i[\mathbf{b}|v_i]$ denote bidder i's utility when the bids are $\mathbf{b} = (b_1, \ldots, b_n)$ and her value is v_i. If bidder i bids $v_i/2$ instead of b_i, her utility will be $v_i/2$ if she wins and 0 otherwise. Thus,

$$u_i\left[\frac{v_i}{2}, \mathbf{b}_{-i}\Big|v_i\right] \geq \frac{v_i}{2}\mathbb{1}_{\{\frac{v_i}{2} > \max_{k \neq i} b_k\}}. \tag{14.18}$$

(14.18) is an inequality only because of the possibility that $v_i/2 = \max_{k \neq i} b_k$ and i wins the auction; i.e., the auctioneer breaks ties in favor of i. It follows that

$$\sum_i u_i\left[\frac{v_i}{2}, \mathbf{b}_{-i}\Big|v_i\right] \geq \sum_i \frac{v_i}{2}\mathbb{1}_{\{\frac{v_i}{2} > \max_{k \neq i} b_k\}} \geq \max_i \frac{v_i}{2} - \max_j b_j. \tag{14.19}$$

---

[15]Social surplus is also called "social welfare" or "efficiency".
[16]See Chapter 8 for an introduction to this concept.

This latter inequality clearly holds if the right-hand side is negative or 0. If it is positive, the inequality follows by considering the summand for which v_i is maximized.

Setting $b_i = \beta_i(V_i)$ in (14.19) and taking expectations, we obtain

$$\sum_i \mathbb{E}\left[u_i\left(\frac{V_i}{2}\Big|V_i\right)\right] \geq \mathbb{E}\left[\frac{\max_i V_i}{2} - \max_j \beta_j(V_j)\right]. \qquad (14.20)$$

Thus, under Bayes-Nash equilibrium bidding, the social surplus satisfies

$$\mathbb{E}\left[V^*\right] = \sum_i \mathbb{E}\left[u_i(V_i|V_i)\right] + \mathbb{E}\left[\max_j \beta_j(V_j)\right] \qquad \text{since the revenue is } \max_j \beta_j(V_j)$$

$$\geq \sum_i \mathbb{E}\left[u_i\left(\frac{V_i}{2}\Big|V_i\right)\right] + \mathbb{E}\left[\max_j \beta_j(V_j)\right] \qquad \text{since } \boldsymbol{\beta} \text{ is a BNE}$$

$$\geq \mathbb{E}\left[\frac{\max_i V_i}{2}\right] \qquad \text{by (14.20).}$$

$\square$

REMARK 14.7.2. This bound of $1/2$ on the price of anarchy can be improved to $1 - 1/e$. See Exercise 14.11 for the derivation of this improved bound.

14.8. The Revelation Principle

In some auctions, the communication between the bidders and the auctioneer is involved; e.g., the English auction, the IPL auction, and the entry-fee auction all involve multiple rounds. In most auction formats we've seen, however, the communication between the bidders and the auctioneer is simple: Each bidder submits a single bid. But even when the communication is restricted to a single bid, as in a first-price sealed-bid auction, determining the equilibrium bid requires each bidder to know the distributions of other bidders' values and might be complicated to compute.

An extremely useful insight, known as the **Revelation Principle**, shows that, for every auction with a Bayes-Nash equilibrium, there is another "equivalent" direct[17] auction in which bidding *truthfully* is a Bayes-Nash equilibrium.

DEFINITION 14.8.1. If bidding truthfully (i.e., $\beta_i(v) = v$ for all v and i) is a Bayes-Nash equilibrium for auction $\mathcal{A}$, then $\mathcal{A}$ is said to be **Bayes-Nash incentive-compatible (BIC)**.

Consider a first-price auction $\mathcal{A}$ in which each bidder's value is drawn from a known prior distribution F. A bidder that is not adept at computing his equilibrium bid might hire a third party to do this for him and submit bids on his behalf.

The Revelation Principle changes this perspective and considers the bidding agents and auction together as a new, more complex, auction $\tilde{\mathcal{A}}$, for which bidding truthfully is an equilibrium.

THEOREM 14.8.2 (**The Revelation Principle**). *Let $\mathcal{A}$ be a direct auction where $\{\beta_i\}_{i=1}^n$ is a Bayes-Nash equilibrium. Recall Definition 14.1.1. Then there is another direct auction $\tilde{\mathcal{A}}$, which is BIC and has the same winners and payments as*

---

[17] A **direct** auction is one in which each bidder submits a single bid to the auctioneer.

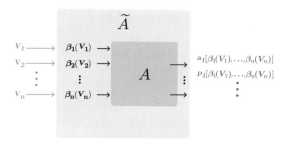

FIGURE 14.7. The figure illustrates the proof of the Revelation Principle. In the auction $\tilde{\mathcal{A}}$, bidders are asked to report their true values. The auction then submits their equilibrium bids for them to the auction $\mathcal{A}$ and then outputs whatever $\mathcal{A}$ would have output on those bids. The auction $\tilde{\mathcal{A}}$ is BIC—it is a Bayes-Nash equilibrium for the bidders to report their values truthfully.

$\mathcal{A}$ *in equilibrium; i.e., for all* $\mathbf{v} = (v_1, \ldots, v_n)$, *if* $b_i = \beta_i(v_i)$ *and* $\mathbf{b} = (b_1, \ldots, b_n)$, *then*

$$\alpha^{\mathcal{A}}[\mathbf{b}] = \alpha^{\tilde{\mathcal{A}}}[\mathbf{v}] \quad and \quad \mathscr{P}^{\mathcal{A}}[\mathbf{b}] = \mathscr{P}^{\tilde{\mathcal{A}}}[\mathbf{v}].$$

PROOF. The auction $\tilde{\mathcal{A}}$ operates as follows: On each input $\mathbf{v}$, $\tilde{\mathcal{A}}$ computes $\boldsymbol{\beta}(\mathbf{v}) = (\beta_1(v_1), \ldots, \beta_n(v_n))$ and then runs $\mathcal{A}$ on $\boldsymbol{\beta}(\mathbf{v})$ to compute the output and payments. (See Figure 14.7.) It is straightforward to check that if $\boldsymbol{\beta}$ is a Bayes-Nash equilibrium in $\mathcal{A}$, then bidding truthfully is a Bayes-Nash equilibrium for $\tilde{\mathcal{A}}$; i.e., $\tilde{\mathcal{A}}$ is BIC. $\qquad\square$

EXAMPLE 14.8.3. Recall Example 14.2.1, a first-price auction with two bidders with U[0, 1] values. An application of the Revelation Principle to this auction yields the following BIC auction: Allocate to the highest bidder and charge him half of his bid.

REMARK 14.8.4. As discussed at the beginning of this section, for some auction formats, the actions of the bidder often go beyond submitting a single bid. The Revelation Principle can be extended to these more general settings. Given any auction format $\mathcal{A}$ (that might involve complex interaction of bidders and auction-eer), an auction $\tilde{\mathcal{A}}$ is constructed as follows: Each bidder truthfully reveals his value to a trusted third party who will calculate and implement his equilibrium strategy for him. If $\tilde{\mathcal{A}}$ includes this third party, then as far as each bidder is concerned, $\tilde{\mathcal{A}}$ is a direct auction which is BIC.

For example, an application of the Revelation Principle to the English auction yields the Vickrey second-price auction. In some online English auctions, the bidder submits to an intermediary (e.g., eBay) a number v which is the maximum price he is willing to pay. The intermediary then bids on his behalf in the actual auction, only increasing his bid when necessary but never bidding more than v.

The primary use of the Revelation Principle is in auction design: To determine which mappings from values to allocations can arise in Bayes-Nash equilibrium, it suffices to consider direct, BIC auctions.

14.9. Myerson's optimal auction

We now consider the design of the **optimal** single-item auction, that is, the auction that maximizes the auctioneer's expected revenue over all auctions, in Bayes-Nash equilibrium. A key assumption here is that the auction designer *knows* the prior distributions from which the bidders' values are drawn.

14.9.1. The optimal auction for a single bidder. Consider a seller with a single item to sell where there is only one potential buyer. Suppose also that the seller knows that the buyer's value V is drawn from a distribution F with support $[0, \infty)$. If the seller offers the item to the buyer at a price of p, the buyer will purchase it if and only if his value V is at least p, which occurs with probability $1 - F(p)$. Thus, the auctioneer's expected revenue will be

$$R(p) := p(1 - F(p)).$$

DEFINITION 14.9.1. The **monopoly reserve price** for the distribution F, denoted by $p^* := p^*(F)$, is defined as the price that maximizes auctioneer revenue; i.e.,

$$p^* = \operatorname{argmax}_p R(p). \tag{14.21}$$

EXAMPLE 14.9.2. When the buyer's value is $\mathrm{U}[0, 1]$, the expected seller revenue for price p is $R(p) = p(1 - p)$, and the monopoly reserve price is $p^* = 1/2$.

THEOREM 14.9.3. *Consider a single buyer with value distribution F in a single-item setting. Suppose that the buyer maximizes his expected utility given his value and the auction format. Then the **maximum** expected revenue that can be achieved by any auction format is $R(p^*)$. This revenue is attained by setting a reserve price of p^*, the monopoly reserve price (which yields a truthful auction).*

PROOF. By the Revelation Principle (Theorem 14.8.2), we only need to consider optimizing over direct BIC auctions. Such an auction is defined by a mapping $\alpha[v]$ that gives the probability of allocation when the reported value is v. Recall that $\alpha[v] = a(v)$ must be increasing and that the expected payment is determined by the allocation probability via part (14.14) of Theorem 14.6.1. (We will fix $p(0) = 0$.)

Any allocation rule $a(v)$ can be implemented by picking $U \sim \mathrm{U}[0, 1]$ and allocating the item if $a(v) \geq U$, i.e., by offering the (random) price $\Psi = \min\{w \mid a(w) \geq U\}$. This auction is truthful, hence BIC. The resulting allocation probability is $a(v)$ and the expected revenue is

$$\mathbb{E}\left[R(\Psi)\right] \leq R(p^*)$$

by (14.21).   □

REMARK 14.9.4. As a consistency check, observe that with the notation of the above proof, the expected payment of a bidder with value v (who buys only if the price Ψ is below his value) is

$$\mathbb{E}\left[\Psi \mathbb{1}_{\Psi \leq v}\right] = \int_0^v \mathbb{P}(w < \Psi \leq v)\, dw = \int_0^v [a(v) - a(w)]\, dw,$$

in agreement with part (14.14) of Theorem 14.6.1.

EXAMPLE 14.9.5. Suppose that the value V of the bidder is drawn from the distribution $F(x) = 1 - 1/x$, for $x \geq 1$. Then $\mathbb{E}[V] = \infty$. However, any price $p > 1$ is accepted with probability $1/p$, resulting in an expected revenue of 1. Thus the maximum expected revenue can be arbitrarily smaller than the expected value.

A digression. We can use the result of Theorem 14.9.3 to bound the expected revenue of a seller in the following extensive-form game.

EXAMPLE 14.9.6 (**The fishmonger's problem**). There is a seller of fish and a buyer who enjoys consuming a fresh fish every day. The buyer has a private value V for each day's fish, drawn from a publicly known distribution $\mathcal{F}$.

However, this value is drawn only once; i.e., the buyer has the same unknown value on all days. Each day, for n days, the seller sets a price for that day's fish, which of course can depend on what happened on previous days. The buyer can then decide whether to buy a fish at that price or to reject it. The goal of the buyer is to maximize his total utility (his value minus the price on each day he buys and 0 on other days), and the goal of the seller is to maximize revenue. How much money can the seller make in n days?

One possible seller strategy is to commit to a daily price equal to the monopoly reserve price p^*, for a total expected revenue of np^* after n days. However, the seller has the freedom to adapt the price each day based on the interaction in prior days and potentially obtain a higher revenue. For example, if the buyer were to buy on day 1 at price ρ, it might make sense for the seller to set a price higher than ρ on day 2, whereas if the buyer rejects this price, it might make sense to lower the price the next day.

More generally, a seller strategy S_I is a binary tree of prices. For instance, if $n = 2$, the seller has to set a single price for day 1 and two prices for day 2, depending on the action of the buyer on day 1. The buyer strategy S_{II} is the best response to the seller strategy and his own value v. Thus, $S_{II}(S_I, v)$ specifies a binary decision (buy or don't buy) at each node in the seller tree S_I; these decisions are chosen to maximize the buyer's utility given v.

CLAIM 14.9.7. *Let $\mathcal{R}_n(S_I, v)$ denote the n-day revenue of the seller when she uses pure strategy S_I, the buyer's value is v, and he uses his best response $S_{II}(S_I, v)$. Then for any S_I,*

$$\mathbb{E}\left[\mathcal{R}_n(S_I, V)\right] \leq nR(p^*).$$

PROOF. The pair of strategies used by the buyer and the seller determine for each possible value v of the buyer a sequence $(\rho_i(v), a_i(v))$, for $1 \leq i \leq n$, where $\rho_i(v)$ is the asking price on day i and $a_i(v) \in \{0, 1\}$ indicates the response of the buyer. Thus,

$$\mathbb{E}\left[\mathcal{R}_n(S_I, V)\right] = \mathbb{E}\left[\sum_{1 \leq i \leq n} a_i(V)\rho_i(V)\right].$$

Now use S_I to construct a single-item, single-round direct auction with $1/n$ of this expected revenue as follows: Ask the buyer to submit his value v. Compute $S_{II}(S_I, v)$ to find $(\rho_i(v), a_i(v))_{i=1}^n$. Finally, pick a uniformly random day i between 1 and n, and sell the item to the buyer (at price $\rho_i(v)$) if and only if $a_i(v) = 1$. The resulting mechanism is BIC since the outcome for the buyer is the result of simulating his best response to S_I given his value. The resulting expected seller revenue is $\mathbb{E}\left[\mathcal{R}_n(S_I, V)\right]/n$ which, by Theorem 14.9.3, is at most $R(p^*)$. $\qquad \square$

14.9.2. A two-bidder special case. Suppose there are two bidders, where bidder 1's value V_1 is known to be exponential with parameter λ_1 and bidder 2's value V_2 is known to be exponential with parameter λ_2. How should an auctioneer facing these two bidders design a BIC auction to maximize his revenue?

Let $\mathcal{A}$ be an auction where truthful bidding ($\beta_i(v) = v$ for all i) is a Bayes-Nash equilibrium, and suppose that its allocation rule is $\boldsymbol{\alpha} : \mathbb{R}^2 \mapsto \mathbb{R}^2$. Recall that $\boldsymbol{\alpha}[\mathbf{b}] := (\alpha_1[\mathbf{b}], \alpha_2[\mathbf{b}])$, where $\alpha_i[\mathbf{b}]$ is the probability[18] that the item is allocated to bidder i on bid vector $\mathbf{b} = (b_1, b_2)$ and $a_i(v_i) = \mathbb{E}\left[\alpha_i(v_i, V_{-i})\right]$.

The goal of the auctioneer is to choose $\boldsymbol{\alpha}[\cdot]$ to maximize

$$\mathbb{E}\left[p_1(V_1) + p_2(V_2)\right].$$

To understand this expression, fix one of the bidders, say i, and let $a_i(v)$, $u_i(v)$, and $p_i(v)$ denote his allocation probability, expected utility and expected payment, respectively, given that $V_i = v_i$ and both bidders are bidding their values. Using condition (14.14) from Theorem 14.6.1[19], we have

$$\mathbb{E}\left[u_i(V_i)\right] = \int_0^\infty \int_0^v a_i(w)\, dw\, \lambda_i e^{-\lambda_i v}\, dv.$$

Reversing the order of integration, we get

$$\mathbb{E}\left[u_i(V_i)\right] = \int_0^\infty a_i(w) \int_w^\infty \lambda_i e^{-\lambda_i v}\, dv\, dw = \int_0^\infty a_i(w) e^{-\lambda_i w}\, dw. \qquad (14.22)$$

Since $u_i(v) = va_i(v) - p_i(v)$, we obtain

$$\mathbb{E}\left[p_i(V_i)\right] = \int_0^\infty va_i(v)\lambda_i e^{-\lambda_i v}\, dv - \int_0^\infty a_i(w) e^{-\lambda_i w}\, dw$$

$$= \int_0^\infty a_i(v)\left[v - \frac{1}{\lambda_i}\right]\lambda_i e^{-\lambda_i v}\, dv. \qquad (14.23)$$

Letting $m_i := 1/\lambda_i$ (the mean of the exponential distribution), we conclude that

$$\mathbb{E}\left[p_1(V_1) + p_2(V_2)\right] = \mathbb{E}\left[a_1(V_1)(V_1 - m_1) + a_2(V_2)(V_2 - m_2)\right]$$

$$= \int_0^\infty \int_0^\infty \left[\alpha_1(\mathbf{v})(v_1 - m_1) + \alpha_2(\mathbf{v})(v_2 - m_2)\right]\lambda_1\lambda_2 e^{-\lambda_1 v_1 - \lambda_2 v_2}\, dv_1\, dv_2. \quad (14.24)$$

Thus, to maximize his expected revenue, the auctioneer should maximize this quantity subject to two constraints: (1) For each $\mathbf{v}$, the item is allocated to at most one bidder, i.e., $\alpha_1(\mathbf{v}) + \alpha_2(\mathbf{v}) \leq 1$, and (2) the auction is BIC.

With only the first constraint in mind, on bid vector $\mathbf{v}$, the auctioneer should never allocate if $v_1 < m_1$ and $v_2 < m_2$, but otherwise, he should allocate to the bidder with the higher value of $v_i - m_i$. (We can ignore ties since they have zero probability.)

It turns out that following this prescription (setting the payment of the winner to be the threshold bid[20] for winning), we obtain a truthful auction, which is therefore BIC. To see this, consider first the case where $\lambda_1 = \lambda_2$. The auction is

---

[18]The randomness here is in the auction itself.

[19]Throughout this section, we assume individual rationality; that is, $p_i(0) = 0$ and hence $u_i(0) = 0$.

[20]The threshold bid is the infimum of the bids a bidder could submit and still win the auction.

then a Vickrey auction with a reserve price of m_1. If $\lambda_1 \neq \lambda_2$, then the allocation rule is

$$\alpha_i(b_1, b_2) = \begin{cases} 1, & \text{if } b_i - m_i > \max(0, b_{-i} - m_{-i}); \\ 0, & \text{otherwise.} \end{cases} \quad (14.25)$$

If bidder 1 wins, he pays $m_1 + \max(0, b_2 - m_2)$, with a similar formula for bidder 2.

EXERCISE 14.e. Show that this auction is truthful; i.e., it is a dominant strategy for each bidder to bid truthfully.

REMARK 14.9.8. Perhaps surprisingly, when $\lambda_1 \neq \lambda_2$, the item might be allocated to the lower bidder. See also Exercise 14.14.

14.9.3. A formula for the expected payment. Fix an allocation rule $\boldsymbol{\alpha}[\cdot]$ and a specific bidder with value V that is drawn from the density $f(\cdot)$. As usual, let $a(v)$, $u(v)$, and $p(v)$ denote his allocation probability, expected utility, and expected payment, respectively, given that $V = v$ and all bidders are bidding truthfully. Using condition (14.14) from Theorem 14.6.1, we have

$$\mathbb{E}\left[u(V)\right] = \int_0^\infty \int_0^v a(w)\, dw\, f(v)\, dv.$$

Reversing the order of integration, we get

$$\mathbb{E}\left[u(V)\right] = \int_0^\infty a(w) \int_w^\infty f(v)\, dv\, dw$$

$$= \int_0^\infty a(w)(1 - F(w))\, dw. \quad (14.26)$$

Thus, since $u(v) = va(v) - p(v)$, we obtain

$$\mathbb{E}\left[p(V)\right] = \int_0^\infty va(v)f(v)\, dv - \int_0^\infty a(w)(1 - F(w))\, dw$$

$$= \int_0^\infty a(v)\left[v - \frac{1 - F(v)}{f(v)}\right] f(v)\, dv. \quad (14.27)$$

REMARK 14.9.9. The quantity in the square brackets in (14.27) is called the bidder's *virtual value*. Contrast the expected payment in (14.27) with the expectation of the *value* allocated to the bidder using $a(\cdot)$, that is,

$$\int_0^\infty a(v)\, v\, f(v)\, dv.$$

The latter would be the revenue of an auctioneer using allocation rule $a(\cdot)$ in a scenario where the buyer could be charged his full value.

The difference between the value and the virtual value captures the auctioneer's loss of revenue that can be ascribed to a buyer with value v, due to the buyer's value being private.

14.9.4. The multibidder case. We now consider the general case of n bidders. The auctioneer knows that bidder i's value V_i is drawn independently from a strictly increasing distribution F_i on $[0, h]$ with density f_i. Let $\mathcal{A}$ be an auction where truthful bidding ($\beta_i(v) = v$ for all i) is a Bayes-Nash equilibrium, and suppose that its allocation rule is $\boldsymbol{\alpha} : \mathbb{R}^n \mapsto \mathbb{R}^n$. Recall that $\boldsymbol{\alpha}[\mathbf{v}] := (\alpha_1[\mathbf{v}], \ldots, \alpha_n[\mathbf{v}])$, where $\alpha_i[\mathbf{v}]$ is the probability[21] that the item is allocated to bidder i on bid[22] vector $\mathbf{v} = (v_1, \ldots, v_n)$ and $a_i(v_i) = \mathbb{E}\left[\alpha_i(v_i, V_{-i})\right]$.

The goal of the auctioneer is to choose $\boldsymbol{\alpha}[\cdot]$ to maximize

$$\mathbb{E}\left[\sum_i p_i(V_i)\right].$$

DEFINITION 14.9.10. For agent i with value v_i drawn from distribution F_i, the **virtual value** of agent i is

$$\psi_i(v_i) := v_i - \frac{1 - F_i(v_i)}{f_i(v_i)}.$$

In the example of §14.9.2, $\psi_i(v_i) = v_i - 1/\lambda_i$.

In §14.9.3, we proved the following proposition:

LEMMA 14.9.11. *The expected payment of agent i in an auction with allocation rule $\boldsymbol{\alpha}(\cdot)$ is*

$$\mathbb{E}\left[p_i(V_i)\right] = \mathbb{E}\left[a_i(V_i)\psi_i(V_i)\right].$$

Summing over all bidders, this means that in any auction, **the expected auctioneer revenue is the expected virtual value of the winning bidder.** Note, however, that the auctioneer directly controls $\boldsymbol{\alpha}(\mathbf{v})$ rather than $a_i(v_i) = \mathbb{E}\left[\boldsymbol{\alpha}(v_i, \mathbf{V}_{-i})\right]$. Expressing the expected revenue in terms of $\boldsymbol{\alpha}(\cdot)$, we obtain

$$\mathbb{E}\left[\sum_i p_i(V_i)\right] = \mathbb{E}\left[\sum_i a_i(V_i)\psi_i(V_i)\right]$$

$$= \int_0^\infty \cdots \int_0^\infty \left[\sum_i \alpha_i(\mathbf{v})\psi_i(v_i)\right] f_1(v_1)\cdots f_n(v_n)\, dv_1 \cdots dv_n.$$
$$(14.28)$$

Clearly, if there is a BIC auction with allocation rule $\boldsymbol{\alpha}[\cdot]$ that maximizes

$$\sum_i \alpha_i(\mathbf{v})\psi_i(v_i) \qquad\qquad (14.29)$$

for every $\mathbf{v}$, then it maximizes expected revenue (14.28). A key constraint on $\boldsymbol{\alpha}[\cdot]$ is that $\sum_i \alpha_i(\mathbf{v}) \leq 1$. To maximize (14.29), if $\max_i \psi_i(v_i) < 0$, then we should set $\alpha_i(\mathbf{v}) = 0$ for all i. Otherwise, we should allocate to the bidder with the highest virtual value (breaking ties by reported value, for instance). This discussion suggests the following auction:

DEFINITION 14.9.12. The Myerson auction for distributions with strictly increasing virtual value functions is defined by the following steps:

(i) Solicit a bid vector $\mathbf{b}$ from the agents.

(ii) Allocate the item to the bidder with the largest virtual value $\psi_i(b_i)$ if positive, and otherwise, do not allocate. That is[23],

---

[21]The randomness here is in the auction itself.

[22]Note that since we are restricting attention to auctions for which truthful bidding is a Bayes-Nash equilibrium, we are *assuming* that $\mathbf{b} = (b_1, \ldots, b_n) = \mathbf{v}$.

[23]Break ties uniformly at random. Ties have 0 probability.

$$\alpha_i(b_i, \mathbf{b}_{-i}) = \begin{cases} 1, & \text{if } \psi(b_i) > \max_{j \neq i} \psi(b_j) \text{ and } \psi(b_i) \geq 0; \\ 0, & \text{otherwise,} \end{cases} \quad (14.30)$$

(iii) If the item is allocated to bidder i, then she is charged her *threshold bid* $t_*(\mathbf{b}_{-i})$, the minimum value she could bid and still win, i.e.,

$$t_*(\mathbf{b}_{-i}) := \min\{b \ : \ \psi_i(b) \geq \max(0, \{\psi_j(b_j)\}_{j \neq i})\}. \quad (14.31)$$

THEOREM 14.9.13. *Suppose that the bidders' values are independent with strictly increasing virtual value functions. Then the Myerson auction is optimal; i.e., it maximizes the expected auctioneer revenue in Bayes-Nash equilibrium. Moreover, bidding truthfully is a dominant strategy.*

PROOF. The truthfulness of the Myerson auction is similar to the truthfulness of the Vickrey auction. Bidder i's utility in the auction is bounded above by $\max(v_i - t_*(\mathbf{b}_{-i}), 0)$, where the first term is the utility for winning and the second term is the utility for losing. This maximum is achieved by bidding truthfully as long as $\psi_i(\cdot)$ is strictly increasing, since $v_i > t_*(\mathbf{b}_{-i})$ if and only if $\psi_i(v_i) > \max(0, \{\psi_j(b_j)\}_{j \neq i})$.

Since the auction is truthful, it is also BIC. Optimality follows from the discussion after (14.28). $\square$

COROLLARY 14.9.14. *The Myerson optimal auction for i.i.d. bidders with strictly increasing virtual value functions is the Vickrey auction with a reserve price of* $\psi^{-1}(0)$.

EXERCISE 14.f. Show that the virtual value function for a uniform distribution is strictly increasing. Use this to conclude that for bidders with i.i.d. $U[0, 1]$ values, the Myerson auction is a Vickrey auction with a reserve price of $1/2$.

REMARK 14.9.15. The fact that truthfulness in the Myerson auction is a dominant strategy means that the bidders do not need to know the prior distributions of other bidders' values. Other BIC auctions with the same allocation rule will also be optimal, but truthfulness need not be a dominant strategy. See Exercise 14.9 for an example.

REMARK 14.9.16. The Myerson auction of Definition 14.9.12 can be generalized to the case where virtual valuations are weakly increasing. Step (i) remains unchanged. In step (ii), a tie-breaking rule is needed. To keep the auction BIC, it is crucial to use a tie-breaking rule that retains the monotonicity of the allocation probabilities $a_i(\cdot)$. Three natural tie-breaking rules are

- break ties by bid (and at random if there is also a tie in the bids);
- break ties according to a predetermined fixed ordering of the bidders, and
- break ties uniformly at random (equivalently, assign a random ranking to the bidders).

The resulting payment in step (iii) is still the threshold bid, the lowest bid the winner could have made without changing the allocation. See Exercise 14.13.

COROLLARY 14.9.17. *Consider n bidders with independent values and strictly increasing virtual value functions. In the class of BIC auctions that always allocate* *the item, i.e., have $\sum_i \alpha_i(\mathbf{b}) = 1$ for all $\mathbf{b}$, the optimal (revenue-maximizing) auction allocates to the bidder with the highest virtual value. If this is bidder i, he is charged $\psi_i^{-1}\left[\max_{j\neq i} \psi_j(b_j)\right]$.*

PROOF. This follows from Lemma 14.9.11 and the proof of truthfulness of the Myerson auction.   □

14.10. Approximately optimal auctions

14.10.1. The advantage of just one more bidder. One of the downsides of implementing the optimal auction is that it requires that the auctioneer know the distributions from which agents' values are drawn (in order to compute the virtual values). The following result shows that in lieu of knowing the distribution from which n i.i.d. bidders are drawn, it suffices to recruit just one more bidder into the auction[24].

THEOREM 14.10.1. *Let F be a distribution for which virtual valuations are increasing. The expected revenue in the optimal auction with n i.i.d. bidders with values drawn from F is upper bounded by the expected revenue in a Vickrey auction with $n+1$ i.i.d. bidders with values drawn from F.*

PROOF. By Corollary 14.9.17, in the i.i.d. setting, the Vickrey auction maximizes expected revenue among BIC auctions that always allocate.

Next, observe that one possible $(n+1)$-bidder auction that always allocates the item consists of, first, running the optimal auction with n bidders and then, if the item is unsold, giving the item to bidder $n+1$ for free.   □

14.10.2. When only the highest bidder can win. Consider the scenario in which an item is being sold by auction to one of n possible buyers whose values are drawn from some joint distribution, known to the auctioneer. We saw that for independent values, the optimal auction might reject the highest bid and allocate to another bidder. In some settings, this is prohibited, and the auctioneer can only allocate the item to the highest bidder (or not at all); e.g., he can run a Vickrey auction with a reserve price. The following auction called the **Lookahead auction** maximizes expected revenue in this class of auctions.

  (i) Solicit bids from the agents. Suppose that agent i submits the highest bid b_i. (If there are ties, pick one of the highest bidders arbitrarily.)
  (ii) Compute the conditional distribution $\tilde{F}_i$ of V_i given the bids $\mathbf{b}_{-i}$ and the event $V_i \geq \max_{j\neq i} b_j$. Let $p_i = p_i(\mathbf{b}_{-i})$ be the price p that maximizes $p(1 - \tilde{F}_i(p))$.
  (iii) Run the optimal single-bidder auction with agent i, using his previous bid b_i and the distribution $\tilde{F}_i$ for his value: This auction sells the item to agent i at price p_i if and only if $b_i \geq p_i$.

REMARK 14.10.2. The Lookahead auction can be implemented even when bidders' values are not independent. The only difference is in the update stage–computing $\tilde{F}_i$ is more involved. See Example 14.10.6.

PROPOSITION 14.10.3. *The Lookahead auction is optimal among truthful auctions that allocate to the highest bidder (if at all).*

---

[24]...under the questionable assumption that this bidder has the same value distribution as the others.

PROOF. Any truthful auction $\mathcal{A}$, after conditioning on b_i being the highest bid and the values $\mathbf{b}_{-i}$, becomes a truthful single bidder auction. Optimizing expected revenue in that auction, by Theorem 14.9.3, yields the Lookahead auction. □

14.10.3. The Lookahead auction is approximately optimal. As we know, the Lookahead auction is not the optimal truthful auction. However, the next theorem shows that it is a factor two approximation.

THEOREM 14.10.4. *The Lookahead (LA) auction yields an expected auctioneer revenue that is at least half that of the optimal truthful and ex-post individually rational auction even when bidders have dependent values.*

PROOF. The expected revenue of an individually rational, truthful auction is the sum of its expected revenue from bidders that are not the highest, say L, plus its expected revenue from the highest bidder, say H.

Assuming truthful bidding, it is immediate that the expected revenue of the LA auction is at least H since it is the *optimal* auction for this bidder conditioned on being highest and conditioned on the other bids. As for the other bidders, since the auction is ex-post individually rational, no bidder can be charged a price higher than her value. Thus, the optimal revenue achievable from these lower bidders is at most the maximum value of bidders in this set, say v^* (since only one item is being sold). But the expected revenue from the highest bidder is at least v^* since one of the possible auctions is to just offer him a price of v^*. Therefore the expected revenue of the LA auction is also at least L. □

REMARK 14.10.5. Note that Theorem 14.10.4 holds for independent values even if virtual valuations are *not* increasing!

EXAMPLE 14.10.6. Two gas fields are being sold as a bundle via a Lookahead auction. X and Y are the profits that bidder 1 can extract from fields 1 and 2. Bidder 2 is more efficient than bidder 1, so he can extract $2X$ from the first but can't reach the second. Thus

$$V_1 = X + Y \quad \text{and} \quad V_2 = 2X.$$

It is known to the auctioneer that X and Y are independent and distributed exponentially with parameter 1.

There are two cases to consider since $\mathbb{P}(X = Y) = 0$.

- $V_2 < V_1$ or equivalently $Y > X$: Given this event and $V_2 = b_2$, the conditional distribution of Y is that of $X + Z$, where Z is an independent exponential with parameter 1. Thus, $\tilde{F}_1$ is the distribution of $b_2 + Z$. The price $p_1 \geq b_2$ that maximizes

$$p \cdot \mathbb{P}(b_2 + Z > p) = p \cdot e^{b_2 - p}$$

  is $\max(b_2, 1)$. Thus, if $b_1 > \max(b_2, 1)$, then in step (iii) the item will be sold to bidder 1 at the price $\max(b_2, 1)$ Notice that if $1 > b_1 > b_2$, then the item will not be sold.
- $V_2 > V_1$ or equivalently $Y < X$: Given this event and $V_1 = b_1$, the conditional distribution of X is uniform[25] on $[b_1/2, b_1]$. Therefore, $\tilde{F}_2$ is

---

[25]Given the sum $X + Y$ of two i.i.d. exponentials, the conditional distribution of each one is uniform on $[0, X + Y]$.

uniform on $[b_1, 2b_1]$. Thus, if $b_2 > b_1$, then the item will be sold to bidder 2 at price b_1, since $p(1 - \tilde{F}_2(p))$ is decreasing on $[b_1, 2b_1]$.

14.11. The plot thickens...

We now briefly consider other settings, where the optimal auctions are surprising or weird.

EXAMPLE 14.11.1. (**Single bidder, two items:**) Suppose the seller has two items and there is a buyer whose private values (V_1, V_2) for the two items are known to be independent samples from distribution F. Suppose further that the buyer's value for getting both items is $V_1 + V_2$. Since the values for the two items are independent, one might think that the seller should sell each of them separately in the optimal way, resulting in twice the expected revenue from a single item. Surprisingly, this is not necessarily the case. Consider the following examples:

(1) Suppose that each V_i is equally likely to be 1 or 2. Then the optimal revenue the seller can get separately from each item is 1: If he sells an item at price 1, the buyer will buy it. If he sells at price 2, the buyer will buy with probability 1/2. Thus, selling separately yields a total expected seller revenue of 2. However, if the seller offers the buyer the bundle of both items at a price of 3, the probability the buyer has $V_1 + V_2 \geq 3$ is 3/4, and so the expected revenue is $3 \cdot 3/4 = 2.25$, more than the optimal revenue from selling the items separately.

(2) On the other hand, suppose that V_i is equally likely to be 0 or 1. Then if each item is sold separately, the optimal price is 1 and the overall expected revenue is $2 \cdot 1/2 = 1$. On the other hand, the optimal bundle price is 1, and the expected revenue is only 3/4.

(3) When V_i is equally likely to be 0, 1, or 2, then selling the items separately yields expected revenue 4/3. That is also the expected revenue from bundling. However, the auction which offers the buyer the choice between any single item at price 2 or the bundle of both items at price 3 obtains an expected revenue of 13/9.

EXAMPLE 14.11.2. **Revenue when values are correlated:** Earlier, we considered the optimal auction when there are two bidders whose values are uniform on $[0, 1]$. In this case, the expected value of the highest bidder is 2/3, and yet the auction which maximizes the seller's revenue obtains an expected revenue of only 5/12. This loss is the "price" the auctioneer has to pay because the values of the bidders are private.[26] However, when bidders' values are correlated, the auctioneer can take advantage of the correlation to extract more revenue, in some cases, the full expected maximum value!

For example, suppose that there are two agents and they each have value either 10 or 100, with the following joint distribution. (The entries in the table are the probabilities of each of the corresponding pairs of values.)

---

[26]This is sometimes called the "information rent" of the bidder.

	$V_2 = 10$	$V_2 = 100$
$V_1 = 10$	1/3	1/6
$V_1 = 100$	1/6	1/3

We consider the following (symmetric) auction, from the perspective of bidder 1: The initial allocation and pricing is determined by a second-price auction yielding

$$u_1(10) = 0, \tag{14.32}$$

$$u_1(100) = \mathbb{P}\left[V_2 = 10 | V_1 = 100\right] \cdot (100 - 10) = \frac{1}{3} \cdot 90 = 30, \tag{14.33}$$

since whenever both bidders are truthful and submit the same bid, they obtain a utility of 0. However, the buyers must commit *up-front* to the following additional rules:

- If $V_2 = 10$, then bidder 1 will receive \$30+ϵ.
- If $V_2 = 100$, then bidder 1 will be charged \$60$-\epsilon$.

(The symmetric rule is applied to bidder 2.) The impact of these rules is to reduce bidder 1's expected utility to 0: If her value is \$10, then the payoff from the extra rules is

$$\mathbb{P}\left[V_2 = 10 | V_1 = 10\right] \cdot (30 + \epsilon) - \mathbb{P}\left[V_2 = 100 | V_1 = 10\right] \cdot (60 - \epsilon)$$

$$= \frac{2}{3} \cdot (30 + \epsilon) - \frac{1}{3} \cdot (60 - \epsilon) = \epsilon, \tag{14.34}$$

whereas if her value is \$100, the payoff from the extra rules is

$$\mathbb{P}\left[V_2 = 10 | V_1 = 100\right] \cdot (30 + \epsilon) - \mathbb{P}\left[V_2 = 100 | V_1 = 100\right] \cdot (60 - \epsilon)$$

$$= \frac{1}{3} \cdot (30 + \epsilon) - \frac{2}{3} \cdot (60 - \epsilon) = \epsilon - 30. \tag{14.35}$$

Combining (14.32) and (14.34) for the case $V_1 = 10$ and combining (14.33) and (14.35) for the case $V_1 = 100$ shows that her combined expected utility from the second-price auction and the extra rules is always ϵ. Since the setting is symmetric, bidder 2's expected utility is also ϵ.

Finally, since

$$2\epsilon = u(V_1) + u(V_2) = \mathbb{E}\left[(V_1 a(V_1) - p_1(V_1)) + (V_2 a(V_2) - p_2(V_2))\right]$$

and the allocation is always to the bidder with the highest value, the auctioneer's expected revenue is

$$\mathbb{E}\left[p_1(V_1) + p_2(V_2)\right] = \mathbb{E}\left[\max(V_1, V_2)\right] - 2\epsilon,$$

essentially the maximum possible (in expectation).

In addition, truth-telling is a Bayes-Nash equilibrium. However, a bidder may be very unhappy with this auction after the fact. For example, if both agents have value 100, they both end up with negative utility ($-60 + \epsilon$)!

REMARK 14.11.3. In this example, the buyer's *ex-interim* expected utility (that is, his expected utility after seeing his value, but before seeing other bidders' values) is positive. However, after the auction concludes, the buyer's utility may be negative. In other words, this auction is not *ex-post individually rational*. Most commonly used auctions are ex-post individually rational.

Notes

There are several excellent texts on auction theory, including the books by Krishna [Kri09], Menezes and Monteiro [MM05], and Milgrom [Mil04]. Other sources that take a more computational perspective are the forthcoming book by Hartline [Har17], the lectures notes by Roughgarden [Rou14] (Lectures 2-6), and Chapters 9 and 13 in [NRTV07].

The first game-theoretic treatment of auctions is due to William Vickrey [Vic61], who analyzed the second-price auction and developed several special cases of the Revenue Equivalence Theorem. These results played an important role in Vickrey's winning the 1996 Nobel Prize, which he shared with James Mirrlees "for their fundamental contributions to the economic theory of incentives under asymmetric information."

William Vickrey   Roger Myerson

The Revenue Equivalence Theorem was proved by Myerson [Mye81] and Riley and Samuelson [RS81]. Myerson's treatment was the most general: He developed the Revelation Principle[27] from §14.8 and optimal (revenue-maximizing) auctions for a number of different settings. For this and "for having laid the foundations of mechanism design theory," Roger Myerson, Leonid Hurwicz, and Eric Maskin won the 2007 Nobel Prize in Economics.

The Revelation Principle also applies to equilibrium concepts other than Bayes-Nash equilibrium. For example, the Revelation Principle for dominant strategies says that if $\mathcal{A}$ is an auction with dominant strategies $\{\beta_i\}_{i=1}^n$, then there is another auction $\tilde{\mathcal{A}}$ for which truth-telling is a dominant strategy and which has the same winner and payments as $\mathcal{A}$. The proof is essentially the same.

In §14.9, we developed Myerson's optimal auction for the case where virtual valuations are (weakly) increasing. Myerson's paper [Mye81] solves the general case. These results were further developed and clarified by Bulow and Roberts [BR89] and Bulow and Klemperer [BK94]. The approximately optimal auction from §14.10.1 is from [BK94]; the proof given here is due to Kirkegaard [Kir06]. The Lookahead auction in §14.10.2 is due to Ronen [Ron01]. The examples in §14.11 are taken from Hart and Nisan [HN12] and Daskalakis et al. [DDT14]. Surprisingly, there are situations where the optimal mechanism may not even be deterministic [HN12, DDT14]. However, Babaioff et al. [BILW14] show that in the setting of a single additive buyer, with a seller with multiple items for

---

[27]According to Myerson [Mye12], the Revelation Principle was independently discovered by others including Dasgupta, Hammond, Maskin, Townsend, Holmstrom, and Rosenthal, building on earlier ideas of Gibbard and Aumann.

sale, if the bidders' valuations for the different items are independent, then the better of selling separately and selling the grand bundle achieves expected revenue within a constant factor of optimal. For other recent developments related to optimal auctions, see e.g., [CS14, Das15, Yao15, RW15, CDW16].

The war of attrition and generalizations thereof have been studied extensively. See, e.g., [Smi74, BK99]. A different type of war of attrition is shown in Figure 14.8.

FIGURE 14.8. Dawkins [Daw06] describes the behavior of emperor penguins in the Antarctic. They have been observed standing on an ice ledge hesitating before diving in to catch their dinner because of the danger of a predator lurking below. Eventually, one of them, the loser of this game, jumps in and (if he's not eaten), the others follow.

Example 14.11.2 is adapted from [Mye81] and [CM88]. Cremer and McLean [CM88] showed that full surplus extraction is possible in a broad range of correlated settings. Uniform price auctions (Example 14.6.5) have been applied to the sale of sports tickets by Baliga and Ely [Tra14]. The generalization to the case where individual bidders demand multiple units is discussed in [Kri09].

Example 14.9.6 and other variants of the fishmonger problem are studied in Hart and Tirole [HT88], as well as [Sch93, DPS15]. These papers focus on the case where the seller cannot commit to future prices.

Exercise 14.18 is from [Kle98]. Theorem 14.7.1 follows [LPL11] and Exercise 14.11 is from [Syr12]. As discussed in the notes of Chapter 8, the price of anarchy in games of incomplete information and auctions has been extensively studied in recent years. For a detailed treatment, see, e.g., [ST13, Rou12, Syr14, HHT14, Rou14, Har17].

Many of the results developed in this chapter, e.g., Myerson's optimal auction apply to settings much more general than single-item auctions, e.g., to the win/lose settings discussed in §15.2. We refer the reader to Klemperer's excellent guide to the literature on auction theory [Kle99] for further details and additional references.

Exercises

14.1. Show that the three-bidder, single-item auction in which the item is allocated to the highest bidder at a price equal to the third highest bid is not truthful.

14.2.   Consider a Vickrey second-price auction with two bidders. Show that for each choice of bidder 1's value v_1 and any possible bid $b_1 \neq v_1$ he might submit, there is a bid by the other bidder that yields bidder 1 strictly less utility than he would have gotten had he bid truthfully.

14.3.   Suppose that each agent's value V_i is drawn independently from the same strictly increasing distribution $F \in [0, h]$. Find the symmetric Bayes-Nash equilibrium bidding strategy in
   - a second-price auction with a reserve price of r,
   - a first-price auction with a reserve price of r,
   - an all-pay auction with a reserve price of r.

14.4.   Consider a descending auction for a single item. The auctioneer starts at a very high price and then lowers the price continuously. The first bidder who indicates that he will accept the current price wins the auction at that price. Show that this auction is equivalent to a first-price auction; i.e., any equilibrium bidding strategy in the first-price auction can be mapped to an equilibrium bidding strategy in this auction and will result in the same allocation and payments.

S 14.5.   Find a symmetric equilibrium in the war of attrition auction discussed in §14.4.3, under the assumption that bids are committed to up-front, rather than in the more natural setting where a player's bid (the decision as to how long to stay in) can be adjusted over the course of the auction.

14.6.   Consider Example 14.5.2 again. Suppose that the bidder's values are independent, but not identically distributed, with $V_i \sim F_i$. Find the revenue-maximizing evaluation fee for the seller (assuming that a second-price auction will be run). The same evaluation fee must be used for all bidders.

14.7.   Prove that the single-item, single-bidder auction described in §14.9.1 is a special case of Myerson's optimal auction.

14.8.   (a) Show that the Gaussian and the equal-revenue ($F(x) = 1 - 1/x$ for $x \geq 1$) distributions have increasing virtual value functions.
   (b) Show that the following distribution does not have an increasing virtual value function: Draw a random variable that is $U[0, 1/2]$ with probability $2/3$ and a random variable that is $U[1/2, 1]$ with probability $1/3$.

14.9.   Consider the following two-bidder, single-item auction: Allocate to the highest bidder if his bid b is at least $1/2$, and if so, charge him $(b/2)+(1/8b)$. Otherwise, don't allocate the item. Show that for two bidders with i.i.d. $U[0, 1]$ values, this auction is BIC.

14.10. Show that the Bayes-Nash price of anarchy for a first-price auction is strictly larger than 1 for two bidders, where bidder 1 has value $V_1 \sim U[0,1]$ and $V_2 \sim U[0,2]$. See §14.7. Hint: Suppose that the higher bidder always wins and then apply the revenue equivalence.

14.11. Under the conditions of Theorem 14.7.1, show that

$$\mathbb{E}\left[V^*\right] \geq \left(1 - \frac{1}{e}\right) \mathbb{E}\left[\max_i V_i\right].$$

See Exercise 8.8 for a hint.

14.12. Show that if the auctioneer has a value of C for the item, i.e., his profit in a single-item auction is the payment he receives minus C (or 0 if he doesn't sell), then with n i.i.d. bidders (with strictly increasing virtual valuation functions), the auction which maximizes his expected profit is Vickrey with a reserve price of $\psi^{-1}(C)$.

S 14.13. Determine the explicit payment rule for the three tie-breaking rules discussed in Remark 14.9.16.

S 14.14. Consider two bidders where bidder 1's value is drawn from an exponential distribution with parameter 1 and bidder 2's value is drawn independently from U[0,1]. What is the Myerson optimal auction in this case? Show that if $(v_1, v_2) = (1.5, 0.8)$, then bidder 2 wins.

14.15. Consider n bidders, where $V_i \sim F_i$, where the V_i's are independent and virtual values are increasing. Let r_i^* be the monopoly reserve price for F_i (recall (14.21)). Show that the following auction obtains at least half the revenue obtained by the optimal truthful and ex-post individually rational auction.
   - Ask the bidders to report their values.
   - If the highest bidder, say bidder 1, reports $b_1 \geq r_1^*$, then he wins the auction at a price equal to $\max(b_2, r_1^*)$, where b_2 is the report of the second-highest bidder. Otherwise, there is no winner.

14.16. Show how to generalize Theorem 14.10.4 to a scenario in which k identical items are being sold.

14.17. Show that if bidders values are i.i.d. from a regular distribution F, then the expected revenue of the first-price auction with reserve price $\psi^{-1}(0)$ is the same as that of the Myerson optimal auction.

14.18. Consider the following game known as the **Wallet Game**: Each of two bidders privately checks how much money she has in her wallet, say s_i for bidder i $(i = 1, 2)$. Suppose an English auction is run where the prize is the combined contents of the two wallets. That is, the price goes up continuously until one of the bidders quits, say at price p. At that point, the remaining bidder pays p and receives $s_1 + s_2$. Find an equilibrium in this game. Can you find an asymmetric equilibrium?

14.19. Show that the lookahead auction does not obtain a better than 2 approximation to the optimal auction.

14.20. Consider an auctioneer selling an item by auction to buyers whose valuations $V_1, V_2, \ldots, V_n$ are drawn from a correlated joint distribution F (not a product distribution). In this case, the characterization of Bayes-Nash equilibrium (Theorem 14.6.1) does not hold. Explain where the proof given there breaks down.

14.21. Prove that Theorem 14.6.1 holds if $\mathcal{A}$ is an auction for selling k items. In fact, prove that it holds in any win/lose setting when each agent's value V_i is drawn independently from F_i. For a definition of win/lose settings, see §15.2.

14.22. Consider an auction in which k identical items are being sold. Each of n bidders is interested in only one of these items. Each bidder's value for the item is drawn independently from the same prior distribution F. Use the result of the previous exercise to derive the optimal auction (i.e., generalize Theorem 14.9.13) for this setting, assuming that $\psi(v) = v - \frac{1-F(v)}{f(v)}$ is strictly increasing.

CHAPTER 15

Truthful auctions in win/lose settings

In a truthful auction, bidders do not need to know anything about their competitors or perform complex calculations to determine their strategy in the auction. In this chapter, we focus exclusively on auctions with this property.

15.1. The second-price auction and beyond

EXAMPLE 15.1.1 (**Spectrum auctions**). The government is running an auction to sell the license for the use of a certain band of electromagnetic spectrum. Its goal is to allocate the spectrum to the company that values it most (rather than maximizing revenue). This value is indicative of how efficiently the company can utilize the bandwidth. One possibility is to run an English auction - we have seen that it is a dominant strategy for bidders to stay in the auction as long as the price is below their value. This ensures that the license is sold to the bidder with the highest value. A key difference between the English auction and the second-price auction is that in the latter, the highest bidder reveals his value. This could be damaging later; e.g., it could lead to a higher reserve price in a subsequent auction. Thus an auction that is truthful in isolation might not be truthful if the same players are likely to participate in a future auction.

FIGURE 15.1. Competing rocket companies.

EXAMPLE 15.1.2 (**Hiring a contractor**). Several rocket companies are competing for a NASA contract to bring an astronaut to the space station. One possibility is for NASA to run a **second-price procurement auction** (where the auctioneer is a buyer instead of a seller): Ask each company to submit a bid, and award the contract to the lowest bidder, but pay him the second-lowest bid. The utility of the winning bidder is the price he is paid minus his cost[1]. Again, it is a dominant strategy for each company to bid its actual cost. Alternatively, to reduce the amount of information the winning company reveals, NASA can run a descending auction: Start from the maximum price NASA is willing to pay and keep reducing the price until exactly one company remains, which is then awarded the contract at that price. It is a dominant strategy for each company to stay in until the price reaches its cost.

EXAMPLE 15.1.3 (**A shared communication channel**). Several users in a large company have data streams to transmit over a shared communication channel. The data stream of each user has a publicly known bandwidth requirement, say w_i, and that user has a private value v_i for getting his data through. If the total capacity of the channel is C, then the set of data streams selected must have a total bandwidth which does not exceed the channel capacity. I.e., only sets of data streams S with $\sum_{i \in S} w_i \le C$ are feasible. Suppose that the company decides to use an auction to ensure that the subset of streams selected has the largest total value. To do so requires incentivizing the users to report their values truthfully.

15.2. Win/lose allocation settings

We now formalize a setting which captures all of these examples.

DEFINITION 15.2.1. a **win/lose allocation problem**[2] is defined by:
- A set U of participants/bidders, where each has a private value v_i for "winning" (being selected) and obtains no value from losing;
- a set of feasible allocations (i.e., possible choices for the set of winning bidders) $\mathcal{L} \subset 2^U$. In a single-item auction, $\mathcal{L}$ contains all subsets of size at most 1 and in the communication channel example, $\mathcal{L} = \{S \subset U \mid \sum_{i \in S} w_i \le C\}$.

A sealed-bid **auction**, or **mechanism**, $\mathcal{A}$ in such a setting asks each bidder to submit a bid b_i. The mechanism then selects (possibly randomly) a single winning set L in $\mathcal{L}$ and a payment vector $(p_i)_{i \in U}$, where p_i is the payment[3] required from bidder i. The mapping from bid vectors to winning sets is called the **allocation rule** and is denoted by $\boldsymbol{\alpha}[\mathbf{b}] = (\alpha_1[\mathbf{b}], \ldots, \alpha_n[\mathbf{b}])$, where $\alpha_i[\mathbf{b}]$ is the probability that bidder i wins when the bid vector is $\mathbf{b} = (b_1, \ldots, b_n)$. (If the auction is deterministic then $\alpha_i[\mathbf{b}] \in \{0, 1\}$.) We denote the **payment rule** of the auction by $\mathbf{p}[\mathbf{b}] = (p_1[\mathbf{b}], \ldots, p_n[\mathbf{b}])$, where $p_i[\mathbf{b}]$ is the expected payment[4] of bidder i when the bid vector is $\mathbf{b}$. (This expectation is taken over the randomness in the auction.)

---

[1] This cost incorporates the bidders' time and effort.

[2] This setting commonly goes under the name "single-parameter problem".

[3] Unless otherwise specified, the payment of each losing bidder is 0.

[4] The quantity $\mathbf{p}[\mathbf{b}] = (p_1[\mathbf{b}], \ldots, p_n[\mathbf{b}])$ was denoted by $\mathscr{P}[\mathbf{b}] = (\mathscr{P}_1[\mathbf{b}], \ldots, \mathscr{P}_n[\mathbf{b}])$ in Definition 14.1.1 of the previous chapter.

Suppose that bidder i has private value v_i for winning. Then his **utility** on bid vector **b** is

$$u_i[\mathbf{b}|v_i] = v_i\alpha_i[\mathbf{b}] - p_i[\mathbf{b}].$$

The **utility of the auctioneer** is his income:

$$\sum_{i \in U} p_i[\mathbf{b}].$$

DEFINITION 15.2.2. We say that a mechanism $\mathcal{M}$ is **truthful** if for every agent, it is a dominant strategy to bid his value. Formally, for all $\mathbf{b}_{-i}$, all i, v_i, and b_i

$$u_i[v_i, \mathbf{b}_{-i}|v_i] \geq u_i[b_i, \mathbf{b}_{-i}|v_i],$$

where $u_i[b_i, \mathbf{b}_{-i}|v_i] = v_i\alpha_i[\mathbf{b}] - p_i[\mathbf{b}]$.

15.3. Social surplus and the VCG mechanism

DEFINITION 15.3.1. Consider an auction with n bidders where the private value of bidder i is v_i. Suppose that the set of winning bidders is L^* and the payment of bidder i is p_i for each i. The **social surplus** of this outcome is the sum of the utilities of all the bidders and the auctioneer; that is,

$$\sum_{i \in U}\left(v_i\mathbb{1}_{\{i \text{ winner}\}} - p_i\right) + \sum_{i \in U} p_i = \sum_{i \in L^*} v_i.$$

(Since the payments are "losses" to the bidders and "gains" to the auctioneer, they cancel out.)

The VCG mechanism for maximizing social surplus in win/lose settings

- Ask each bidder to report his private value v_i (which he may or may not report truthfully). Assume that bidder i reports b_i.
- Choose as the winning set a feasible $L \in \mathcal{L}$ that maximizes $b(L)$, where $b(L) = \sum_{j \in L} b_j$. Call this winning set L^*.
- To compute payments, let $\mathcal{L}_i^+ = \{S | S \cup \{i\} \in \mathcal{L}\}$. Then i only wins if his bid b_i satisfies

$$b_i + \max_{L \in \mathcal{L}_i^+} b(L) \geq \max_{L \in \mathcal{L}^-} b(L), \tag{15.1}$$

where $\mathcal{L}^-$ is the collection of sets in $\mathcal{L}$ that do *not* contain i. His payment is his **threshold bid**, the minimum b_i for which (15.1) holds; i.e.,

$$p_i = \max_{L \in \mathcal{L}^-} b(L) - \max_{L \in \mathcal{L}_i^+} b(L). \tag{15.2}$$

The payment given by (15.2) precisely captures the **externality** imposed by bidder i, i.e., the reduction in total (reported) values obtained by the other bidders due to i's presence in the auction.

THEOREM 15.3.2. *The VCG mechanism is truthful. Moreover, with truthful bidding, $0 \leq p_i \leq v_i$ for all i and therefore the auction is individually rational.*

PROOF. Fix any set of bids $\mathbf{b}_{-i}$ of all bidders but i and note that p_i is determined by $\mathbf{b}_{-i}$. Whatever b_i is, player i's utility is at most $v_i - p_i$ if $v_i > p_i$ and at most 0 if $v_i \leq p_i$. Bidding truthfully guarantees a bidder i this maximum utility of $\max(0, v_i - p_i) \geq 0$. $\square$

REMARK 15.3.3. Note that if payments were instead equal to $p_i' = p_i + h_i(\mathbf{b}_{-i})$, for any function $h_i(\cdot)$, the mechanism would still be truthful. However, it would no longer be true that each bidder would be guaranteed nonnegative utility.

EXERCISE 15.a. Check that the Vickrey second-price auction and the Vickrey k-unit auction are both special cases of the VCG mechanism.

REMARK 15.3.4. Since the VCG auction incentivizes truth-telling in dominant strategies, henceforth we **assume** that bidders bid truthfully. Therefore, we will not refer to the bids using the notation $\mathbf{b} = (b_1, \ldots, b_n)$, but rather will assume they are $\mathbf{v} = (v_1, \ldots, v_n)$.

15.4. Applications

15.4.1. Shared communication channel, revisited.
See Figure 15.2 for an example of the application of VCG to choosing a feasible winning set of maximum total value.

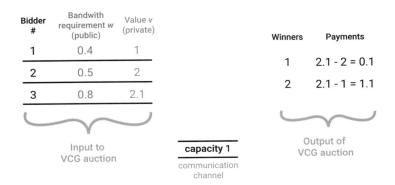

FIGURE 15.2. This figure illustrates the execution of the VCG algorithm on Example 15.1.3 (a shared communication channel) when $C = 1$ and there are three bidders with the given values and weights. In this example, $\mathcal{L} = \{\{1, 2\}, \{1\}, \{2\}, \{3\}, \emptyset\}$. With the given values, the winning set selected is $\{1, 2\}$. To compute, for example, bidder 1's payment, we observe that without bidder 1, the winning set is $\{3\}$ for a value of 2.1. Therefore the loss of value to other bidders due to bidder 1's presence is $2.1 - 2$.

15.4.2. Spanning tree auctions.
Netflix wishes to set up a distribution network for its streaming video. The links that can be used for streaming form a graph, and each link is owned by a different service provider. Netflix must purchase the use of a set of links that will enable it to reach all nodes in the graph. For each link ℓ, the owner incurs a private *cost* $c_\ell \in [0, C]$ for transmitting the Netflix data. Netflix runs an auction to select the spanning tree of minimum cost in the graph (owners of the selected links will be the "winners"). In this setting, the feasible sets $\mathcal{L}$ are the spanning trees of the graph.

This is a social surplus maximization problem, with $v_e := -c_e$. The VCG mechanism for buying a minimum spanning tree (MST) is the following:

- Ask each bidder (link owner) ℓ to report his (private) cost c_ℓ.

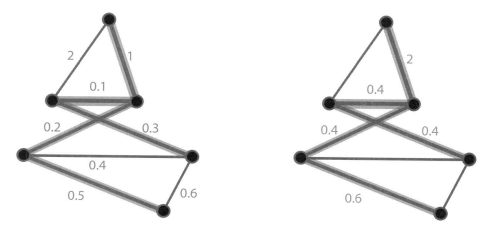

FIGURE 15.3. This figure shows the outcome and payments in a spanning tree auction. The blue labels on edges in the left figure are the costs of the edges. The minimum spanning tree consists of the gray highlighted edges. The payments are shown in pink on the right side of the figure.

- Choose the MST T^* with respect to the reported costs.
- Pay each winning link owner ℓ his threshold bid, which is C if removing it disconnects the graph, and otherwise

(cost of MST with ℓ deleted) $-$ (cost of MST with ℓ contracted),

which is always at most C. (This is the minimum cost the owner of ℓ can report and still be part of the MST.) Contracting an edge (i, j) consists of identifying its endpoints, thereby creating a new merged node ij, and replacing all edges (i, k) and (j, k), $k \neq i, j$, with an edge (ij, k).

See Figure 15.3 for an example.

In fact, the VCG auction for this example can be implemented as a "descending auction". It works as follows: Initialize the asking price for each link to C.

- Reduce the asking price on each link uniformly (e.g., at time t, the asking price is $C - t$) until some link owner declares the price unacceptable and withdraws from the auction.
- If at this point, there is a link ℓ whose removal would disconnect the graph, buy it at the current price and contract ℓ.

See Figure 15.4 for a sample execution of the descending auction.

15.4.3. Public project. Suppose that the government is trying to decide whether or not to build a new library which will cost C dollars. Each person in society has his own value v_i for this library, that is, how much it is worth to that person to have the library built. A possible goal for the government is to make sure that the library is built if the population's total value for it is at least C dollars, and not otherwise. How can the government incentivize the members of society to truthfully report their values?

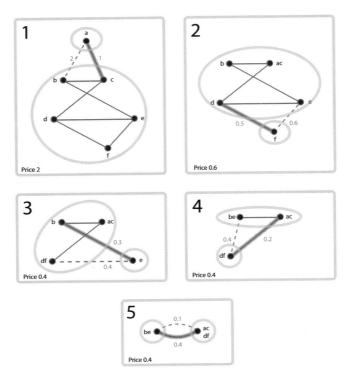

FIGURE 15.4. This figure shows the evolution of the descending auction for purchasing a spanning tree on the example shown in Figure 15.3. In this example $C > 2$. When the price reaches 2 (upper left), edge (a, b) is ready to drop out. The consecutive figures (top to bottom, and within each row, left to right) show the points at which edges are selected and the corresponding payments.

The social surplus in this setting is 0 if the library isn't built and $\sum_i v_i - C$ if the library is built. We apply VCG to maximize social surplus:

- Ask each person i to report his value v_i for the library.
- Build the library if and only if $\sum_i v_i \geq C$.
- If the library is built, person i pays his threshold bid: If $\sum_{j \neq i} v_j \geq C$, he pays nothing. Otherwise, he pays

$$p_i = C - \sum_{j \neq i} v_j.$$

In practice, this scheme is not implemented as it suffers from several problems.

- The government might not recover its cost. For example, if there are n people and each has value $C/(n-1)$, then all payments will be 0, but the library will be built. Unfortunately, this is inevitable; no truthful mechanism can simultaneously balance the budget and maximize social surplus. See the chapter notes.
- It is deeply susceptible to collusion. If two people report that their values are C, then the library will be built and nobody will pay anything.

- Our technical definition of "value" is the amount a person is willing to pay for an item. It is not appropriate in this example. Indeed, the library may be more valuable to someone who cannot pay for it than to someone who can (e.g., has the resources to buy books). In this case, there is discord between the intuitive meaning of social surplus and the technical definition.

15.4.3.1. *VCG might not be envy-free.*

DEFINITION 15.4.1. We say that a truthful mechanism in a win/lose setting is **envy-free** if, when bidding truthfully, each bidder prefers his own allocation and payment to that of any other bidder. That is, for every $\mathbf{v}$ and i, we have

$$\alpha_i[\mathbf{v}]v_i - p_i[\mathbf{v}] \geq \alpha_j[\mathbf{v}]v_i - p_j[\mathbf{v}].$$

EXAMPLE 15.4.2. Suppose that the government is trying to decide whether or not to build a bridge from the mainland to a big island A or to a small island B. The cost of building a bridge is \$90 million. Island A has a million people, and each one values the bridge at \$100. Island B has five billionaires, and each one values the bridge at \$30 million. Running VCG will result in a bridge to island B and payments of 0. In this case, the people on island A will envy the outcome for the people on island B.

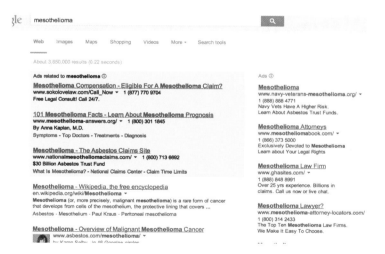

FIGURE 15.5. A typical page of search results: some organic and some sponsored. If you click on a sponsored search result, the associated entity pays the search engine some amount of money. Some of the most expensive keywords relate to lawyers, insurance, and mortgages, and have a cost per click upwards of \$40. Clicks on sponsored search slots for other keywords can go for as low as a penny.

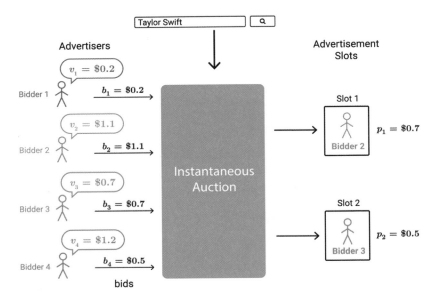

FIGURE 15.6. Behind the scenes when you do a query for a keyword, like "Taylor Swift", in a search engine: At that moment, some of the advertisers who have previously bid on this keyword participate in an instantaneous auction that determines which sponsored search slot, if any, they are allocated to and how much they will have to pay the search engine in the event of a user click. Notice that they may or may not bid truthfully. In this example, the highest bidder's ad (in this case, bidder 2) is allocated to the top slot, and the price p_1 he is charged per click is the bid of the second-highest bidder (bidder 3).

15.5. Sponsored search auctions, GSP, and VCG

What most people don't realize is that all that money comes in pennies at a time.[5]   Hal Varian, Google chief economist

EXAMPLE 15.5.1. **Sponsored search auctions:** An individual performing a search, say for the term "Hawaii timeshare", in a search engine receives a page of results containing the links the search engine has deemed relevant to the search, together with *sponsored links*, i.e., advertisements. These links might lead to the webpages of hotels and companies selling timeshares in Hawaii. To have their ads shown in these slots, these companies participate in an instantaneous auction.

In this auction, each interested advertiser reports[6] a bid b_i representing the maximum he is willing to pay when a searcher clicks on his ad. The search engine, based on these bids, decides which ad to place in each slot and what price to charge the associated advertiser in the event of a user click.

---

[5]Google's revenue in 2015 was approximately $74,500,000,000.

[6]Usually this bid is submitted in advance.

Suppose there are k ad slots[7], with publicly known **clickthrough rates** $c_1 \geq c_2 \geq \cdots \geq c_k \geq 0$. The clickthrough rate of a slot is the probability that a user viewing the webpage will click on an ad in that slot. If bidder i has value v_i per click, then the expected value he obtains from having his ad assigned to slot j is $v_i c_j$. In this setting, the social surplus of the allocation which assigns slot j to bidder π_j is $\sum_{j=1}^{k} v_{\pi_j} c_j$.

This is not formally a win/lose auction (because of the clickthrough rates), but the VCG mechanism readily extends to this case: The social surplus maximizing allocation is selected, and the price a bidder pays is the externality his presence imposes on others. Specifically:

- Each bidder is asked to submit a bid b_i representing the maximum he is willing to pay per click.
- The bidders are reordered so that their bids satisfy $b_1 \geq b_2 \geq \ldots$, and slot i is allocated to bidder i for $1 \leq i \leq k$.
- The participation of bidder i pushes each bidder $j > i$ from slot $j - 1$ to slot j (with the convention that $c_{k+1} = 0$). Thus, i's participation imposes an expected cost of $b_j(c_{j-1} - c_j)$ on bidder j in one search (assuming that b_j is j's value for a click). The auctioneer then charges bidder i a price of $p_i(\mathbf{b})$ per click, chosen so that his expected payment in one search equals the total externality he imposes on other bidders; i.e.,

$$c_i p_i(\mathbf{b}) = \sum_{j=i+1}^{k+1} b_j(c_{j-1} - c_j). \tag{15.3}$$

In other words, bidder i's payment per click is then

$$p_i(\mathbf{b}) := \sum_{j=i+1}^{k+1} b_j \frac{c_{j-1} - c_j}{c_i}. \tag{15.4}$$

Figure 15.6 shows the timing of events in a sponsored search auction. Figure 15.7 shows the allocation, payments, and advertiser utilities that result from maximizing social surplus for a two-slot example, and Figure 15.8 shows the allocation and payments as a function of a particular bidder's bid.

15.5.1. Another view of the VCG auction for sponsored search. First suppose that all clickthrough rates are equal; e.g., $c_i = 1$, for $1 \leq i \leq k$. Then we have a simple k-item Vickrey auction (the items are slots) where bidder i's value for winning slot i is v_i and the threshold bid for each bidder is the kth highest of the other bids. Thus, the payment of each winning bidder is the $(k + 1)^{\text{st}}$ highest bid b_{k+1} and the utility of winning bidder i is $v_i - b_{k+1}$. Since the payment of a winner doesn't depend on his bid, the auction is also truthful. Also, with truthful bidding the auction is **envy-free**: A winner would not want to lose, and vice versa.

In the general case, the VCG auction $\mathcal{A}$ can be thought of as a "sum" of k different auctions, where the ℓth auction $\mathcal{A}_\ell$, for $1 \leq \ell \leq k$, is an ℓ-unit auction and the values and payments are scaled by $(c_\ell - c_{\ell+1})$. Thus, in $\mathcal{A}_\ell$, the value to bidder i of winning is $(c_\ell - c_{\ell+1})v_i$, his payment is $(c_\ell - c_{\ell+1})b_{\ell+1}$, and his utility is $(c_\ell - c_{\ell+1})(v_i - b_{\ell+1})$.

---

[7]The model we consider here greatly simplifies reality.

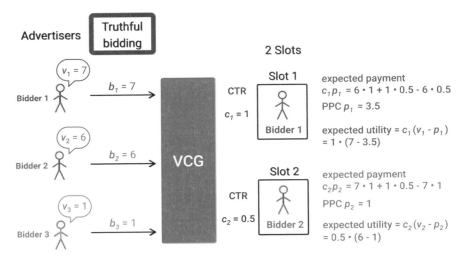

FIGURE 15.7. VCG on sponsored search example: An advertiser's expected value for a slot is her value per click times the clickthrough rate of the slot. For example, bidder 2's expected value for slot 1 is 6, and her expected value for slot 2 is $6 \cdot 0.5 = 3$. Her expected payment is the value other players obtain if she wasn't there ($7 \cdot 1 + 1 \cdot 0.5$) (since bidder 3 would get the second slot in her absence) minus the value the other players get when she is present ($7 \cdot 1$). Her expected payment is the price-per-click (PPC) times the clickthrough rate.

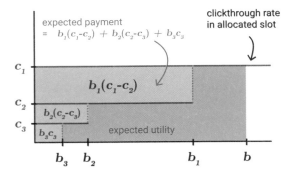

FIGURE 15.8. Suppose that in an ad auction there are 4 bidders, and 3 slots, with clickthrough rates $c_1 > c_2 > c_3$. The first three have submitted bids $b_1 > b_2 > b_3$. The figure shows the allocation and payment for the fourth bidder as a function of his bid b. The piecewise linear black curve shows the probability that the fourth bidder gets a click as a function of his bid. The blue area is his expected VCG payment when he bids $b > b_1$. The payment per click for this bidder is the blue area divided by the click through rate of the slot he is assigned to, in this case c_1. His expected utility for this outcome is the purple area. As indicated below the graph, if he bids between b_{i+1} and b_i, he gets slot $i + 1$. If he bids below b_3, he gets no slot.

If bidder i submits the jth highest bid b_j to the VCG auction $\mathcal{A}$ and wins slot j, the expected value he will obtain is

$$c_j v_i = \sum_{\ell=j}^{k} (c_\ell - c_{\ell+1}) v_i,$$

where $c_{k+1} = 0$. The right-hand side is also the sum of the values he obtains from bidding b_j in each of the auctions $\mathcal{A}_1, \ldots, \mathcal{A}_k$, assuming that the other bidders bid as they did in $\mathcal{A}$; in this case, he will win in $\mathcal{A}_j, \ldots, \mathcal{A}_k$.

Similarly, his expected payment in $\mathcal{A}$ is the sum of the corresponding payments in these auctions; that is,

$$c_j p_j(\mathbf{b}) = \sum_{\ell=j}^{k} (c_\ell - c_{\ell+1}) b_{\ell+1},$$

where b_ℓ is the ℓth largest bid.

LEMMA 15.5.2. *The VCG auction for sponsored search auctions is truthful. Moreover, if bidder i bids truthfully, then he does not envy any other bidder j; i.e.,*

$$c_i(v_i - p_i) \geq c_j(v_i - p_j).$$

(We take $c_j = p_j = 0$ if $j > k$.)

PROOF. The utility of a bidder i in the sponsored search auction $\mathcal{A}$ is the sum of his utilities in the k auctions $\mathcal{A}_1, \ldots, \mathcal{A}_k$. Since bidding truthfully maximizes his utility in each auction separately, it also maximizes his utility in the combined auction. Similarly, since he is not envious of any other bidder j in any of the auctions $\mathcal{A}_1, \ldots, \mathcal{A}_k$, he is not envious in the combined auction.   □

15.5.2. Generalized second-price mechanism. Search engines and other Internet companies run millions of auctions every second. Some companies use VCG, e.g., Facebook, but there is another format that is more popular, known as **generalized second-price (GSP)** mechanism, so named because it generalizes the Vickrey second-price auction.

A simplified version of the GSP auction works as follows:

- Each advertiser interested in bidding on keyword K submits a bid b_i, indicating the price he is willing to pay per click.
- Each ad is ranked according to its bid b_i and ads allocated to slots in this order.
- Each winning advertiser pays the minimum bid needed to win the allocated slot. For example, if the advertisers are indexed according to the slot they are assigned to, with advertiser 1 assigned to the highest slot (slot 1), then advertiser i's payment p_i is

$$p_i = b_{i+1}.$$

(Without loss of generality, there are more advertisers than slots. If not, add dummy advertisers with bids of value 0.)

When there is only one slot, GSP is the same as a second-price auction. However, when there is more than one slot, GSP is no longer truthful, as Figure 15.9 shows.

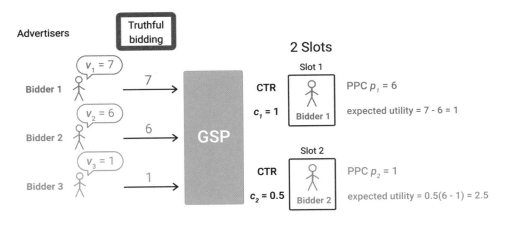

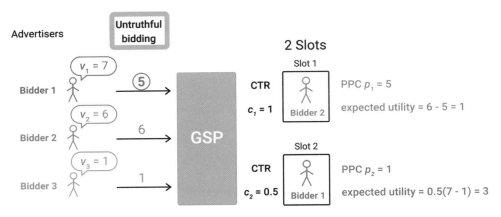

FIGURE 15.9. The top of the figure shows an execution of GSP when advertisers bid truthfully. Bidding truthfully is not an equilibrium though. For
example, if bidder 1 reduces his bid to 5, as shown in the bottom figure, then
he gets allocated to the second slot instead of the first, but his utility is higher.

Although GSP is not truthful, one of its Nash equilibria precisely corresponds
to the outcome of VCG in the following sense.

LEMMA 15.5.3. *Consider n competing advertisers with values v_i sorted so that*
$v_1 \geq v_2 \geq \cdots \geq v_n$. *Assuming truthful bidding in the VCG auction, from* (15.4) *we*
have that bidder i's price-per-click is

$$p_i^{\text{VCG}} = \sum_{j=i+1}^{k+1} v_j \frac{c_{j-1} - c_j}{c_i}. \tag{15.5}$$

Then, in GSP, it is a Nash equilibrium for these advertisers to bid $(b_1, \ldots, b_n)$
where $b_1 > p_1^{\text{VCG}}$ and $b_i = p_{i-1}^{\text{VCG}}$ for $i \geq 2$.

REMARK 15.5.4. For an example of this bidding strategy, see Figure 15.10.

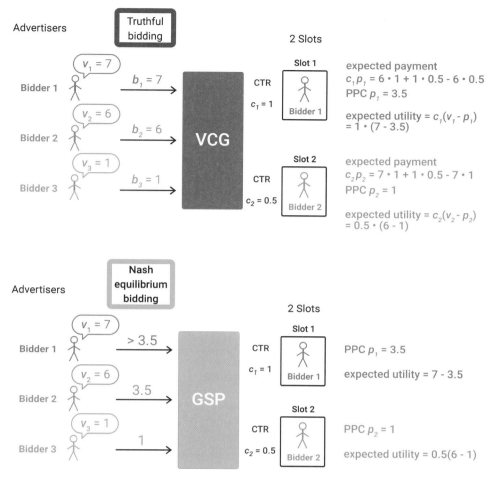

FIGURE 15.10. The top figure illustrates the allocation and payments using VCG. The bottom figure illustrates the allocation and payments in GSP when bidders bid so as to obtain the same allocation and payments. Bidding this way is a Nash equilibrium for GSP.

PROOF. If the advertisers bid $(b_1, \ldots, b_n)$, then the allocation and payments in GSP are exactly those of VCG. Moreover, if each bidder $\ell \neq i$ bids b_ℓ, then bidding b_i is a best response for bidder i: If he bids $b_i' \in (b_{i-1}, b_{i+1})$, then his utility will not change. Otherwise, if he bids b_i' and that yields him a different slot j, then it will be at a price p_j^{VCG}, which cannot increase his utility by the envy-free property of VCG (Lemma 15.5.2). $\qquad\square$

15.5.2.1. *Advertisers differ.* In sponsored search auctions carried out by search engines, they take into account the fact that users prefer certain advertisers to others. To model this, let f_i be an appeal factor for advertiser i that affects the probability that a user will click on an ad from that advertiser. Specifically, we assume that the clickthrough rate of advertiser i in slot j has the form $f_i \cdot c_j$. In

GSP, bidders are first reordered so that $b_1 f_1 \geq b_2 f_2 \geq \cdots \geq b_n f_n$. Bidder i is then assigned to slot i, and he is charged a price-per-click which is the minimum bid required to retain his slot:

$$p_i = \frac{b_{i+1} f_{i+1}}{f_i}.$$

See the exercises for further details.

15.6. Back to revenue maximization

EXAMPLE 15.6.1. **Trips to the moon**: A billionaire is considering selling tours to the moon. The cost of building a rocket is T. There are n_0 people that have declared an interest in the trip. The billionaire wishes to set prices that will recover his cost but does not have good information about the distribution of values the people (bidders) have. Therefore he runs the following auction: Let n_1 be the number of bidders willing to pay T/n_0. If $n_1 = n_0$, the auction ends with the sale of a ticket to each of the n_0 bidders at price T/n_0. If $n_1 < n_0$, let n_2 be the number of bidders willing to pay T/n_1. Iterating, if $n_{j+1} = n_j$ for some j, then the auction terminates with a sale to each of these n_j bidders at a price of T/n_j. Otherwise, some n_j is 0 and the auction terminates with no sale.

EXERCISE 15.b. Show that it is a dominant strategy to be truthful in the auction of Example 15.6.1. Also, show that if the bidders are truthful, the auction finds the largest set of bidders that can share the target cost T equally, if there is one.

15.6.1. Revenue maximization without priors. In the previous chapter, we assumed that the auctioneer knew the prior distributions from which bidders' values are drawn. In this section, we present an auction that is guaranteed to achieve high revenue without any prior information about the bidders. We do this in the context of a *digital goods auction*. These are auctions[8] to sell digital goods such as mp3s, digital video, pay-per view TV, etc. A characteristic feature of digital goods is that the cost of reproducing the items is negligible and therefore the auctioneer effectively has an unlimited supply of the items.

For digital goods auctions, the VCG mechanism allocates to all of the bidders and charges them all nothing! Thus, while VCG perfectly maximizes social surplus, it can be disastrous when the goal is to maximize revenue. We present a truthful auction that does much better.

DEFINITION 15.6.2. The **optimal fixed-price revenue** that can be obtained from bidders with bid vector $\mathbf{b} = (b_1, b_2, \ldots, b_n)$ is

$$\mathcal{R}^*(\mathbf{b}) = \max_p \; p \cdot |\{i \; : \; b_i \geq p\}|. \tag{15.6}$$

Let $\mathsf{p}^*(\mathbf{b})$ denote the (smallest) **optimal fixed price**, i.e., the smallest p where the maximum in (15.6) is attained.

Equivalently, $\mathcal{R}^*(\mathbf{b})$ can be defined as follows: Reorder the bidders so that $b_1 \geq b_2 \geq \cdots \geq b_n$. Then

$$\mathcal{R}^*(\mathbf{b}) = \max_{k \geq 1} \; k \cdot b_k. \tag{15.7}$$

---

[8]Economists call this the "monopoly pricing problem with constant marginal cost".

If the auctioneer knew the true values $\mathbf{v} = (v_1, \ldots, v_n)$ of the agents, he could easily obtain a revenue of $\mathcal{R}^*(\mathbf{v})$ by setting the price to $\mathsf{p}^*(\mathbf{v})$. But the auction which asks the agents to submit bids and then sets the price to $\mathsf{p}^*(\mathbf{b})$ is not truthful. In fact, no auction can approximate $\mathcal{R}^*(\mathbf{v})$ for all $\mathbf{v}$.

CLAIM 15.6.3. *Fix $\delta > 0$. No (randomized) auction can guarantee expected revenue which is at least $\delta \cdot \mathcal{R}^*(\mathbf{v})$ for all $\mathbf{v}$.*

PROOF. Let $n = 1$, and suppose that V_1 is drawn from the distribution $F(x) = 1 - 1/x$, for $x > 1$. Then by Theorem 14.9.3, no auction can obtain revenue more than 1. On the other hand, $\mathbb{E}[\mathcal{R}^*(V_1)] = \infty$. □

The difficulty in the claim arose from a single high bidder. This motivates the goal of designing an auction that achieves a constant fraction of the revenue

$$\mathcal{R}_2^*(\mathbf{b}) := \max_{k \geq 2} \; k \cdot b_k.$$

Since there are no value distributions assumed, we seek to maximize revenue in mechanisms that admit dominant strategies. By a version of the Revelation Principle (see Notes of Chapter 14), it suffices to consider truthful mechanisms. To obtain this, we can ensure that each agent is offered a price which does not depend on her own bid. The following auction is a natural candidate.

The deterministic optimal price auction (DOP):
> For each bidder i, compute $t_i = \mathsf{p}^*(\mathbf{b}_{-i})$, the optimal fixed price for the remaining bidders, and use that as the threshold bid for bidder i.

Unfortunately, this auction does not work well, as the following example shows.

EXAMPLE 15.6.4. Consider a group of bidders of which 11 bidders have value 100 and 1,001 bidders have value 1. The best fixed price is 100 - at that price 11 items can be sold for a total revenue of \$1,100. (The only plausible alternative is to sell to all 1,012 bidders at price \$1, which would result in a lower revenue.)

However, if we run the DOP auction on this bid vector, then for each bidder of value 100, the threshold price that will be used is \$1, whereas for each bidder of value 1, the threshold price is \$100, for a total revenue of only \$11!

In fact, the DOP auction can obtain arbitrarily poor revenue compared to $\mathcal{R}_2^*(\mathbf{v})$. The key to overcoming this problem is to use randomization.

15.6.2. Revenue extraction. A key ingredient in the auction we will develop is the notion of a revenue extractor (discussed in the context of trips to the moon, Example 15.6.1).

DEFINITION 15.6.5 (**A revenue extractor**). The revenue extractor $\mathsf{pe_T}(\mathbf{b})$ with target revenue T sells to the largest set of k bidders that can equally share the cost T and charges each T/k. If there is no such set, the revenue is \$0.

Using the ascending auction procedure discussed in Example 15.6.1, we obtain the following:

LEMMA 15.6.6. *The revenue extractor pe_T is truthful and guarantees a revenue of T on any $\mathbf{b}$ such that $\mathcal{R}^*(\mathbf{b}) \geq T$.*

See Exercise 15.8 for an alternative implementation and proof.

15.6.3. An approximately optimal auction.

DEFINITION 15.6.7 (RSRE). The **random sampling revenue extraction auction** (RSRE) works as follows:

(1) Randomly partition the bids **b** into two groups by flipping a fair coin for each bidder and assigning her bid to $\mathbf{b}'$ or $\mathbf{b}''$.
(2) Compute the optimal fixed-price revenue $T' := \mathcal{R}^*(\mathbf{b}')$ and $T'' := \mathcal{R}^*(\mathbf{b}'')$.
(3) Run the revenue extractors: $\mathsf{pe}_{T'}$ on $\mathbf{b}''$ and $\mathsf{pe}_{T''}$ on $\mathbf{b}'$. Thus, the target revenue for $\mathbf{b}''$ is determined by $\mathbf{b}'$ and vice versa.

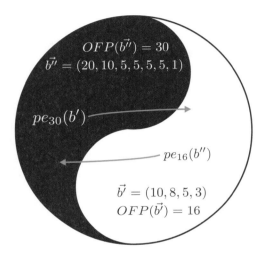

FIGURE 15.11. This figure illustrates a possible execution of the RSRE auction when the entire set of bids is $(20, 10, 10, 8, 5, 5, 5, 5, 5, 3, 1)$. Running the revenue extractor $\mathsf{pe}_{30}(\mathbf{b}')$ will not sell to anyone. Running the revenue extractor $\mathsf{pe}_{16}(\mathbf{b}'')$ will sell to the top six bidders at a price of $16/6$.

REMARK 15.6.8. It seems more natural to compute the price $p^*(\mathbf{b}')$ and use that as a threshold bid for the bidders corresponding to $\mathbf{b}''$ and vice versa. The analysis of this auction, known as the Random Sampling Optimal Price (RSOP) auction, is more delicate. See the notes.

THEOREM 15.6.9. *The random sampling revenue extraction (RSRE) auction is truthful and for all bid vectors* **b**, *the expected revenue of RSRE is at least* $\mathcal{R}_2^*(\mathbf{b})/4$. *Thus, if bidders are truthful, this auction extracts at least* $\mathcal{R}_2^*(\mathbf{v})/4$ *in expectation.*

PROOF. The RSRE auction is truthful since it is simply randomizing over truthful auctions, one for each possible partition of the bids. (Note that any target revenue used in step (3) of the auction is independent of the bids to which it is applied.) So we only have to lower bound the revenue obtained by RSRE on each input **b**. The crucial observation is that for any particular partition of the bids, the revenue of RSRE is at least $\min(T', T'')$. Indeed, if, say, $T' \leq T''$, then $\mathcal{R}_2^*(\mathbf{b}'') = T''$ is large enough to ensure that $\mathsf{pe}_{T'}(\mathbf{b}'')$ will extract a revenue of T'.

Thus, we just need to analyze $E(\min(T', T''))$. Assume that $\mathcal{R}_2^*(\mathbf{b}) = k\mathsf{p}^*$ has $k \geq 2$ winners at price p^*. Of these k winners, suppose that k' are in $\mathbf{b}'$ and k'' are in $\mathbf{b}''$. Thus, $T' \geq k'\mathsf{p}^*$ and $T'' \geq k''\mathsf{p}^*$. Therefore,

$$\frac{\mathbb{E}\left[\min(T', T'')\right]}{k\mathsf{p}^*} \geq \frac{\mathbb{E}\left[\min(k'\mathsf{p}^*, k''\mathsf{p}^*)\right]}{k\mathsf{p}^*} = \frac{\mathbb{E}\left[\min(k', k'')\right]}{k} \geq \frac{1}{4}$$

by Claim 15.6.10. $\square$

CLAIM 15.6.10. *If $k \geq 2$ and $X \sim \mathrm{Bin}(k, \frac{1}{2})$, then $\mathbb{E}\left[\min(X, k - X)\right] \geq k/4$.*

PROOF. For $k = 2, 3$, it is easy to check that the claim holds with equality. Suppose $k \geq 4$. Then $\min(X, k - X) = \frac{k}{2} - |X - \frac{k}{2}|$ and, e.g., by Appendix C, (12),

$$\mathbb{E}[|X - k/2|] \leq \sqrt{\mathrm{Var}X} = \sqrt{\frac{k}{4}} \leq \frac{1}{4}.$$ $\square$

REMARK 15.6.11. Notice that if the bidders actually had values i.i.d. from a distribution F, then the optimal auction would be to offer each bidder the price p that maximizes $p(1 - F(p))$. Thus, the optimal auction would in fact be a fixed-price auction.

Notes

The VCG mechanism is named for Vickrey [Vic61], Clarke [Cla71], and Groves [Gro79]. The most general version of VCG we present in this book is in Chapter 16. See the chapter notes there for more on VCG, the related theory, and its applications. In this chapter, we have focused on truthful *single-parameter* mechanism design, wherein each bidder's value for an allocation depends on only a single private real parameter v_i; e.g., the bidder has value v_i for winning and value 0 for losing.

The descending auction for buying a minimum spanning tree is due to Bikhchandani et al. [BdVSV11]. In §15.4.3, we observed that for public projects, the VCG mechanism does not achieve *budget balance*. That is, the mechanism designer (the government) did not necessarily recover the cost of building the library. Green and Laffont [GL77] showed that no truthful mechanism can simultaneously balance the budget and maximize social surplus.

FIGURE 15.12. Envy

Note that Definition 15.4.1 doesn't capture all types of envy; e.g., the possibility of one agent envying another agent's value function is not addressed. For instance, consider two diners in a restaurant. One orders the $100 lobster. The second cannot afford that and orders rice and beans for $10. By our formal definition, this situation is envy-free, but clearly it is not.

The model and results on sponsored search auctions are due to Edelman, Ostrovsky, and Schwarz [EOS07], and Varian [Var07]. Sponsored search auctions date back to the early 1990s: The idea of allowing advertisers to bid for keywords and charging them per click was introduced by Overture (then GoTo) in 1994. In this early incarnation, the mechanism was "first-price": An advertiser whose ad was clicked on paid the search engine his bid. In the late 1990s, Yahoo! and MSN implemented the Overture scheme as well. However, the use of first-price auctions was observed to be unstable, with advertisers needing to constantly update their bids to avoid paying more than their value. In 2002, Google adopted the generalized second price mechanism. Yahoo!, MSN, and Microsoft followed suit. For a brief history of sponsored search as well as results and research directions related to sponsored search, see [JM08] and Chapter 28 in [NRTV07].

Facebook is one of the only companies selling ads online via auctions that uses the VCG mechanism [Met15]. Possible reasons that VCG isn't widely used are (a) it is relatively complicated for the advertisers to understand and (b) optimizing social surplus may not be the objective. The obvious goal for the search engines is to maximize long-term profit. However, it is not clear what function of the current auction is the best proxy for that. Besides profit and social surplus, other key parameters are user and advertiser satisfaction.

In the previous chapter, we assumed the auctioneer knew the prior distributions from which agents' values are drawn. An attractive alternative is to design an auction whose performance does not depend on knowledge of priors. Indeed, where do priors come from? Typically, they come from previous similar interactions. This is problematic when markets are small, when interactions are novel or when priors change over time. Another difficulty is that agents may alter their actions to bias the auctioneer's beliefs about future priors in their favor. This is the motivation for the material in §15.6.

The first prior-free digital goods auctions and the notion of prior-free analysis of auctions were introduced by Goldberg, Hartline, and Wright [GHW01]. Their paper shows that any deterministic truthful auction that treats the bidders symmetrically will fail to consistently obtain a constant fraction of the optimal fixed-price revenue, thus motivating the need for randomization. (Aggarwal et al. [AFG+11] showed that this barrier can be circumvented using an auction that treats the bidders asymmetrically.)

The RSOP auction (see Remark 15.6.8) proposed by [GHW01] was first analyzed in [GHK+06]; its analysis was improved in a series of papers culminating in [AMS09]. The random sampling revenue extraction auction presented here is due to Fiat, Goldberg, Hartline, and Karlin [FGHK02]. The key building block, revenue extraction, is due to Moulin and Shenker [MS01].

Stronger positive results and further applications of the prior-free framework for revenue maximization and cost minimization are surveyed in Chapter 13 of [NRTV07] and in Chapter 7 of [Har17]. The strongest positive results known for prior-free digital goods auctions are due to Chen, Gravin, and Lu [CGL14]. They showed that the lower bound of 2.42 on the competitive ratio of any digital goods auction, proved in [GHKS04], is tight.

The problem considered in Exercise 15.1 was studied in [MS83]. Exercise 15.3 is from [Nis99], Exercise 15.2 is from [NR01], and Exercise 15.10 is the analogue of Theorem 14.6.1, which is due to [Mye81]. See also [GL77].

Exercises

15.1.   A seller has an item that he values at $v_s \in [0,1]$ and a buyer has a value $v_b \in [0,1]$ for this same item. Consider designing a mechanism for determining whether or not the seller should transfer the item to the buyer.

Social surplus is maximized by transferring the item if $v_b \geq v_s$ and not transferring it otherwise. How do the VCG payments depend on the values v_s and v_b?

15.2. Consider a communication network, where each link is owned by a different company (henceforth, bidder). Each bidder's private information is the cost of routing data along that link. A procurement auction to buy the use of a path between two specific nodes s and t is run: Each company submits a bid representing the minimum price that company is willing to be paid for the use of its link. The auction consists of choosing the minimum cost path between s and t and then paying each bidder (link owner) along that path the maximum amount he could have bid while still being part of the minimum cost path.
- Show that this auction is truthful.
- Construct an example in which the amount paid by the auctioneer is $\Omega(n)$ times as large as the actual cost of the shortest path. Here n is the number of links in the network.

15.3. There are n computers connected in a line (with computer i, with $1 < i < n$, connected to computers $i - 1$ and $i + 1$). Each computer has private value v_i for executing a job; however, a computer can only successfully execute the job with access to both of its neighboring links. Thus, no two adjacent computers can execute jobs. Consider the following protocol:
- In the first, left-to-right, phase, each computer places a bid r_i for the link on its right, where $r_1 = v_1$ and $r_i = \max(v_i - r_{i-1}, 0)$ for $1 < i < n$.
- In the next, right-to-left, phase, each computer places a bid ℓ_i for the link on its left, where $\ell_n = v_n$ and $\ell_i = \max(v_i - \ell_{i+1}, 0)$.

Computer i can execute its task if and only if $\ell_i > r_{i-1}$ and $r_i \geq \ell_{i+1}$. In this case, it "wins" both links, and its payment is $r_{i-1} + \ell_{i+1}$.
(a) Show that this mechanism maximizes social surplus; that is, the set of computers selected has the maximum total value among all subsets that do not contain any adjacent pairs.
(b) Show that, under the assumption that any misreports by the computers are consistent with having a single value v_i', it is in each computer's best interest to report truthfully.
(c) Show that if the computer can misreport arbitrarily in the two phases, then it is no longer always a dominant strategy to report truthfully; however, reporting truthfully is a Nash equilibrium.

15.4. Consider the descending price auction described in §15.4.2. Prove that the outcome and payments are the same as those of the VCG auction and that it is a dominant strategy for each edge (bidder) to stay in the auction as long as the current price is above his cost.

15.5.   Consider a search engine selling advertising slots on one of its pages. There are three advertising slots with publicly known clickthrough rates (probability that an individual viewing the webpage will click on the ad) of 0.08, 0.03, and 0.01, respectively, and four advertisers whose values per click are 10, 8, 2, and 1, respectively. Assume that the expected value for an advertiser to have his ad shown in a particular slot is his value times the clickthrough rate. What is the allocation and what are the payments if the search engine runs VCG? How about GSP?

15.6.   Consider the following model for a keyword auction, slightly more general than the version considered in the text. There are k slots and n bidders. Each bidder has a private value v_i per click and a publicly known quality q_i. The quality measures the ad-bidder (ad) dependent probability of being clicked on. Assume that the slots have clickthrough rates $c_1 \geq c_2 \geq \cdots \geq c_k$ and that the expected value to bidder i to be shown in slot j is $q_i c_j v_i$. Determine the VCG allocation and per click payments for this setting. (Note that the expected utility of bidder i if allocated slot j at price-per-click of p_{ij} is $q_i c_j (v_i - p_{ij})$.)

15.7.   Consider the GSP auction in the setting of the previous exercise, where bidders are assigned to slots in decreasing order of $b_i q_i$, where b_i is the advertiser i's bid, with a bidder's payment being the minimum bid he could make to retain his slot. Prove an analogue of Lemma 15.5.3 for this setting.

15.8.   Consider the following implementation of the profit extractor pe_T from §15.6.2: Given the reported bids, for each bidder i separately, the auctioneer pretends that bidder's bid is ∞ and then, using bid vector $(\infty, \mathbf{b}_{-i})$, determines the largest k such that k bidders can pay T/k. This price is then offered to bidder i who accepts if $b_i \geq T/k$ and rejects otherwise. With this formulation it is clear that the auction is truthful. However, it is less obvious that this implements the same outcome. Show that it does.

15.9.   Prove the exact formula for RSRE

$$E(\min(k', k'')) = \sum_{0 \leq i \leq k} \min(i, k-i) \binom{k}{i} 2^{-k}$$
$$= k \left(\frac{1}{2} - \binom{k-1}{\lfloor \frac{k}{2} \rfloor} 2^{-k} \right)$$
$$\geq \frac{k}{4}.$$

15.10. Let $\mathcal{A}$ be an auction for a win/lose setting defined by U and $\mathcal{L}$, and suppose that $\alpha_i[\mathbf{b}]$ is the probability that bidder i wins when the bids are $\mathbf{b} = (b_1, \ldots, b_n)$. (This expectation is taken solely over the randomness in the auction.) Assume that, for each bidder, $v_i \in [0, \infty)$. Prove that it is a dominant strategy in $\mathcal{A}$ for bidder i to bid truthfully if and only if, for any bids $\mathbf{b}_{-i}$ of the other bidders, the following holds:

(a) The expected allocation $\alpha_i[v_i, \mathbf{b}_{-i}]$ is (weakly) increasing in v_i.

(b) The expected payment of bidder i is determined by the expected allocation up to an additive constant:

$$p_i[v_i, \mathbf{b}_{-i}] = v_i \cdot \alpha_i[v_i, \mathbf{b}_{-i}] - \int_0^{v_i} \alpha_i[z, \mathbf{b}_{-i}]dz + p_i[0, \mathbf{b}_{-i}].$$

Hint: The proof is analogous to that of Theorem 14.6.1.

15.11. Show that the allocation and payment rule of the VCG mechanism for maximizing social surplus in win/lose settings satisfies conditions (a) and (b) of the previous exercise.

15.12. Generalize the result of Exercise 15.10 to the following setting:

- Each bidder has a private value $v_i \geq 0$.
- Each outcome is a vector $\mathbf{q} = (q_1, \ldots, q_n)$, where q_i is the "quantity" allocated to bidder i. (In a win/lose setting, each q_i is either 0 or 1.) Denote by $\mathcal{Q}$ the set of feasible outcomes (generalizing $\mathcal{L}$ from win/lose settings).
- A bidder with value v_i who receives an allocation of q_i and is charged p_i obtains a utility of $v_i q_i - p_i$.

Now let $\mathcal{A}$ be an auction for an allocation problem defined by U and $\mathcal{Q}$, and suppose that $\alpha_i[\mathbf{b}]$ is the expected quantity allocated to bidder i when the bids are $\mathbf{b} = (b_1, \ldots, b_n)$. (This expectation is taken solely over the randomness in the auction.) Assume that, for each bidder, $v_i \in [0, \infty)$. Show that it is a dominant strategy in $\mathcal{A}$ for bidder i to bid truthfully if and only if, for any bids $\mathbf{b}_{-i}$ of the other bidders, the following holds:

(a) The expected allocation $\alpha_i[v_i, \mathbf{b}_{-i}]$ is (weakly) increasing in v_i.

(b) The expected payment of bidder i is determined by the expected allocation up to an additive constant:

$$p_i[v_i, \mathbf{b}_{-i}] = v_i \cdot \alpha_i[v_i, \mathbf{b}_{-i}] - \int_0^{v_i} \alpha_i[z, \mathbf{b}_{-i}]dz + p_i[0, \mathbf{b}_{-i}].$$

15.13. Consider a single-item auction, but with the following alternative bidder model. Suppose that each of n bidders has a signal, say s_i for bidder i, and suppose that agent i's value $v_i := v_i(s_1, \ldots, s_n)$ is a function of all the signals. This captures scenarios where each bidder has different information about the item being auctioned and weighs all of these signals, each in his or her own way.

Suppose that there are two bidders, where $v_1(s_1, s_2) = s_1$ and $v_2(s_1, s_2) = s_1^2$. Assume that $s_1 \in [0, \infty)$. Show that there is no truthful social surplus maximizing auction for this example. (A social surplus maximizing auction must allocate the item to the bidder with the higher value, so to bidder 2 when $s_1 \geq 1$ and to bidder 1 when $s_1 < 1$.)

CHAPTER 16

VCG and scoring rules

In the previous chapter we studied the design of truthful auctions in a setting where each bidder was either a winner or a loser in the auction. In this chapter, we explore the more general setting of **mechanism design**. **Mechanisms** allow for the space of outcomes to be richer than, say, simply allocating items to bidders. The goal of the mechanism designer is to design the game (i.e., the mechanism) so that, in equilibrium, desirable outcomes are achieved.

16.1. Examples

EXAMPLE 16.1.1 (**Spectrum auctions**). In a spectrum auction, the government allocates licenses for the use of some band of electromagnetic spectrum in a certain geographic area. The participants in the auction are cell phone companies who need such licenses to operate. Each company has a value for each combination of licenses. The government wishes to design a procedure for allocating and pricing the licenses that maximizes the cumulative value of the outcome to all participants. What procedure should be used?

FIGURE 16.1. A spectrum auction.

EXAMPLE 16.1.2 (**Building roads**). The state is trying to determine which roads to build to connect a new city C to cities A and B (which already have a road between them). The options are to build a road from A to C or a road from B to C, both roads, or neither. Each road will cost the state $10 million to build. Each city obtains a certain economic/social benefit for each outcome. For example, city A might obtain a $5 million benefit from the creation of a road to city C, but no real benefit from the creation of a road between B and C. City C, on the other hand, currently disconnected from the others, obtains a significant benefit

($9 million) from the creation of each road, but the marginal benefit of adding a second connection is not as great as the benefit of creating a first connection. The following table summarizes these values (in millions of dollars), and the cost to the state for each option. The final row is the social surplus, the sum of the values to the three cities plus the value to the state.

	road A-C	road B-C	both	none
city A	5	0	5	0
city B	0	5	5	0
city C	9	9	15	0
state	-10	-10	-20	0
social surplus	4	4	5	0

The state's goal might be to choose the option that maximizes social surplus, which, for these numbers, is the creation of both roads. However, these numbers are reported to the state by the cities themselves, who may have an incentive to exaggerate their values, so that their preferred option will be selected. Thus, the state would like to employ a mechanism that incentivizes truth-telling.

16.2. Social surplus maximization and the general VCG mechanism

A mechanism $\mathcal{M}$ selects an outcome from a set of possible outcomes $\mathcal{A}$, based on inputs from a set of agents. Each agent i has a **valuation function** $v_i : \mathcal{A} \to \mathbb{R}$ that maps the possible outcomes $\mathcal{A}$ to nonnegative real numbers. The quantity $v_i(a)$ represents the value[1] that i assigns to outcome $a \in \mathcal{A}$, measured in a common currency, such as dollars. We denote by $V(a)$ the value the (mechanism) designer has for outcome a. Given the reported valuation functions, the mechanism selects an outcome and a set of payments, one per agent. We have seen the following definition several times; we repeat it in the context of this more general setting.

DEFINITION 16.2.1. We say that a mechanism $\mathcal{M}$ is **truthful** if, for each agent i, each valuation function $v_i(\cdot)$, and each possible report $\mathbf{b}_{-i}$ of the other agents, it is a dominant strategy for agent i to report his valuation truthfully. Formally, for all $\mathbf{b}_{-i}$, all i, $v_i(\cdot)$, and $b_i(\cdot)$,

$$u_i[v_i, \mathbf{b}_{-i}|v_i] \geq u_i[b_i, \mathbf{b}_{-i}|v_i],$$

where i's utility[2] $u_i[b_i, \mathbf{b}_{-i}|v_i] = v_i(a(\mathbf{b})) - p_i(\mathbf{b})$. Here $\mathbf{b} = (b_1(\cdot), b_2(\cdot), \dots, b_n(\cdot))$, $a(\mathbf{b}) \in \mathcal{A}$ is the outcome selected by $\mathcal{M}$ on input $\mathbf{b}$, and $p_i(\mathbf{b})$ is the payment of agent i.

---

[1] See discussion in §15.4.3.
[2] This is called *quasilinear* utility and is applicable when the value is measured in the same units as the payments.

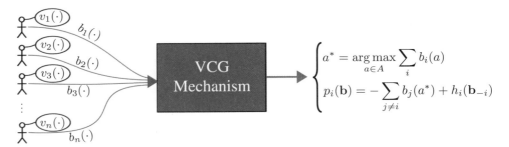

FIGURE 16.2. A depiction of the VCG mechanism for the setting where $V(a) = 0$ for all $a \in A$. The outcome selected is $a^*(\mathbf{b})$, and the payment of agent i is $p_i(\mathbf{b})$. Note that Theorem 16.2.6 holds for any choice of functions $h_i(\mathbf{b}_{-i})$ as long as it is independent of $b_i(\cdot)$. In Definition 16.2.3, we take $h_i(\mathbf{b}_{-i}) = \max_a \sum_{j \neq i} b_j(a)$. This choice guarantees that $u_i[v_i, \mathbf{b}_{-i}|v_i] \geq 0$ for all v_i, so long as all $v_i(a) \geq 0$ for all a and i.

DEFINITION 16.2.2. The **social surplus** of an outcome a is $\sum_i v_i(a) + V(a)$. The **reported social surplus** of an outcome a is $\sum_i b_i(a) + V(a)$.

DEFINITION 16.2.3. The **Vickrey-Clarke-Groves (VCG) mechanism**, illustrated in Figure 16.2, works as follows: Each agent is asked to report his valuation function $v_i(\cdot)$ and submits a function $b_i(\cdot)$ (which may or may not equal $v_i(\cdot)$). Write $\mathbf{b} = (b_1(\cdot), \ldots, b_n(\cdot))$. The outcome selected is

$$a^* := a^*(\mathbf{b}) = \mathrm{argmax}_a \left(\sum_j b_j(a) + V(a) \right),$$

breaking ties arbitrarily. The payment $p_i(\mathbf{b})$ agent i makes is the loss his presence causes others (with respect to the reported bids); formally,

$$p_i(\mathbf{b}) = \max_a \left(\sum_{j \neq i} b_j(a) + V(a) \right) - \left(\sum_{j \neq i} b_j(a^*) + V(a^*) \right). \tag{16.1}$$

The first term is the total reported value the other agents would obtain if i was absent, and the term being subtracted is the total reported value the others obtain when i is present.

EXAMPLE 16.2.4. Consider the outcome and payments for the VCG mechanism on Example 16.1.2, assuming that the cities report truthfully.

	road A-C	road B-C	both	none
City A	5	0	5	0
City B	0	5	5	0
City C	9	9	15	0
state	−10	−10	−20	0
social surplus	4	4	5	0
surplus without A	−1	4	0	0
surplus without C	−5	−5	−10	0

As we saw before, the social surplus maximizing outcome would be to build both roads, yielding a surplus of 5. What about the payments using VCG? For city A, the surplus attained by others in A's absence is 4 (road B-C only would be built), whereas with city A, the surplus attained by others is 0, and therefore city A's payment, the harm its presence causes others, is 4. By symmetry, B's payment is the same. For city C, the surplus attained by others in C's absence is 0, whereas the surplus attained by others in C's presence is −10, and therefore C's payment is 10. Notice that the total payment is 18, whereas the state spends 20.

EXAMPLE 16.2.5 (**Employee housing**). A university owns a number of homes and plans to lease them to employees. They choose the allocation and pricing by running a VCG auction. A set of n employees, each interested in leasing at most one house, participates in the auction. The ith employee has value v_{ij} for a yearly lease of the jth house. Figure 16.3 shows an example.

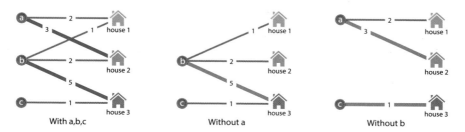

FIGURE 16.3. The label on the edge from i on the left to j on the right is the value v_{ij} that employee i has for a yearly lease of house j (say in thousands of dollars). The VCG mechanism allocates according to purple shaded edges. The payment of bidder a is 0 since in his absence house 3 is still allocated to bidder b. The payment of bidder b is 1 since in his absence the allocation is as follows: house 2 to bidder a and house 3 to bidder c, and therefore the externality he imposes is 1.

THEOREM 16.2.6. *VCG is a truthful mechanism for maximizing social surplus.*

PROOF. We prove the theorem assuming that the designer's value $V(a)$ for all outcomes a is 0. See Exercise 16.1 for the general case. Fix the reports $\mathbf{b}_{-i}$ of all agents except agent i. Suppose that agent i's true valuation function is $v_i(\cdot)$, but he reports $b_i(\cdot)$ instead. Then the outcome is

$$a^* = \text{argmax}_a \sum_j b_j(a)$$

and his payment is

$$p_i(\mathbf{b}) = \max_a \sum_{j \neq i} b_j(a) - \sum_{j \neq i} b_j(a^*) = C - \sum_{j \neq i} b_j(a^*).$$

where $C := \max_a \sum_{j \neq i} b_j(a)$ is a constant that agent i's report has no influence on. Thus, agent i's utility is

$$u_i(\mathbf{b}|v_i) = v_i(a^*) - p_i(\mathbf{b}) = v_i(a^*) + \sum_{j \neq i} b_j(a^*) - C. \tag{16.2}$$

On the other hand, if he were to report his true valuation function $v_i(\cdot)$, the outcome would be

$$a' = \text{argmax}_a \left(v_i(a) + \sum_{j \neq i} b_j(a) \right),$$

and thus, by (16.2), his utility would be

$$u_i[v_i, \mathbf{b}_{-i}|v_i] = v_i(a') + \sum_{j \neq i} b_j(a') - C \geq v_i(a^*) + \sum_{j \neq i} b_j(a^*) - C.$$

In other words, $u_i[\mathbf{b}|v_i] \leq u_i[v_i, \mathbf{b}_{-i}|v_i]$ for every $\mathbf{b}$, $v_i(\cdot)$, and i. □

REMARK 16.2.7. It follows from the proof of Theorem 16.2.6 that, if all the bids are truthful, then bidder i's utility is $u_i[\mathbf{v}|v_i] = \max_{a \in A} \sum_i v_i(a) + h_i(\mathbf{v}_{-i})$, which is the same, up to a constant, as the objective function optimized by the mechanism designer.

The following example illustrates a few of the deficiencies of the VCG mechanism.

EXAMPLE 16.2.8 (**Spectrum auctions, revisited**). Company A has recently entered the market and needs two licenses in order to operate efficiently enough to compete with the established companies. Thus, A has no value for a single license but values a pair of licenses at $1 billion. Companies B and C are already well established and only seek to expand capacity. Thus, each one needs just one license and values that license at $1 billion.

Suppose the government runs a VCG auction to sell two licenses. If only companies A and B compete in the auction, the government revenue is $1 billion (either A or B can win). However, if A, B, and C all compete, then companies B and C will each receive a license but pay nothing. Thus, VCG revenue is not necessarily monotonic in participation or bidder values.

A variant on this same setting illustrates another problem with the VCG mechanism that we saw earlier: susceptibility to collusion. Suppose that company A's preferences are as above and companies B and C still only need one license each, but now they only value a license at $25 million. In this case, if companies B and

C bid honestly, they lose the auction. However, if they collude and each bid $1 billion, they both win at a price of $0.

16.3. Scoring rules

16.3.1. Keeping the meteorologist honest. On the morning of each day $t = 1, 2, \ldots$, a meteorologist M reports an estimate q_t of the probability that it will rain that day. With some effort, he can determine the true probability p_t that it will rain that day, given the current atmospheric conditions, and set $q_t = p_t$. Or, with no effort, he could just guess this probability. How can his employer motivate him to put in the effort and report p_t?

A first idea is to pay M, at the end of the tth day, the amount q_t (or some dollar multiple of that amount) if it rains and $1 - q_t$ if it shines. If $p_t = p > \frac{1}{2}$ and M reports truthfully, then his expected payoff is $p^2 + (1 - p)^2$, while if he were to report $q_t = 1$, then his expected payoff would be $p = p^2 + p(1 - p)$, which is higher.

Another idea is to pay M an amount that depends on how **calibrated** his forecast is. Suppose that M reports q_t values on a scale of $\frac{1}{10}$, so that he has nine choices[3], namely $\{k/10 : k \in \{1, \ldots, 9\}\}$. When a year has gone by, the days of that year may be divided into nine types according to the q_t value that the weatherman declared. Suppose there are n_k days that the predicted value q_t is $\frac{k}{10}$, while in fact it rained on r_k of these n_k days. Then, the forecast is calibrated if r_k/n_k is close to $k/10$ for all k. Thus, we might want to penalize the weatherman according to the squared error of his predictions; i.e.,

$$\sum_{k=1}^{9} \left(\frac{r_k}{n_k} - \frac{k}{10} \right)^2.$$

Unfortunately, it is easy to be calibrated without being accurate: If the true probability of rain is 90% for half of the days each year and 10% the rest and M reports 50% every day, then he is calibrated for the year.

16.3.2. A solution. Suppose that the employer pays $s_1(q_t)$ to M if it rains and $s_2(1 - q_t)$ if it shines on day t. If $p_t = p$ and $q_t = q$, then the expected payment made to M on day t is

$$g_p(q) := p s_1(q) + (1 - p) s_2(1 - q).$$

The employer's goal is to choose the functions $s_1, s_2 : [0, 1] \to \mathbb{R}$ so that

$$g_p(p) \geq g_p(q) \quad \text{for all } q \in [0, 1] \setminus \{p\}. \tag{16.3}$$

In this case, the pair $s_1(\cdot), s_2(\cdot)$ is called a **proper scoring rule.** Suppose these functions are differentiable. Then we must have

$$g_p'(p) = p s_1'(p) - (1 - p) s_2'(1 - p) = 0$$

for all p. Define

$$h(p) := p s_1'(p) = (1 - p) s_2'(1 - p). \tag{16.4}$$

Thus

$$g_p'(q) = p s_1'(q) - (1 - p) s_2'(1 - q) = h(q) \left(\frac{p}{q} - \frac{1 - p}{1 - q} \right). \tag{16.5}$$

---

[3]M's instruments are not precise enough to yield complete certainty.

PROPOSITION 16.3.1. *A pair of smooth[4] functions s_1, s_2 define a proper scoring rule if and only if there is a continuous $h : (0,1) \to [0, \infty)$ such that for all p*

$$s_1'(p) = \frac{h(p)}{p} \quad and \quad s_2'(p) = \frac{h(1-p)}{p}. \tag{16.6}$$

PROOF. If s_1, s_2 satisfy (16.6) and $h(\cdot)$ is nonnegative, then $g_p(q)$ is indeed maximized at p by (16.5), since $p/q - (1-p)/(1-q)$ is positive for $q < p$ and negative for $q > p$.

Conversely, if the smooth functions s_1, s_2 define a proper scoring rule, then h defined by (16.4) is continuous. If there is some p for which $h(p) < 0$, then $g_p'(q) > 0$ for $q \downarrow p$, and the scoring rule is not proper at p. □

REMARK 16.3.2. If $h(p) > 0$ for all p, then the inequality in (16.3) is strict and the scoring rule is called **strictly proper**.

For example, we can apply Proposition 16.3.1 to obtain the following two well known scoring rules:

- Letting $h(p) = 1$, we get the **logarithmic scoring rule**:

$$s_i(p) = \log p.$$

- Letting $h(p) = 2p(1-p)$, we get the **quadratic scoring rule**:

$$s_i(p) = \int_0^p 2(1-x)\,dx = 2p - p^2.$$

16.3.3. A characterization of scoring rules*. We extend the previous discussion to the case where the prediction is not binary.

DEFINITION 16.3.3. Consider a forecaster whose job is to assign probabilities to n possible outcomes. Denote by Δ_n° the open simplex consisting of strictly positive probability vectors. A **scoring rule** $\mathbf{s} : \Delta_n^\circ \to \mathbb{R}^n$ specifies the score/reward the forecaster will receive as a function of his prediction[5] and the outcome. That is, if $\mathbf{s}(\mathbf{q}) = (s_1(\mathbf{q}), \dots, s_n(\mathbf{q}))$, then $s_i(\mathbf{q})$ is the reward for report $\mathbf{q} = (q_1, \dots, q_n)$ if the ith outcome occurs. The scoring rule is **proper** if for all $\mathbf{p}$ the function

$$g_{\mathbf{p}}(\mathbf{q}) = \sum_i p_i s_i(\mathbf{q})$$

is maximized at $\mathbf{q} = \mathbf{p}$. Thus, a scoring rule is proper if a forecaster who believes the probability distribution over outcomes is $\mathbf{p}$ maximizes his expected reward by reporting this distribution truthfully. (If the maximum is attained only at $\mathbf{p}$, then $\mathbf{s}$ is called a **strictly proper** scoring rule.)

In order to characterize proper scoring rules, we recall the following definition[6].

---

[4] A function is smooth if it has a continuous derivative.

[5] We've restricted predictions to the interior of the simplex because one of the most important scoring rules, the logarithmic scoring rule, becomes $-\infty$ at the boundary of the simplex.

[6] See Appendix C for a review of basic properties of convex functions.

DEFINITION 16.3.4. Let $K \subset \mathbb{R}^n$ be a convex set. A vector $\mathbf{v} \in \mathbb{R}^n$ is a **subgradient** of a convex function $f : K \to \mathbb{R}$ at $\mathbf{y} \in K$ if for all $\mathbf{x} \in K$

$$f(\mathbf{x}) \geq f(\mathbf{y}) + \mathbf{v} \cdot (\mathbf{x} - \mathbf{y}).$$

If f extends to a differentiable, convex function on an open neighborhood of $K \subset \mathbb{R}^n$, then $\nabla f(\mathbf{y})$ is a subgradient at $\mathbf{y}$ for every $\mathbf{y} \in K$.

THEOREM 16.3.5. *Let* $\mathbf{s} : \Delta_n^{\circ} \to \mathbb{R}^n$. *Then* $\mathbf{s}(\cdot)$ *is a proper scoring rule if and only if there is a convex function* $f : \Delta_n^{\circ} \to \mathbb{R}$ *such that for all* $\mathbf{q} \in \Delta_n^{\circ}$ *there is a subgradient* $\mathbf{v_q}$ *of* f *at* $\mathbf{q}$ *satisfying*

$$s_i(\mathbf{q}) = f(\mathbf{q}) + (\mathbf{e}_i - \mathbf{q}) \cdot \mathbf{v_q} \qquad \forall i. \tag{16.7}$$

PROOF. Suppose that S is a proper scoring rule. Let

$$f(\mathbf{p}) := \mathbf{p} \cdot \mathbf{s}(\mathbf{p}) = \sup_{\mathbf{q} \in \Delta_n^{\circ}} \mathbf{p} \cdot \mathbf{s}(\mathbf{q}).$$

Since $f(\cdot)$ is the supremum of linear functions, it is convex. Next, we fix $\mathbf{q}$. Then for all $\mathbf{p}$

$$f(\mathbf{p}) \geq \mathbf{p} \cdot \mathbf{s}(\mathbf{q}) = f(\mathbf{q}) + (\mathbf{p} - \mathbf{q}) \cdot \mathbf{s}(\mathbf{q}).$$

Thus, $\mathbf{s}(\mathbf{q})$ is a subgradient at $\mathbf{q}$ and (16.7) holds for $\mathbf{v_q} = \mathbf{s}(\mathbf{q})$.

For the converse, suppose that $\mathbf{s}(\cdot)$ is of the form (16.7) for some convex function f. We observe that

$$\mathbf{p} \cdot \mathbf{s}(\mathbf{q}) = \sum_i p_i \left[f(\mathbf{q}) + (\mathbf{e}_i - \mathbf{q}) \cdot \mathbf{v_q} \right] = f(\mathbf{q}) + (\mathbf{p} - \mathbf{q}) \cdot \mathbf{v_q}$$

$$= f(\mathbf{p}) - \left[f(\mathbf{p}) - f(\mathbf{q}) - (\mathbf{p} - \mathbf{q}) \cdot \mathbf{v_q} \right].$$

In particular, $\mathbf{p} \cdot \mathbf{s}(\mathbf{p}) = f(\mathbf{p})$. Since $\mathbf{v_q}$ is a subgradient, the quantity inside the square braces is nonnegative for all $\mathbf{q}$, so $\mathbf{p} \cdot \mathbf{s}(\mathbf{q}) \leq f(\mathbf{p}) = \mathbf{p} \cdot \mathbf{s}(\mathbf{p})$. $\qquad \square$

REMARK 16.3.6. The proof implies that $\mathbf{s}(\cdot)$ is a strictly proper scoring rule if and only if there is a strictly convex function $f(\cdot)$ for which (16.7) holds.

COROLLARY 16.3.7. *Using the recipe of the previous theorem, we obtain proper scoring rules as follows:*

- *The **quadratic scoring rule** is*

$$s_i(\mathbf{q}) = 2q_i - \sum_j q_j^2.$$

  This is a scoring rule obtained using

$$f(\mathbf{q}) = \sum_i q_i^2.$$

- *The **logarithmic scoring rule** is*

$$s_i(\mathbf{q}) = \log(q_i).$$

  This is the scoring rule obtained using

$$f_L(\mathbf{q}) = \sum_i q_i \log q_i.$$

EXERCISE 16.a. Prove Corollary 16.3.7.

Notes

The VCG mechanism is named for Vickrey [Vic61], Clarke [Cla71], and Groves [Gro79]. Vickrey's work focused on single-item auctions and multiunit auction with down-sloping valuations, Clarke studied the public project problem, and Groves presented the general formulation of VCG. Clarke proposed the "Clarke pivot rule" - the specific constant added to payments that guarantees individually rational mechanisms. This rule enables the interpretation of the payments as the externalities imposed on other bidders.

"The lovely but lonely Vickrey auction" by Ausubel and Milgrom [AM06] describes the features and deficiencies of VCG and discusses why it is used infrequently in practice. For more on VCG, the related theory, and its many applications, see [MCWJ95, Kri09, Mil04, NRTV07, Har17, Rou13, Rou14].

In some settings, e.g., combinatorial auctions, computing the social surplus maximizing outcome is intractable. Unfortunately, the VCG payment scheme does not remain truthful when an approximate social surplus maximizing outcome is selected (see, e.g., [LOS02, NR01, NR07]).

Consider the setting of §16.2 and let f be a function from valuation functions to outcomes $\mathcal{A}$. We say that the function f can be *truthfully implemented* if there is a payment rule from reported valuation functions to outcomes such that for every i, $v_i(\cdot)$, $b_i(\cdot)$, and $\mathbf{b}_{-i}(\cdot)$,

$$v_i(f(v_i, \mathbf{b}_{-i})) - p(v_i, \mathbf{b}_{-i}) \geq v_i(f(b_i, \mathbf{b}_{-i})) - p(b_i, \mathbf{b}_{-i}).$$

In other words, the payment rule incentivizes truth-telling. We've already seen that the social surplus function, $f = \mathrm{argmax}_{a \in \mathcal{A}} v_i(a)$, can be truthfully implemented. Affine maximization, i.e., $f = \mathrm{argmax}_{a \in \mathcal{A}'} \sum_i (c_i v_i(a) + \gamma_a)$, where $\mathcal{A}' \subseteq \mathcal{A}$, can also be truthfully implemented. (See Exercise 16.3.) In single-parameter domains (such as the win/lose settings discussed in Chapter 15), where the bidder's private information is captured by a single number, any monotone function can be truthfully implemented. (See, e.g., [NRTV07].) In contrast, when valuation functions $v_i : \mathcal{A} \to \mathbb{R}$ can be arbitrary, Roberts [Rob79] has shown that the *only* functions that can be implemented by a truthful mechanism are affine maximizers. For other results on this and related topics, see, e.g., [DHM79, GL77, Mye81, GMV04, SY05, AK08].

As we have discussed, mechanism design is concerned with the design of protocols (and auctions) so that rational participants, motivated solely by their self-interest, will end up achieving the designer's goal. Traditional applications of mechanism design include writing insurance contracts, regulating public utilities and constructing the tax code. Modern applications include scheduling tasks in the cloud, routing traffic in a network, allocating resources in a distributed system, buying and selling goods in online marketplaces, etc. Moreover, some of these problems are being solved on a grand scale, with mechanisms that are implemented in software (and sometimes hardware). Even the bidding is often done by software agents. For this reason, efficient computability has become important.

The objective of understanding to what extent appropriate incentives and computational efficiency can be achieved simultaneously was brought to the fore in a paper by Noam Nisan and Amir Ronen, who shared the 2012 Gödel Prize[7] for "laying the foundations of algorithmic game theory."

For further background on algorithmic game theory, with emphasis on its computational aspects, see [Pap01, NRTV07, Har17, Rou13, Rou14] and the survey chapter by Nisan in [YZ15].

The first scoring rule is due to Brier [Bri50], who discussed, in an equivalent form, the quadratic scoring rule. Selten [Sel98] provides an axiomatic characterization. The

---

[7]The Gödel Prize is an annual prize for outstanding papers in theoretical computer science. As discussed in Chapter 8, Nisan and Ronen shared the prize with Koutsoupias, Papadimitriou, Roughgarden, and Tardos.

Noam Nisan

logarithmic scoring rule was first introduced by Toda [Tod63] and further developed by Winkler and Murphy [WM68, Win69]. Theorem 16.3.5 is adapted from Gneiting and Raftery [GR07].

In the text, we argued that the weather forecaster M can achieve calibration just by knowing the overall percentage of rainy days. In fact, M can be well–calibrated without any information about the weather, as was first shown by Foster and Vohra [FV99]. See also [HMC00a].

Another interesting application of scoring rules is in prediction markets. See, e.g., [Han12, Han03, CP10].

Exercises

16.1.   Generalize the VCG mechanism to handle the case where the designer has a value $V(a)$ for each outcome $a \in \mathcal{A}$. Specifically, suppose the outcome selected maximizes social surplus (see Definition 16.2.2). Show that the payment scheme proposed in (16.1) incentivizes truthful reporting.

16.2.   Consider two bidders A and B and two items a and b. Bidder A's value for item a is 3, for item b is 2, and for both is 5. Bidder B's value for item a is 2, and for item b is 2, and for both is 4. Suppose that two English auctions are run simultaneously for the two items. Show that truthful bidding is not a dominant strategy in this auction.

16.3.   Consider the setting discussed in §16.2. Fix a subset A' of the outcomes $\mathcal{A}$, a set of numbers $\{\gamma_a | a \in A'\}$, and c_i for each $1 \leq i \leq n$. Design a truthful mechanism that takes as input the agents' reported valuation functions, and chooses as output an outcome $a \in A'$ that maximizes $\gamma_a + \sum_i c_i v_i(a)$. In other words, come up with a payment scheme that incentivizes truthful reporting.

16.4.   Let f be a mapping from reported valuation functions $\mathbf{v} = (v_1(\cdot), \ldots, v_n(\cdot))$ to outcomes in some set $\mathcal{A}$. Suppose that $\mathcal{M}$ is a direct revelation mechanism that on input $\mathbf{v}$ chooses outcome $f(\mathbf{v})$ and sets payments $p_i(\mathbf{v})$ for each agent. Show that M is truthful if and only if the following conditions hold for each i and $\mathbf{v}_{-i}$:

- The payment doesn't depend on v_i, but only on the alternative selected. That is, for fixed $\mathbf{v}_{-i}$ and for all $v_i(\cdot)$ such that $f(v_i, \mathbf{v}_{-i}) = a$, the payment p_i is the same. Thus, for all v_i with $f(v_i, \mathbf{v}_{-i}) = a$, it holds that $p(v_i, \mathbf{v}_{-i}) = p_a$.
- For each $\mathbf{v}_{-i}$, let $\mathcal{A}(\mathbf{v}_{-i})$ be the range of $f(\cdot, \mathbf{v}_{-i})$. Then $f(v_i, \mathbf{v}_{-i}) \in \operatorname{argmax}_{a \in \mathcal{A}(\mathbf{v}_{-i})}(v_i(a) - p_a)$.

16.5.   (a) Show that the scoring rule $s_i^\alpha(\mathbf{p}) = \alpha p_i^{\alpha-1} - (\alpha - 1)\sum_{j=1}^n p_j^\alpha$ is proper for any $\alpha > 1$.

(b) Show that these scoring rules interpolate between the logarithmic and the quadratic scoring rules, by showing that

$$\lim_{\alpha \to 1} \frac{s_i^\alpha(\mathbf{p}) - 1}{\alpha - 1} = \log p_i.$$

16.6.   Show that the scoring rule

$$s_i(\mathbf{p}) = \frac{p_i}{\|\mathbf{p}\|}$$

is proper.

CHAPTER 17

Matching markets

17.1. Maximum weighted matching

Consider a seller with n different items for sale and n buyers, each interested in buying at most one of these items. Suppose that the seller prices item j at p_j, so the price vector is $\mathbf{p} = (p_1, \ldots, p_n)$. Buyer i's value for item j is $v_{ij} \geq 0$; i.e., buyer i would only be willing to buy item j if $p_j \leq v_{ij}$, and, in this case, his utility for item j is $v_{ij} - p_j$. Given these prices, each buyer i has a set of **preferred items**:

$$D_i(\mathbf{p}) = \{j \mid \forall k \ v_{ij} - p_j \geq v_{ik} - p_k \ \text{and} \ v_{ij} \geq p_j\}. \tag{17.1}$$

If each buyer with nonempty $D_i(\mathbf{p})$ selects one of his preferred items, a conflict could arise between buyers who select the same item.

We will show that there is a price vector $\mathbf{p}^*$ and a corresponding perfect matching M between buyers and items in which each buyer is matched to one of his preferred items (so there is no conflict). Moreover, this matching M maximizes the **social surplus**:

$$\sum_i (v_{iM(i)} - p^*_{M(i)}) + \sum_j p^*_j = \sum_i v_{iM(i)}.$$

This will follow from a generalization of König's Lemma (Lemma 3.2.5).

THEOREM 17.1.1. *Given a nonnegative matrix $V = (v_{ij})_{n \times n}$, let*

$$K := \big\{ (\mathbf{u}, \mathbf{p}) \in \mathbb{R}^n \times \mathbb{R}^n : \quad u_i, p_j \geq 0 \ \text{and} \ u_i + p_j \geq v_{ij} \ \forall i, j \big\}.$$

Then

$$\min_{(\mathbf{u}, \mathbf{p}) \in K} \Big\{ \sum_i u_i + \sum_j p_j \Big\} = \max_{matchings \ M} \Big\{ \sum_i v_{i,M(i)} \Big\}. \tag{17.2}$$

REMARKS 17.1.2.

(1) The permutation maximizing the right-hand side is called a **maximum weight matching**, and the pair $(\mathbf{u}, \mathbf{p})$ minimizing the left-hand side is called a **minimum (fractional) cover.** The special case when the entries of $V, \mathbf{u}$, and $\mathbf{p}$ are all 0/1 reduces to Lemma 3.2.5.

(2) Let $v = \max v_{ij}$. If $(\mathbf{u}, \mathbf{p}) \in K$, then replacing every $u_i > v$ and $p_j > v$ by v yields a pair $(\tilde{\mathbf{u}}, \tilde{\mathbf{p}}) \in K$ with a smaller sum. Therefore, the minimization in (17.2) can be restricted to the set $K' := K \cap \{(\mathbf{u}, \mathbf{p}) : u_i, p_j \leq v \ \forall i, j\}$. This means that a minimum cover exists because the continuous map

$$F(\mathbf{u}, \mathbf{p}) := \sum_{i=1}^n (u_i + p_i),$$

defined on the closed and bounded set K' does indeed attain its infimum.

PROOF OF THEOREM 17.1.1. Suppose the left-hand side of (17.2) is minimized at some $(\mathbf{u}^*, \mathbf{p}^*) \in K$. By the definition of K, the following holds for all i and j:

$$u_i^* + p_j^* \geq v_{ij} \quad \text{and therefore} \quad F(\mathbf{u}^*, \mathbf{p}^*) \geq \sum_i v_{i,M(i)} \quad \forall \text{ matchings } M.$$

To prove the equality in (17.2), consider the bipartite graph G between the rows and the columns, in which

$$(i, j) \text{ is an edge iff } u_i^* + p_j^* = v_{ij}.$$

(In the context of the example at the beginning of this section, an edge (i, j) means that at the prices $\mathbf{p}^*$, item j is one of buyer i's preferred items.)

If there is a perfect matching M^* in G, we are done. If not, then by Hall's Theorem (3.2.2), there is a subset S of rows such that $|N(S)| < |S|$, where $N(S)$ is the set of neighbors of S in G; we will show this contradicts the definition of $(\mathbf{u}^*, \mathbf{p}^*)$. Let

$$u_i := u_i^* - \delta \cdot \mathbf{1}_{\{i \in S\}} \quad \text{and} \quad p_j := p_j^* + \delta \cdot \mathbf{1}_{\{j \in N(S)\}} \quad \forall i, j,$$

where

$$\delta = \min_{i \in S, j \notin N(S)} u_i^* + p_j^* - v_{ij} > 0.$$

(See Figure 17.1.)

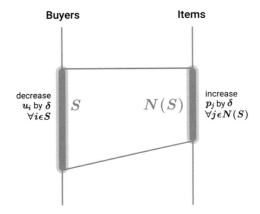

FIGURE 17.1. The figure illustrates the first part of the argument used to show that there must be a perfect matching.

Then $u_i + p_j \geq v_{ij}$ holds for all (i, j) and, since $|N(S)| < |S|$, we have $F(\mathbf{u}, \mathbf{p}) < F(\mathbf{u}^*, \mathbf{p}^*)$. However, some u_i might be negative. To rectify this, let $\epsilon = \min_j p_j$, and define

$$\tilde{u}_i := u_i + \epsilon \quad \forall i \quad \text{and} \quad \tilde{p}_j := p_j - \epsilon \quad \forall j.$$

Then $\tilde{u}_i + \tilde{p}_j \geq v_{ij}$ holds for all (i, j). Since $\tilde{p}_j = 0$ for some j and $v_{ij} \geq 0$, it follows that $\tilde{u}_i \geq 0$ for all i. Thus, $(\tilde{\mathbf{u}}, \tilde{\mathbf{p}}) \in K$. Clearly, $F(\tilde{\mathbf{u}}, \tilde{\mathbf{p}}) = F(\mathbf{u}, \mathbf{p}) < F(\mathbf{u}^*, \mathbf{p}^*)$, contradicting the minimality of $F(\mathbf{u}^*, \mathbf{p}^*)$. $\square$ $\square$

REMARK 17.1.3. The assumption that V is a square matrix can easily be relaxed. If V has fewer rows than columns, then adding rows of zeros reduces to the square case, similarly for columns.

17.2. Envy-free prices

Consider again n buyers and a seller with n items for sale, where v_{ij} is buyer i's value for item j.

DEFINITION 17.2.1. Given the values $V = (v_{ij})$, $1 \le i, j \le n$, and a set of prices $\mathbf{p} = (p_1, \dots, p_n)$, the **demand graph** $D(\mathbf{p})$ is the bipartite graph with an edge between a buyer i and an item j if, at the prices $\mathbf{p}$, item j is one of i's most preferred items; that is,

$$\forall k \quad v_{ij} - p_j \ge v_{ik} - p_k \text{ and } v_{ij} \ge p_j.$$

We say that a price vector $\mathbf{p}$ is **envy-free** if there exists a perfect matching in the demand graph $D(\mathbf{p})$.

LEMMA 17.2.2. *Let $V = (v_{ij})_{n \times n}$, $\mathbf{u}, \mathbf{p} \in \mathbb{R}^n$, all nonnegative, and let M be a perfect matching from $[n]$ to $[n]$. The following are equivalent:*

    (i) *$(\mathbf{u}, \mathbf{p})$ is a minimum cover of V and M is a maximum weight matching for V.*

    (ii) *The prices $\mathbf{p}$ are envy-free prices, M is contained in the demand graph $D(\mathbf{p})$, and $u_i = v_{iM(i)} - p_{M(i)}$.*

PROOF. (i)→(ii): In a minimum cover $(\mathbf{u}, \mathbf{p})$

$$u_i = \max_\ell v_{i\ell} - p_\ell. \tag{17.3}$$

Given a minimum cover $(\mathbf{u}, \mathbf{p})$, define, as we did earlier, a graph G where (i, j) is an edge if and only if $u_i + p_j = v_{ij}$. By (17.3), G is the demand graph for $\mathbf{p}$. By Theorem 17.1.1, the weight of a maximum weight matching in V is $\sum_i u_i + \sum_j p_j$. Therefore

$$\sum_i v_{iM(i)} = \sum_i u_i + \sum_j p_j = \sum_i (u_i + p_{M_i}).$$

Since $v_{iM(i)} \le u_i + p_{M(i)}$ for each i and since we have equality on the sum, these must be equalities for each i, so M is contained in $G = D(\mathbf{p})$.

(ii)→(i): To verify that $(\mathbf{u}, \mathbf{p})$ is a cover, observe that for all i, k, we have $u_i = v_{iM(i)} - p_{M(i)} \ge v_{ik} - p_k$ since M is contained in the demand graph. The equality $\sum_i u_i + p_{M(i)} = \sum_i v_{iM(i)}$ together with Theorem 17.1.1 implies that $(\mathbf{u}, \mathbf{p})$ is a minimum cover and M is a maximum weight matching. $\qquad\square$

COROLLARY 17.2.3. *Let $\mathbf{p}$ be an envy-free pricing for V and let M be a perfect matching of buyers to items. Then M is a maximum weight matching for V if and only if it is contained in $D(\mathbf{p})$.*

17.2.1. Highest and lowest envy-free prices.

LEMMA 17.2.4. *The envy-free price vectors for $V = (v_{ij})_{n \times n}$ form a **lattice**: Let $\mathbf{p}$ and $\mathbf{q}$ be two vectors of envy-free prices. Then, defining*

$$a \wedge b := \min(a, b) \quad and \quad a \vee b := \max(a, b),$$

the two price vectors

$$\mathbf{p} \wedge \mathbf{q} = (p_1 \wedge q_1, \dots, p_n \wedge q_n) \quad and \quad \mathbf{p} \vee \mathbf{q} = (p_1 \vee q_1, \dots, p_n \vee q_n)$$

are also envy-free.

PROOF. It follows from Lemma 17.2.2 that $\mathbf{p}$ is envy-free iff there is a nonnegative vector $\mathbf{u}$ such that $(\mathbf{u}, \mathbf{p})$ is a minimum cover of V. That is, $u_i + p_j \geq v_{ij}$ for all (i, j) with equality along every edge (i, j) that is in some maximum weight matching. Thus, it suffices to prove the following:

$$\text{If } (\mathbf{u}, \mathbf{p}) \text{ and } (\tilde{\mathbf{u}}, \tilde{\mathbf{p}}) \text{ are minimum covers, then so is } (\mathbf{u} \vee \tilde{\mathbf{u}}, \mathbf{p} \wedge \tilde{\mathbf{p}}). \quad (17.4)$$

Fix an edge (i, j). Without loss of generality $p_j = p_j \wedge \tilde{p}_j$, so

$$(u_i \vee \tilde{u}_i) + (p_j \wedge \tilde{p}_j) \geq u_i + p_j \geq v_{ij}.$$

Moreover, if (i, j) is in a maximum weight matching, then since $u_i + p_j = \tilde{u}_i + \tilde{p}_j = v_{ij}$, the assumption $p_j \leq \tilde{p}_j$ implies that $u_i \geq \tilde{u}_i$, and hence

$$(u_i \vee \tilde{u}_i) + (p_j \wedge \tilde{p}_j) = u_i + p_j = v_{ij}.$$

Switching the roles of $\mathbf{u}$ and $\mathbf{p}$ in (17.4) shows that if $(\mathbf{u}, \mathbf{p})$ and $(\tilde{\mathbf{u}}, \tilde{\mathbf{p}})$ are minimum covers, then so is $(\mathbf{u} \wedge \tilde{\mathbf{u}}, \mathbf{p} \vee \tilde{\mathbf{p}})$. □

COROLLARY 17.2.5. *Let $\mathbf{p}$ minimize $\sum_j p_j$ among all envy-free price vectors for V. Then:*

(i) *Every envy-free price vector $\mathbf{q}$ satisfies $p_i \leq q_i$ for all i.*
(ii) $\min_j p_j = 0$.

PROOF. (i) If not, then $\mathbf{p} \wedge \mathbf{q}$ has lower sum than $\mathbf{p}$, a contradiction.
(ii) Let $(\mathbf{u}, \mathbf{p})$ be a minimum cover. If $\epsilon = \min_j p_j > 0$, then subtracting ϵ from each p_j and adding ϵ to each u_i yields a minimum cover with lower prices. □

We refer to the vector $\mathbf{p}$ in this corollary as the **lowest envy-free price vector**. The next theorem gives a formula for these prices and the corresponding utilities.

THEOREM 17.2.6. *Given an $n \times n$ nonnegative valuation matrix V, let M^V be a maximum weight matching and let $\|M^V\|$ be its weight; that is, $\|M^V\| = \sum_i v_{iM^V(i)}$. Write V_{-i} for the matrix obtained by replacing row i of V by $\mathbf{0}$. Then the lowest envy-free price vector $\mathbf{p}$ for V and the corresponding utility vector $\mathbf{u}$ are given by*

$$M^V(i) = j \implies p_j = \|M^{V_{-i}}\| - (\|M^V\| - v_{ij}), \quad (17.5)$$

$$u_i = \|M^V\| - \|M^{V_{-i}}\| \quad \forall i. \quad (17.6)$$

REMARK 17.2.7. To interpret (17.5), observe that $\|M^{V_{-i}}\|$ is the maximum weight matching of all agents besides i and $\|M^V\| - v_{ij}$ is the total value received by these agents in the maximum matching that includes i. The difference in (17.5) is the externality i's presence imposes on the other agents, which is the price charged to i by the VCG mechanism (as discussed in §16.2).

PROOF. Let $\mathbf{p} := \underline{\mathbf{p}}$ be the lowest envy-free price vector for V, and let $\mathbf{u} := \bar{\mathbf{u}}$ be the corresponding utility vector. We know that

$$\sum_k (u_k + p_k) = \|M^V\|. \quad (17.7)$$

If we could find another matrix for which $\mathbf{p}$ is still envy-free, buyer i has utility 0 and all other utilities are unchanged, then applying (17.7) to that matrix in place of V would yield a formula for u_i. A natural way to reduce buyer i's utility is to zero out his valuations. Below, we will prove the following:

CLAIM 17.2.8. $((0, \mathbf{u}_{-i}), \mathbf{p})$ *is a minimum cover for* V_{-i} *(though* $\mathbf{p}$ *may no longer be the lowest envy-free price vector for* V_{-i}*).*

With this claim in hand, we obtain

$$\sum_k (u_k + p_k) - u_i = \|M^{V_{-i}}\|. \tag{17.8}$$

Subtracting (17.8) from (17.7) yields (17.6):

$$u_i = \|M^V\| - \|M^{V_{-i}}\|.$$

Thus, if $M^V(i) = j$, then

$$p_j = v_{ij} - u_i = \|M^{V_{-i}}\| - (\|M^V\| - v_{ij}). \qquad \square$$

PROOF OF CLAIM 17.2.8. Clearly $((0, \mathbf{u}_{-i}), \mathbf{p})$ is a cover for V_{-i}. To prove that it is a minimum cover, by Lemma 17.2.2 it suffices to show that there is a perfect matching in the demand graph $D'(\mathbf{p})$ for V_{-i}. Observe that the only edges that are changed relative to the demand graph $D(\mathbf{p})$ for V are those incident to i. Furthermore, edge (i, j) is in $D'(\mathbf{p})$ if and only if $p_j = 0$.

Suppose that there is no perfect matching in $D'(\mathbf{p})$. By Hall's Theorem (3.2.2), there is a subset S of items such that its set of neighbors T in $D'(\mathbf{p})$ has $|T| < |S|$. Since there is a perfect matching in $D(\mathbf{p})$ and the only buyer whose edges changed is i, it must be that $i \notin T$, and therefore for all $k \in S$, we have $p_k > 0$. But this means that, for small enough ϵ, the vector $(\mathbf{u}', \mathbf{p}')$ with

$$p'_k := p_k - \epsilon \cdot \mathbf{1}_{\{k \in S\}} \quad \text{and} \quad u'_\ell := u_\ell + \epsilon \cdot \mathbf{1}_{\{\ell \in T \cup \{i\}\}} \quad \forall k, \ell$$

is a minimum cover for V, a contradiction to the choice of $\mathbf{p}$ as the lowest envy-free price vector. (See Figure 17.2.) $\qquad \square$

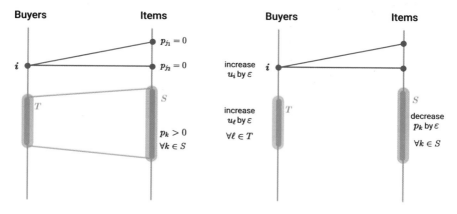

FIGURE 17.2. The modified utility and price vector shown in the right figure form a minimum cover for V with prices that have lower sum then $\mathbf{p}$, the presumed minimum envy-free price vector. This is a contradiction.

Employing the symmetry between $\mathbf{u}$ and $\mathbf{p}$, we obtain

COROLLARY 17.2.9. *Under the hypotheses of Theorem 17.2.6, the highest envy-free price vector* $\mathbf{p}$ *for* V *and the corresponding utility vector* $\mathbf{u}$ *are given by*

$$M^V(i) = j \implies u_i = \|M^{V^{-j}}\| - (\|M^V\| - v_{ij}), \qquad (17.9)$$

$$p_j = \|M^V\| - \|M^{V^{-j}}\| \quad \forall j, \qquad (17.10)$$

where V^{-j} *is the matrix obtained by replacing column* j *of* V *by* $\mathbf{0}$.

17.2.2. Seller valuations and unbalanced markets. In the previous sections, we implicitly assumed that the seller had value 0 for the item being sold since the minimal acceptable price for an item was 0. The general case where the seller has value s_j for item j, i.e., would not accept a price below s_j, can be reduced to this special case by replacing each buyer valuation v_{ij} by $\max(v_{ij} - s_j, 0)$. (If buyer i is matched to item j with $s_j > v_{ij}$, it must be at a price of 0, and no actual transaction takes place.)

Also, we assumed that the number of buyers and items was equal. If there are n buyers but $k < n$ items for sale, we can add $n - k$ dummy items of value 0 to all buyers. Similarly, if there are $j < n$ buyers but n items, we can add $n - j$ dummy buyers with value 0 for all items. In the latter case, any item that is matched to a dummy buyer (i.e., is unsold) must have price 0.

17.3. Envy-free division of rent

Consider a group of n people that would like to find an n-room house to rent. Suppose that person i assigns value v_{ij} to the jth room. Is there an envy-free way[1] to partition the rent R, accounting for the different valuations $V = (v_{ij})$?

FIGURE 17.3. Three roommates need to decide who will get each room and how much of the rent each person will pay.

Clearly, if the rent is higher than the weight of every matching in V, then it does not admit fair division. However, even if the rent is too low, there may be a

---

[1]That is, a vector $p_1, \ldots, p_n$ of envy-free prices such that $\sum_j p_j = R$.

problem. For example, suppose $v_{i1} > v_{i2} + R$ for $i = 1, 2$. Then whichever of renter 1 and 2 is not assigned to room 1 will envy the occupant of that room.

Theorem 17.2.6 and Corollary 17.2.9 provide an approach for computing the minimum and maximum envy-free rent: Use (17.5) and (17.10) to determine the lowest envy-free rent $\underline{R} = \sum_j \underline{p}_j$ and the highest envy-free rent $\overline{R} = \sum_j \overline{p}_j$. By Lemma 17.2.2, the set of envy-free prices is a convex set. Thus, for any $R = \alpha\underline{R} + (1 - \alpha)\overline{R}$, with $0 < \alpha < 1$, envy-free prices are obtained by setting

$$p_j = \alpha\underline{p}_j + (1 - \alpha)\overline{p}_j.$$

In the next section, we present an algorithm for finding the value of a maximum weighted matching in a graph. This can be used to compute (17.5) and (17.10).

17.4. Finding maximum matchings via ascending auctions

Consider an auction with n bidders and n items, where each bidder wants at most one item, and the valuations of the bidders are given by a valuation matrix $V = (v_{ij})$.

- Fix the minimum bid increment $\delta = 1/(n + 1)$.
- Initialize the prices $\mathbf{p}$ of all items to 0 and set the matching M of bidders to items to be empty.
- As long as M is not perfect:
  - one unmatched bidder i selects an item j in his demand set

  $$D_i(\mathbf{p}) := \{j \mid v_{ij} - p_j \geq v_{ik} - p_k \quad \forall k \quad \text{and } v_{ij} \geq p_j\}$$

  and bids $p_j + \delta$ on it.
  (We will see that the demand set $D_i(\mathbf{p})$ is nonempty.)
  - If j is unmatched, then $M(i) := j$; otherwise, say $M(\ell) = j$, remove (ℓ, j) from the matching and add (i, j), so that $M(i) := j$.
  - Increase p_j by δ.

THEOREM 17.4.1. *Suppose that the elements of the valuation matrix $V = (v_{ij})$ are integers. Then the above auction terminates with a maximum weight matching M, and the final prices $\mathbf{p}$ satisfy*

$$M(i) = j \quad \Longrightarrow \quad v_{ij} - p_j \geq v_{ik} - p_k - \delta \quad \forall k. \tag{17.11}$$

PROOF. From the moment an item is matched, it stays matched forever. Also, until an item is matched, its price is 0. Therefore, as long as a bidder is unmatched, there must be an unmatched item, so his demand set is nonempty. Moreover, all items in $D_i(\mathbf{p})$ have price at most $\max_j v_{ij}$. Since no item's price can exceed $\max_{ij} v_{ij} + \delta$ and some price increases by δ each round, the auction must terminate. When it terminates, the matching is perfect.

Suppose that in the final matching, $M(i) = j$. After i bid on j for the last time, p_j was increased by δ, so (17.11) held. Later, prices of other items were only increased, so the final prices also satisfy (17.11).

Finally, let M^* be any other perfect matching. By (17.11)

$$\sum_i (v_{iM(i)} - p_{M(i)}) \geq \sum_i (v_{iM^*(i)} - p_{M^*(i)} - \delta);$$

i.e.,

$$\sum_i v_{iM(i)} \geq \sum_i v_{iM^*(i)} - n\delta = \sum_i v_{iM^*(i)} - \frac{n}{n+1}.$$

Since the weight of any perfect matching is an integer, M must be a maximum weight matching. □

EXERCISE 17.a. Argue that the number of times the main loop is executed in the above algorithm is at most $\frac{1}{\delta} \sum_i \max_j(v_{ij} + \delta)$.

17.5. Matching buyers and sellers

The previous theorems can be interpreted in a different context known as the **assignment game**: Consider n buyers and n sellers, e.g., homeowners. (The case where the number of buyers differs from the number of sellers can be handled as in §17.2.2.) Let v_{ij} denote the value buyer i assigns to house j. We assume that each house has value 0 to its owner if unsold. (We consider the case where the value is positive at the end of the section.) If seller j sells the house to buyer i at a price of p_j, then the buyer's **utility** is $u_i = v_{ij} - p_j$.

DEFINITION 17.5.1. An **outcome** $(M, \mathbf{u}, \mathbf{p})$ of the assignment game is a matching M between buyers and sellers and a partition (u_i, p_j) of the value v_{ij} on every matched edge; i.e., $u_i + p_j = v_{ij}$, where $u_i, p_j \geq 0$ for all i, j. If buyer i is unmatched, we set $u_i = 0$. Similarly, $p_j = 0$ if seller j is unmatched.

We say the outcome is **stable**[2] if $u_i + p_j \geq v_{ij}$ for all i, j.

By Theorem 17.1.1 we have

PROPOSITION 17.5.2. *An outcome $(M, \mathbf{u}, \mathbf{p})$ is stable if and only if M is a maximum weight matching for V and $(\mathbf{u}, \mathbf{p})$ is a minimum cover for V. In particular, every maximum weight matching supports a stable outcome.*

DEFINITION 17.5.3. Let $(M, \mathbf{u}, \mathbf{p})$ be an outcome of the assignment game. Define the **excess** β_i of buyer i to be the difference between his utility and his **best outside option**[3]; i.e., (denoting $x_+ := \max(x, 0)$),

$$\beta_i := u_i - \max_k\{(v_{ik} - p_k)_+ \; : \; (i, k) \notin M\}.$$

Similarly, the **excess** s_j of seller j is

$$s_j := p_j - \max_\ell\{(v_{\ell,j} - u_\ell)_+ \; : \; (\ell, j) \notin M\}.$$

The outcome is **balanced** if it is stable and, for every matched edge (i, j), we have $\beta_i = s_j$.

REMARK 17.5.4. Observe that an outcome is stable if and only if all excesses are nonnegative. In fact, it suffices for all buyer (or all seller) excesses to be nonnegative since in an unstable pair both buyer and seller will have a negative excess. If an outcome is balanced, then every matched pair has reached the Nash bargaining solution between them. (See Exercise 12.b.)

---

[2]In other words, an outcome is unstable if there is an unmatched pair (k, ℓ) with value $v_{k\ell} > u_k + p_\ell$.

[3]The best outside option is the maximum utility buyer i could obtain by matching with a different seller and paying him his current price.

THEOREM 17.5.5. *Every assignment game has a balanced outcome. Moreover, the following process converges to a balanced outcome: Start with the minimum cover* $(\mathbf{u}, \mathbf{p})$ *where* $\mathbf{p}$ *is the vector of lowest envy-free prices and a maximum weight matching* M. *Repeatedly pick an edge in* M *to balance, ensuring that every edge in* M *is picked infinitely often.*

We will need the following lemma.

LEMMA 17.5.6. *Let* $(M, \mathbf{u}, \mathbf{p})$ *be a stable outcome with* $\beta_i \geq s_j \geq 0$ *for every* $(i, j) \in M$. *Pick a pair* $(i, j) \in M$ *with* $\beta_i > s_j$ *and balance the pair by performing the update*

$$u_i' := u_i - \frac{\beta_i - s_j}{2} \quad and \quad p_j' := p_j + \frac{\beta_i - s_j}{2},$$

leaving all other utilities and profits unchanged. Then the new outcome is stable and has excesses $\beta_i' = s_j'$ *and* $\beta_k' \geq s_\ell' \geq 0$ *for all* $(k, \ell) \in M$.

PROOF. By construction, after the update, we have $\beta_i' = s_j'$. Next, consider any other matched pair (k, ℓ). Since $p_j' > p_j$, buyer k's best outside option is no better than it was before, so $\beta_k' \geq \beta_k$. Similarly, since $u_i' < u_i$, seller ℓ's best outside option is at least what it was before, so $s_\ell' \leq s_\ell$.

The updated outcome is still stable since all buyer excesses are nonnegative (see Remark 17.5.4): The excess of buyer i is at least half of what it was before and the excesses of other buyers have not decreased. $\qquad\square$

PROOF OF THEOREM 17.5.5. Let M be a maximum weight matching. Start with the stable outcome defined by lowest envy-free prices $\{p_j^0\}$ and corresponding utilities $\{u_i^0\}$. The corresponding initial buyer excess is $\beta_i^0 \geq 0$ for all i, and the initial seller excess is $s_j^0 = 0$ for all j. (If $s_j^0 = \epsilon > 0$, then decreasing p_j by ϵ and increasing u_i by ϵ is still stable and has lower prices.)

Fix a sequence $\{(i_t, j_t)\}_{t \geq 1}$ of matched pairs. At time $t \geq 1$, balance the edge (i_t, j_t). This yields a new price vector $\mathbf{p}^t$ and a new utility vector $\mathbf{u}^t$. Using the lemma and induction, we conclude that for all t, the outcome $(M, \mathbf{u}^t, \mathbf{p}^t)$ is stable with $\beta_i^t \geq s_j^t$ for all matched pairs (i, j).

Moreover, $u_i^t \leq u_i^{t-1}$ and $p_j^t \geq p_j^{t-1}$ (with equality for $i \neq i_t$ and $j \neq j_t$). Thus, $\{u_i^t\}_{t \geq 1}$ decreases to a limit u_i for all i and $\{p_j^t\}_{t \geq 1}$ increases to a limit p_j for all j.

We claim that if every edge $(i, j) \in M$ occurs infinitely often in $\{(i_t, j_t)\}_{t \geq 1}$, then the limiting $(\mathbf{u}, \mathbf{p})$ is balanced. Indeed, let $(i, j) \in M$. Observe that β_i and s_j are continuous functions of $(\mathbf{u}, \mathbf{p})$. Then $\mathbf{u}^t \to \mathbf{u}$ and $\mathbf{p}^t \to \mathbf{p}$ imply that $\beta_i^t \to \beta_i$ and $s_j^t \to s_j$. Since $\beta_i^t = s_j^t$ after every rebalancing of (i, j), i.e., when $(i_t, j_t) = (i, j)$, we deduce that $\beta_i = s_j$. $\qquad\square$

REMARK 17.5.7. In general, balanced outcomes are not unique.

17.5.1. Positive seller values. Suppose seller j has value $h_j \geq 0$ for his house. We may reduce this case to the case $h_j = 0$ for all j as follows. Let $\tilde{v}_{ij}$ denote the value buyer i assigns to house j. Then $v_{ij} := (\tilde{v}_{ij} - h_j)_+$ is the **surplus** generated by the sale of house j to buyer i. (If $\tilde{v}_{ij} < h_j$, then there is no sale and $v_{ij} = 0$.) If seller j sells the house to buyer i at a price of $h_j + p_j$, then the seller's **profit** is p_j and the buyer's **utility** is $u_i = \tilde{v}_{ij} - (p_j + h_j) = v_{ij} - p_j$. The previous discussion goes through with surpluses replacing values and profits replacing prices.

17.6. Application to weighted hide-and-seek games

In this section, we show how to apply Theorem 17.1.1 to solve the following game.

EXAMPLE 17.6.1 (**Generalized Hide and Seek**). We consider a weighted version of the Hide and Seek game from §3.2. Let H be an $n \times n$ nonnegative matrix. Player II, the robber, chooses a pair (i,j) with $h_{ij} > 0$, and player I, the cop, chooses a row i or a column j. The value $h_{ij} > 0$ represents the payoff to the cop if the robber hides at location (i,j) and the cop chooses row i or column j. (E.g., certain safehouses might be safer than others, and h_{ij} could represent the probability the cop actually catches the robber if she chooses either i or j when he is hiding at (i,j).) The game is zero-sum.

Consider the following class of player II strategies: Player II first chooses a fixed permutation M of the set $\{1, \ldots, n\}$ and then hides at location $(i, M(i))$ with a probability q_i that he chooses. If $h_{iM(i)}$ is 0, then q_i is 0. For example, if $n = 5$ and the fixed permutation M is $3, 1, 4, 2, 5$, then the following matrix gives the probability of player II hiding in different locations:

$$
\begin{array}{ccccc}
0 & 0 & q_1 & 0 & 0 \\[1em]
q_2 & 0 & 0 & 0 & 0 \\[1em]
0 & 0 & 0 & q_3 & 0 \\[1em]
0 & q_4 & 0 & 0 & 0 \\[1em]
0 & 0 & 0 & 0 & q_5
\end{array}
$$

Given a permutation M and a robber strategy defined by the probability vector $(q_1, \ldots, q_n)$, the payoff to the cop if she chooses row i is $q_i h_{iM(i)}$, and her payoff if she chooses column j is $q_{M^{-1}(j)} h_{M^{-1}(j),j}$. Obviously, against this robber strategy, the cop will never choose a row i with $h_{iM(i)} = 0$ or a column j with $h_{M^{-1}(j),j} = 0$ since these would yield her a payoff of 0. Let

$$
v_{ij} = \begin{cases} \frac{1}{h_{i,j}} & \text{if } h_{ij} > 0, \\ 0 & \text{otherwise.} \end{cases}
$$

To equalize the payoffs to the cop on rows with $h_{iM(i)} > 0$ and columns with $h_{M^{-1}(j),j} > 0$ (recall Proposition 2.5.3), the robber can choose

$$
q_i = \frac{v_{i,M(i)}}{V_M} \quad \text{where} \quad V_M = \sum_{i=1}^{n} v_{i,M(i)}.
$$

With this choice, each row or column with positive expected payoff yields an expected payoff of $1/V_M$.

If the robber restricts himself to this class of strategies, then to minimize his expected payment to the cop, he should choose the permutation M^* that minimizes $1/V_M$, i.e., maximizes V_M. We will show that doing this is optimal for him not just within this restricted class of strategies but in general.

To see this, observe that by Theorem 17.1.1, there is a cover $(\mathbf{u}^*, \mathbf{p}^*)$, such that $u_i^* + p_j^* \geq v_{ij}$, for which

$$\sum_{i=1}^{n} \left(u_i^* + p_i^* \right) = \max_M V_M = V_{M^*}. \tag{17.12}$$

Now suppose that the cop assigns row i probability u_i^*/V_{M^*} and column j probability p_j^*/V_{M^*} for all i, j. Against this strategy, if the robber chooses some (i, j) (necessarily with $h_{ij} > 0$), then the payoff will be

$$\frac{(u_i^* + p_j^*)}{V_{M^*}} h_{ij} \geq \frac{v_{ij} h_{ij}}{V_{M^*}} = \frac{1}{V_{M^*}}.$$

We deduce that the cop can guarantee herself a payoff of at least $1/V_{M^*}$, whereas the permutation strategy for the robber described above guarantees that the cop's expected payoff is at most $1/V_{M^*}$. Consequently, this pair of strategies is optimal.

EXAMPLE 17.6.2. Consider the Generalized Hide and Seek game with probabilities given by the following matrix:

$$\begin{pmatrix} 1 & 1/2 \\ 1/3 & 1/5 \end{pmatrix}.$$

This means that the matrix V is equal to

$$\begin{pmatrix} 1 & 2 \\ 3 & 5 \end{pmatrix}.$$

For this matrix, the maximum matching M^* is given by the identity permutation with $V_{M^*} = 6$. This matrix has a minimum cover $\mathbf{u} = (1, 4)$ and $\mathbf{p} = (0, 1)$. Thus, an optimal strategy for the robber is to hide at location $(1, 1)$ with probability $1/6$ and location $(2, 2)$ with probability $5/6$. An optimal strategy for the cop is to choose row 1 with probability $1/6$, row 2 with probability $2/3$, and column 2 with probability $1/6$. The value of the game is $1/6$.

Notes

The assignment problem (a.k.a. maximum weighted bipartite matching) has a long and glorious history which is described in detail in Section 17.5 of Schrijver [Sch03]. An important early reference not mentioned there is Jacobi [Jac90b]. Theorem 17.1.1 was proved by Egerváry [Ege31] and led to the development of an algorithm for the assignment problem by Kuhn [Kuh55]. He called it the "Hungarian method" since the ideas were implicit in earlier papers by König [Kön31] and Egerváry. Munkres [Mun57] later sharpened the analysis of the Hungarian method, proving that the algorithm is strongly polynomial, i.e., the number of steps to execute the algorithm does not depend on the weights, rather is polynomial in $n \times m$. For more about matching theory, including more efficient algorithms, see, e.g., Lovász and Plummer [LP09] and Schrijver [Sch03].

John von Neumann [vN53] considered the Generalized Hide and Seek game discussed in §17.6, and used the Minimax Theorem applied to that game to reprove Theorem 17.1.1; see Exercise 17.4.

The assignment game as described in §17.5 was introduced by Shapley and Shubik [SS71]. They considered a cooperative game, where the payoff $v(S)$ for a set of agents is $v(S) := \max_{M_S} \sum_{(i,j) \in M_S} v_{ij}$ (the maximum is over matchings M_S of agents in S). Using the natural linear programming formulation, they proved that the core of this game, i.e., the set of stable outcomes, is nonempty, thereby proving Proposition 17.5.2. They also showed that the core of this game coincides with the set of *competitive equilibria*. (A competitive equilibrium, also known as a Walrasian equilibrium, is a vector of prices and an allocation of items to agents such that (a) each agent gets an item in his demand set and (b) if an item is not allocated to an agent, then its price is 0.) Furthermore, they proved the lattice property of outcomes (Lemma 17.2.4), thereby showing that there are two extreme points in the core, one best for buyers and one best for sellers.

Demange [Dem82] and Leonard [Leo83] proved Theorem 17.2.6 and showed that the mechanism that elicits buyer valuations and sets prices as prescribed is truthful. This is the VCG mechanism applied to the setting of unit-demand agents.

The auction algorithm for finding a maximum matching from §17.4 follows Demange, Gale, and Sotomayor [DGS86], who considered a number of auction mechanisms, building on work of Crawford and Knoer [CK81].

Gabrielle Demange   David Gale   Marilda Sotomayor

There are striking analogies between some of the results in this chapter and those on stable marriage, including the lattice structure, deferred-acceptance algorithms and resilience to manipulation by the proposers. (However, if the matrix V is used to indicate preferences as discussed in §10.3.1, then the unique stable matching is not necessarily a maximum weight matching. See Exercise 10.5.) See Roth and Sotomayor [RS92] for more on this.

Balanced outcomes were first studied by Sharon Rochford [Roc84] who proved Theorem 17.5.5 via Brouwer's Fixed-Point Theorem (Theorem 5.1.2). The same notion in general graphs was studied by Kleinberg and Tardos [KT08], who showed that whenever a stable outcome exists, there is also a balanced outcome. They also presented a polynomial time algorithm for finding a balanced outcome when one exists. Azar et al. [ABC+09] showed that the dynamics described in Theorem 17.5.5 converge to a balanced outcome starting from any stable outcome. Kanoria et al. [KBB+11] prove convergence to a balanced outcome for a different dynamics in which the matching changes over time.

The rent-division problem discussed in §17.3 was considered by Francis Su [Su99], who showed that there is always an envy-free partition of rent under the "Miserly Tenants" assumption: *No tenant will choose the most expensive room if there is a free room available.* This assumption is not always reasonable; e.g., consider a house with one excellent room that costs ϵ and 10 free closets. Other papers on fair rent division include [ADG91, ASÜ04]. (We have been unable to verify the results in the latter paper.)

Exercises

17.1.   Given the values $V = (v_{ij})$, $1 \leq i, j \leq n$, let G be a bipartite graph whose
         edge set is the union of all maximum weight matchings for V. Show that
         every perfect matching in G is a maximum weight matching for V.

17.2.   Use Brouwer's Fixed-Point Theorem (Theorem 5.1.2) to give a shorter
         proof of the existence of balanced outcomes (Theorem 17.5.5). Hint: Order
         the edges and balance them one by one.

17.3.   If $A = (a_{ij})_{n \times n}$ is a nonnegative matrix with row and column sums at most
         1, then there is a doubly stochastic matrix $S_{n \times n}$ with $A \leq S$ entrywise.
         (A doubly stochastic matrix is a nonnegative matrix with row and column
         sums equal to 1.)

17.4.   Let $V = (v_{ij})_{n \times n}$ be a nonnegative matrix. Consider the Generalized Hide
         and Seek game where $h_{ij} = v_{ij}^{-1}$ (or 0 if $v_{ij} = 0$). Prove Theorem 17.1.1 as
         follows:
         (a) Prove that

$$\min_{(\mathbf{u}, \mathbf{p}) \in K} \left\{ \sum_i u_i + \sum_j p_j \right\} = \frac{1}{w}$$

         where w is the value of the game by showing that $(\mathbf{u}, \mathbf{p})$ is a cop
         strategy ensuring her a payoff at least γ if and only if $\frac{1}{\gamma}(\mathbf{u}, \mathbf{p})$ is a
         cover of V.
         (b) Given a robber optimal strategy (q_{ij}) (where all $q_{ij} \geq 0$, $\sum_{i,j} q_{ij} = 1$,
         and $q_{ij} = 0$ if $v_{ij} = 0$), verify that the the $n \times n$ matrix

$$A = (a_{ij}) := \left(\frac{q_{ij} h_{ij}}{w} \right)$$

         has row and column sums at most 1. Clearly $w \sum_{i,j} a_{ij} v_{ij} = 1$. Then
         apply the previous exercise and the Birkhoff–von Neumann Theorem
         (Exercise 3.4) to complete the proof.

CHAPTER 18

Adaptive decision making

Suppose that two players are playing multiple rounds of the same game. How would they adapt their strategies to the outcomes of previous rounds? This fits into the broader framework of adaptive decision making which we develop next and later apply to games. We start with a very simple setting.

18.1. Binary prediction with expert advice and a perfect expert

EXAMPLE 18.1.1. [**Predicting the stock market**] Consider a trader trying to predict whether the stock market will go up or down each day. Each morning, for T days, he solicits the opinions of n experts, who each make up/down predictions. Based on their predictions, the trader makes a choice between up and down and buys or sells accordingly.

In this section, we assume that at least one of the experts is perfect, that is, predicts correctly every day, but the trader doesn't know which one it is. What should the trader do to minimize the number of mistakes he makes in T days?

First approach—follow the majority of leaders:

On any given day, the experts who have never made a mistake are called *leaders*. By following the majority opinion among the leaders, the trader is guaranteed never to make more than $\log_2 n$ mistakes: Each mistake the trader makes eliminates at least half of leaders and, obviously, never eliminates the perfect expert.

This analysis is tight when the minority is right each time and has size nearly equal to that of the majority.

Second approach—follow a random leader (FRL):

Perhaps surprisingly, following a random leader yields a slightly better guarantee: For any n, the number of mistakes made by the trader is at most $H_n - 1$ in expectation.[1] We verify this by induction on the number of leaders. The case of a single leader is clear. Consider the first day on which some number of experts, say $k > 0$, make a mistake. Then by the induction hypothesis, the expected number of mistakes the trader ever makes is at most

$$\frac{k}{n} + H_{n-k} - 1 \leq H_n - 1.$$

This analysis is tight for $T \geq n$. Suppose that for $1 \leq i < n$, on day i, only expert i makes a mistake. Then the probability that the trader makes a mistake that day is $1/(n-i+1)$. Thus, the expected number of mistakes he makes is $H_n - 1$.

---

[1] $H_n := \sum_{i=1}^{n} \frac{1}{i} \in (\ln n, \ln n + 1)$.

REMARK 18.1.2. We can think of this setting as an extensive-form zero-sum game between an adversary and a trader. The adversary chooses the daily advice of the experts and the actual outcome on each day t, and the trader chooses a prediction each day based on the experts' advice. In this game, the adversary seeks to maximize his gain, the number of mistakes the trader makes.

Next we derive a lower bound on the expected number of mistakes made by *any* trader algorithm, by presenting a strategy for the adversary.

PROPOSITION 18.1.3. *In the setting of Example 18.1.1 with at least one perfect expert, there is an adversary strategy that causes any trader algorithm to incur at least $\lfloor \log_2 n \rfloor / 2 \geq \lfloor \log_4 n \rfloor$ mistakes in expectation.*

PROOF. Let $2^k \leq n < 2^{k+1}$. Let E_0 denote the set consisting of the first 2^k experts, and let E_t be the subset of experts in E_{t-1} that predicted correctly on day t.

Now suppose that for each $t \in \{1, \ldots, k\}$, on day t half of the experts in E_{t-1} predict up and half predict down, and the rest of the experts predict arbitrarily. Suppose also that the truth is equally likely to be up or down. Then no matter how the trader chooses up or down, with probability $1/2$, he makes a mistake. Thus, in the first k days, any trader algorithm makes $k/2$ mistakes in expectation. Moreover, after $k = \lfloor \log_2 n \rfloor$ days, there is still a perfect expert, and the expected number of mistakes made by the trader is $\lfloor \log_2 n \rfloor / 2 \geq \lfloor \log_4 n \rfloor$.   □

To prove a matching upper bound, we will take a middle road between following the majority of the leaders (ignoring the minority) and FRL (which weights the minority too highly).

Third approach—a function of majority size:
Given any function $p : [1/2, 1] \to [1/2, 1]$, consider the trader algorithm A_p: When the leaders are split on their advice in proportion $(x, 1-x)$ with $x \geq 1/2$, follow the majority with probability $p(x)$.

If $p(x) = 1$ for all $x > 1/2$, we get the deterministic majority vote, while if $p(x) = x$, we get FRL.

What is the largest $a > 1$ for which we can prove an upper bound of $\log_a n$ on the expected number of mistakes? To do so, by induction, we need to verify two inequalities for all $x \in [1/2, 1]$:

$$\log_a(nx) + 1 - p(x) \leq \log_a n, \tag{18.1}$$
$$\log_a(n(1-x)) + p(x) \leq \log_a n. \tag{18.2}$$

The left-hand side of (18.1) is an upper bound on the expected mistakes of A_p assuming the majority is right (using the induction hypothesis) and the left-hand side of (18.2) is an upper bound on the expected mistakes of A_p assuming the minority is right.

Adding these inequalities and setting $x = 1/2$ gives $2\log_a(1/2) + 1 \leq 0$; that is, $a \leq 4$. We already know this since $\lfloor \log_4 n \rfloor$ is a lower bound for the worst case performance. Setting $a = 4$, the two required inequalities become

$$p(x) \geq 1 + \log_4 x,$$
$$p(x) \leq -\log_4(1-x).$$

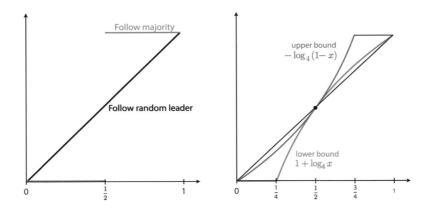

FIGURE 18.1. The function $p(x)$ represents the probability that the trader follows the prediction of a proportion x of leaders. Thus, $p(x) = x$ for FRL. The right figure shows the upper and lower bounds for optimal choices of $p(x)$.

We can easily satisfy both of these inequalities, e.g., by taking $p(x) = 1 + \log_4 x$, since $x(1 - x) \leq 1/4$. Since $p(1/2) = 1/2$ and $p(1) = 1$, concavity of the logarithm implies that $p(x) > x$ for $x \in (1/2, 1)$.

THEOREM 18.1.4. *In the binary prediction problem with n experts including a perfect expert, consider the trader algorithm A_p that follows the majority of leaders with probability $p(x) = 1 + \log_4 x$ when that majority comprises a fraction x of leaders. Then for any horizon T, the expected number of mistakes made by A_p is at most $\log_4 n$.*

Comparing this to Proposition 18.1.3 shows that when n is a power of two, we have identified minimax optimal strategies for the trader and the adversary.

FIGURE 18.2. Expert advice.

18.2. Nobody is perfect

Unfortunately, the assumption that there is a perfect expert is unrealistic. In the setting of Example 18.1.1, let L_i^t be the *cumulative loss* (i.e., total number of mistakes) incurred by expert i on the first t days. Denote

$$L_*^t = \min_i L_i^t \quad \text{and} \quad S_j = \{t \mid L_*^t = j\};$$

i.e., S_j is the time period during which the best expert has made j mistakes.

It is natural (but far from optimal) to apply the approach of the previous section: Suppose that, for each t, on day $t+1$ the trader follows the majority opinion of the *leaders*, i.e., those experts with $L_i^t = L_*^t$. Then during S_j, the trader's loss is at most $\log_2 n + 1$ (by the discussion of the case where there is a perfect expert). Thus, for any number T of days, the trader's loss is bounded by $(\log_2 n + 1)(L_*^T + 1)$. Similarly, the expected loss of FRL is at most $\mathrm{H}_n(L_*^T + 1)$ and the expected loss of the third approach above is at most $(\log_4 n + 1)(L_*^T + 1)$.

REMARK 18.2.1. There is an adversary strategy that ensures that any trader algorithm that only uses the advice of the leading experts will incur an expected loss that is at least $\lfloor \log_4(n) \rfloor L_*^T$ in T steps. See Exercise 18.1.

18.2.1. Weighted majority. There are strategies that guarantee the trader an asymptotic loss that is at most twice that of the best expert. One such strategy is based on weighted majority, where the weight assigned to an expert is decreased by a factor of $1 - \epsilon$ each time he makes a mistake.

Weighted Majority Algorithm

Fix $\epsilon \in [0, 1]$. On each day t, associate a weight w_i^t with each expert i.
Initially, set $w_i^0 = 1$ for all i.
Each day t, follow the **weighted majority** opinion: Let U_t be the set of experts predicting up on day t, and D_t the set predicting down. Predict "up" on day t if

$$W_U(t-1) := \sum_{i \in U_t} w_i^{t-1} \geq W_D(t-1) := \sum_{i \in D_t} w_i^{t-1}$$

and "down" otherwise.
At the end of day t, for each i such that expert i predicted incorrectly on day t, set

$$w_i^t = (1 - \epsilon) w_i^{t-1}. \tag{18.3}$$

Thus, $w_i^t = (1 - \epsilon)^{L_i^t}$, where L_i^t is the number of mistakes made by expert i in the first t days.

For the analysis of this algorithm, we will use the following facts.

LEMMA 18.2.2. *Let $\epsilon \in [0, 1/2]$. Then $\epsilon \leq -\ln(1 - \epsilon) \leq \epsilon + \epsilon^2$.*

PROOF. Taylor expansion gives

$$-\ln(1 - \epsilon) = \sum_{k \geq 1} \frac{\epsilon^k}{k} \geq \epsilon.$$

On the other hand,

$$\sum_{k \geq 1} \frac{\epsilon^k}{k} \leq \epsilon + \frac{\epsilon^2}{2} \sum_{k=0}^{\infty} \epsilon^k \leq \epsilon + \epsilon^2$$

since $\epsilon \leq 1/2$. □

THEOREM 18.2.3. *Suppose that there are n experts. Let $L(T)$ be the number of mistakes made by the Weighted Majority Algorithm in T steps with $\epsilon \leq \frac{1}{2}$, and let L_i^T be the number of mistakes made by expert i in T steps. Then for any sequence of up/down outcomes and for* every *expert i, we have*

$$L(T) \leq 2(1 + \epsilon)L_i^T + \frac{2 \ln n}{\epsilon}. \tag{18.4}$$

PROOF. Let $W(t) = \sum_i w_i^t$ be the total weight on all the experts after the tth day. If the algorithm incurs a loss on the tth day, say by predicting up instead of correctly predicting down, then $W_U(t - 1) \geq \frac{1}{2}W(t - 1)$. But in that case

$$W(t) \leq W_D(t - 1) + (1 - \epsilon)W_U(t - 1) \leq \left(1 - \frac{\epsilon}{2}\right)W(t - 1).$$

Thus, after a total loss of $L := L(T)$,

$$W(T) \leq \left(1 - \frac{\epsilon}{2}\right)^L W(0) = \left(1 - \frac{\epsilon}{2}\right)^L n.$$

Now consider expert i who incurs a loss of $L_i := L_i^T$. His weight at the end is

$$w_i^T = (1 - \epsilon)^{L_i},$$

which is at most $W(T)$. Thus

$$(1 - \epsilon)^{L_i} \leq \left(1 - \frac{\epsilon}{2}\right)^L n.$$

Taking logs and negating, we have

$$-L_i \ln(1 - \epsilon) \geq -L \ln\left(1 - \frac{\epsilon}{2}\right) - \ln n. \tag{18.5}$$

Applying Lemma 18.2.2, we obtain that for $\epsilon \in (0, 1/2]$,

$$L_i(\epsilon + \epsilon^2) \geq L\frac{\epsilon}{2} - \ln n$$

or

$$L(T) \leq 2(1 + \epsilon)L_i^T + \frac{2 \ln n}{\epsilon}. \qquad\qquad □$$

REMARKS 18.2.4.

(1) It follows from (18.5) that for all $\epsilon \in (0, 1]$

$$L(T) \leq \frac{|\ln(1 - \epsilon)|L_i^T + \ln n}{|\ln(1 - \frac{\epsilon}{2})|}. \tag{18.6}$$

If we know in advance that there is an expert with $L_i^T = 0$, then letting $\epsilon \uparrow 1$ recovers the result of the first approach to Example 18.1.1.

(2) There are cases where the Weighted Majority Algorithm incurs at least twice the loss of the best expert. In fact, this holds for every deterministic algorithm. See Exercise 18.3.

18.3. Multiple choices and varying costs

"I hear the voices, and I read the front page, and I know the speculation. But I'm the decider, and I decide what is best."
George W. Bush

In the previous section, the decision maker used the advice of n experts to choose between two options, and the cost of any mistake was the same. We saw that a simple deterministic algorithm could guarantee that the number of mistakes was not much more than twice that of any expert. One drawback of the Weighted Majority Algorithm is that it treats a majority of 51% with the same reverence as a majority of 99%. With careful randomization, we can avoid this pitfall and show that the decision maker can do almost as well as the best expert.

In this section, the decider faces multiple options, e.g., which stock to buy, rather than just up or down, now with varying losses. We will refer to the options of the decider as *actions*: This covers the task of prediction with expert advice, as the ith action could be "follow the advice of expert i".

EXAMPLE 18.3.1 (**Route-picking**). Each day you choose one of a set of n routes from your house to work. Your goal is to minimize your travel time. However, traffic is unpredictable, and you do not know in advance how long each route will take. Once you choose your route, you incur a loss equal to the latency on the route you selected. This continues for T days. For $1 \leq i \leq n$, let L_i^T be the total latency you would have incurred over the T days if you had taken route i every day. Can we find a strategy for choosing a route each day such that the total latency incurred is close to $\min_i L_i^T$?

The following setup captures the stock market and route-picking examples above and many others.

DEFINITION 18.3.2 (**Sequential adaptive decision making**). On day t, a decider $\mathcal{D}$ chooses a probability distribution $\mathbf{p}^t = (p_1^t, \ldots, p_n^t)$ over a set of n actions, e.g., stocks to own or routes to drive. (The choice of $\mathbf{p}^t$ can depend on the history, i.e., the prior losses of each action and prior actions taken by $\mathcal{D}$.) The losses $\boldsymbol{\ell}^t = (\ell_1^t, \ldots, \ell_n^t) \in [0,1]^n$ of each action on day t are then revealed.

Given the history, $\mathcal{D}$'s expected loss on day t is $\mathbf{p}^t \cdot \boldsymbol{\ell}^t = \sum_{i=1}^n p_i^t \ell_i^t$. The total expected loss $\mathcal{D}$ incurs in T days is

$$\overline{L}_{\mathcal{D}}^T = \sum_{t=1}^T \mathbf{p}^t \cdot \boldsymbol{\ell}^t.$$

(See the chapter notes for a more precise interpretation of $\overline{L}_{\mathcal{D}}^T$ in the case where losses depend on prior actions taken by $\mathcal{D}$.)

REMARK 18.3.3. In stock-picking examples, $\mathcal{D}$ could have a fraction p_i^t of his portfolio in stock i instead of randomizing.

DEFINITION 18.3.4. The *regret of a decider $\mathcal{D}$ in T steps against loss sequence* $\boldsymbol{\ell} = \{\boldsymbol{\ell}^t\}_{t=1}^T$ is defined as the difference between the total expected loss of the decider and the total loss of the best single action; that is,

$$\mathcal{R}_T(\mathcal{D}, \boldsymbol{\ell}) := \overline{L}_{\mathcal{D}}^T - \min_i L_i^T,$$

where $L_i^T = \sum_{t=1}^T \ell_i^t$. We define the **regret of a decider** $\mathcal{D}$ as

$$\mathcal{R}_T(\mathcal{D}) := \max_{\boldsymbol{\ell}} \mathcal{R}_T(\mathcal{D}, \boldsymbol{\ell}). \tag{18.7}$$

Perhaps surprisingly, there exist algorithms with regret that is **sublinear in** T; i.e., the average regret per day tends to 0. We will see one below.

18.3.1. Discussion. Let $\boldsymbol{\ell} = \{\boldsymbol{\ell}^t\}_{t=1}^T$ be a sequence of loss vectors. A natural goal for a decision-making algorithm $\mathcal{D}$ is to minimize its worst case loss, i.e., $\max_{\boldsymbol{\ell}} \overline{L}_{\mathcal{D}}^T$. But this is a dubious measure of the quality of the algorithm since on a worst-case sequence there may be nothing any decider can do. This motivates evaluating $\mathcal{D}$ by its performance gap

$$\max_{\boldsymbol{\ell}} (\overline{L}_{\mathcal{D}}^T - \mathcal{B}(\boldsymbol{\ell})),$$

where $\mathcal{B}(\boldsymbol{\ell})$ is a benchmark loss for $\boldsymbol{\ell}$. The most obvious choice for the benchmark is $\mathcal{B}^* = \sum_{t=1}^T \min \ell_i^t$, but this is too ambitious: E.g., if $n = 2$, the losses of the first expert $\{\ell_1^t\}_{t=1}^T$ are independent unbiased bits, and $\ell_2^t = 1 - \ell_1^t$, then $\mathbb{E}\left[\overline{L}_{\mathcal{D}}^T - \mathcal{B}^*\right] = T/2$ since $\mathcal{B}^* = 0$. Instead, in the definition of regret, we employ the benchmark $\mathcal{B}(\boldsymbol{\ell}) = \min_i L_i^T$. At first sight, this benchmark looks weak since why should choosing the same action every day be a reasonable option? We give two answers: (1) Often there really is a best action and the goal of the decision algorithm is to learn its identity without losing too much in the process. (2) Alternative decision algorithms (e.g., use action 1 on odd days, action 2 on even days, except if one action has more than double the cumulative loss of the other) can be considered experts and incorporated into the model as additional actions. We show below that the regret of any decision algorithm is at most $\sqrt{T \log n/2}$ when there are n actions to choose from. Note that this grows linearly in T if n is exponential in T. To see that this is unavoidable, recall that in the binary prediction setting, if we include 2^T experts making all possible predictions, one of them will make no mistakes, and we already know that for this case, any decision algorithm will incur worst-case regret at least $T/2$.

18.3.2. The Multiplicative Weights Algorithm. We now present an algorithm for adaptive decision making, with regret that is $o(T)$ as $T \to \infty$. The algorithm is a randomized variant of the Weighted Majority Algorithm; it uses the weights in that algorithm as probabilities. The algorithm and its analysis in Theorem 18.3.7 deal with the case where the decider incurs losses only.

Multiplicative Weights Algorithm (MW)

Fix $\epsilon < 1/2$ and n possible actions.
On each day t, associate a weight w_i^t with the ith action.
Initially, $w_i^0 = 1$ for all i.
On day t, use the mixed strategy $\mathbf{p}^t$, where

$$p_i^t = \frac{w_i^{t-1}}{\sum_k w_k^{t-1}}.$$

For each action i, with $1 \le i \le n$, observe the loss $\ell_i^t \in [0,1]$ and update the weight w_i^t as follows:

$$w_i^t = w_i^{t-1} \exp(-\epsilon \ell_i^t). \tag{18.8}$$

In the next proof we will use the following lemma.

LEMMA 18.3.5 (Hoeffding Lemma). *Suppose that X is a random variable with distribution F such that $a \leq X \leq a + 1$ for some $a \leq 0$ and $\mathbb{E}[X] = 0$. Then for all $\lambda \in \mathbb{R}$,*

$$\mathbb{E}\left[e^{\lambda X}\right] \leq e^{\lambda^2/8}.$$

For a proof, see Appendix B.2.1. For the reader's convenience, we prove here the following slightly weaker version.

LEMMA 18.3.6. *Let X be a random variable with $\mathbb{E}[X] = 0$ and $|X| \leq 1$. Then for all $\lambda \in \mathbb{R}$,*

$$\mathbb{E}\left[e^{\lambda X}\right] \leq e^{\lambda^2/2}.$$

PROOF. By convexity of the function $f(x) = e^{\lambda x}$, we have

$$e^{\lambda x} \leq \frac{(1+x)e^\lambda + (1-x)e^{-\lambda}}{2} = \ell(x)$$

for $x \in [-1, 1]$. See Figure 18.3.

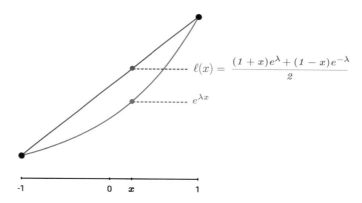

FIGURE 18.3. The blue curve is $e^{\lambda x}$. The purple curve is a line between $e^{-\lambda}$ and e^λ.

Thus, since $|X| \leq 1$ and $\mathbb{E}[X] = 0$, we have

$$\mathbb{E}\left[e^{\lambda X}\right] \leq \mathbb{E}[\ell(X)] = \frac{e^\lambda + e^{-\lambda}}{2} = \sum_{k=0}^{\infty} \frac{\lambda^{2k}}{(2k)!}$$

$$\leq \sum_{k=0}^{\infty} \frac{\lambda^{2k}}{2^k k!} = e^{\lambda^2/2}. \qquad \square$$

THEOREM 18.3.7. *Consider the Multiplicative Weights Algorithm with n actions. Define*

$$\overline{L}_{\text{MW}}^T := \sum_{t=1}^{T} \mathbf{p}^t \cdot \boldsymbol{\ell}^t,$$

where $\boldsymbol{\ell}^t \in [0,1]^n$. Then, for every loss sequence $\{\boldsymbol{\ell}^t\}_{t=1}^T$ and every action i, we have

$$\overline{L}_{\text{MW}}^T \leq L_i^T + \frac{T\epsilon}{8} + \frac{\log n}{\epsilon},$$

where $L_i^T = \sum_{t=1}^T \ell_i^t$. In particular, taking $\epsilon = \sqrt{\frac{8 \log n}{T}}$, we obtain that for all i,

$$\overline{L}_{\mathrm{MW}}^T \le L_i^T + \sqrt{\frac{1}{2} T \log n};$$

i.e., the regret $\mathcal{R}_T(\mathrm{MW}, \boldsymbol{\ell})$ is at most $\sqrt{\frac{1}{2} T \log n}$.

PROOF. Let $W^t = \sum_{1 \le i \le n} w_i^t = \sum_{1 \le i \le n} w_i^{t-1} \exp(-\epsilon \ell_i^t)$. Then

$$\frac{W^t}{W^{t-1}} = \sum_i \frac{w_i^{t-1}}{W^{t-1}} \exp(-\epsilon \ell_i^t) = \sum_i p_i^t \exp(-\epsilon \ell_i^t) = \mathbb{E}\left[e^{-\epsilon X_t}\right], \qquad (18.9)$$

where X_t is the loss the algorithm incurs at time t; i.e.,

$$\mathbb{P}\left[X_t = \ell_i^t\right] = p_i^t.$$

Let

$$\overline{\ell}^t := \mathbb{E}\left[X_t\right] = \mathbf{p}^t \cdot \boldsymbol{\ell}^t.$$

By Hoeffding's Lemma (Lemma 18.3.5), we have

$$\mathbb{E}\left[e^{-\epsilon X_t}\right] = e^{-\epsilon \overline{\ell}^t} \mathbb{E}\left[e^{-\epsilon(X_t - \overline{\ell}^t)}\right] \le e^{-\epsilon \overline{\ell}^t} e^{\epsilon^2/8},$$

so plugging back into (18.9), we obtain

$$W^t \le e^{-\epsilon \overline{\ell}^t} e^{\epsilon^2/8} W^{t-1} \quad \text{and thus} \quad W^T \le e^{-\epsilon \overline{L}^T} e^{T \epsilon^2/8} n.$$

On the other hand,

$$W^T \ge w_i^T = e^{-\epsilon L_i^T},$$

so combining these two inequalities, we obtain

$$e^{-\epsilon L_i^T} \le e^{-\epsilon \overline{L}^T} e^{T \epsilon^2/8} n.$$

Taking logs, we obtain

$$\overline{L}_{\mathrm{MW}}^T \le L_i^T + \frac{T \epsilon}{8} + \frac{\log n}{\epsilon}.$$

$\square$

The following proposition shows that the bound of Theorem 18.3.7 is asymptotically optimal as T and n go to infinity.

PROPOSITION 18.3.8. *Consider a loss sequence $\boldsymbol{\ell}$ in which all losses are independent and equally likely to be 0 or 1. Then for any decision algorithm $\mathcal{D}$, its expected regret satisfies*

$$\mathbb{E}\left[\mathcal{R}_T(\mathcal{D}, \boldsymbol{\ell})\right] = \frac{\gamma_n}{2} \sqrt{T} \cdot (1 + o(1)) \qquad \text{as } T \to \infty \qquad (18.10)$$

where

$$\gamma_n = \mathbb{E}\left[\max_{1 \le i \le n} Y_i\right] \quad \text{and} \quad Y_i \sim N(0, 1).$$

Moreover,

$$\gamma_n = \sqrt{2 \log n} \, (1 + o(1)) \qquad \text{as} \quad n \to \infty. \qquad (18.11)$$

PROOF. Clearly, for any $\mathcal{D}$, the decider's expected loss is $T/2$. Expert i's loss, L_i^T, is binomial with parameters T and $1/2$, and thus by the Central Limit Theorem $\frac{L_i^T - T/2}{\sqrt{T/4}}$ converges in law to a normal $(0,1)$ random variable Y_i. Let $L_*^T = \min_i L_i^T$. Then as $T \to \infty$,

$$\frac{\mathbb{E}\left[L_*^T - T/2\right]}{\sqrt{T/4}} \to \mathbb{E}\left[\min_{1 \leq i \leq n} Y_i\right] = -\gamma_n,$$

which proves (18.10). See Exercise 18.7 for the proof of (18.11). $\qquad\square$

18.3.3. Gains. Consider a setting with gains instead of losses, where $\mathbf{g}^t = (g_1^t, \ldots, g_n^t)$ are the gains of the n experts on day t. As in the setting of losses, let

$$\overline{G}_{\mathcal{D}}^T = \sum_{t=1}^{T} \mathbf{p}^t \cdot \mathbf{g}^t$$

be the expected gain of decider $\mathcal{D}$, and let $G_i^T := \sum_{t=1}^{T} g_i^t$ be the total gain of expert i over the T days.

COROLLARY 18.3.9. *In the setup of Definition 18.3.2, suppose that for all t, $\max_i g_i^t - \min_j g_j^t \leq \Gamma$. Then the Multiplicative Weights Algorithm can be adapted to the gains setting and yields regret at most $\Gamma\sqrt{\frac{T \log n}{2}}$; i.e.*

$$\overline{G}_{\mathrm{MW}}^T \geq G_j^T - \Gamma\sqrt{\frac{T \log n}{2}}$$

for all j.

PROOF. Let $g_{\max}^t = \max_k g_k^t$. Run the Multiplicative Weights Algorithm using the (relative) losses

$$\ell_i^t = \frac{1}{\Gamma}(g_{\max}^t - g_i^t).$$

By Theorem 18.3.7, we have for all actions j that

$$\frac{1}{\Gamma}\sum_t \sum_i p_i^t(g_{\max}^t - g_i^t) \leq \sqrt{\frac{T \log n}{2}} + \frac{1}{\Gamma}\sum_t (g_{\max}^t - g_j^t)$$

and the corollary follows. $\qquad\square$

18.4. Using adaptive decision making to play zero-sum games

Consider a two-person zero-sum game with payoff matrix $A = \{a_{ij}\}$. Suppose T rounds of this game are played. We can apply the Multiplicative Weights Algorithm to the decision-making process of player II. In round t, he chooses a mixed strategy $\mathbf{p}^t$; i.e., column j is assigned probability p_j^t. Knowing $\mathbf{p}^t$ and the history of play, player I chooses a row i_t. The loss of action j in round t is $\ell_j^t = a_{i_t j}$, and the total loss of action j in T rounds is $L_j^T = \sum_{t=1}^{T} a_{i_t j}$.

The next proposition bounds the total loss $\overline{L}_{\mathrm{MW}}^T = \sum_{t=1}^{T}(A\mathbf{p}^t)_{i_t}$ of player II.

PROPOSITION 18.4.1. *Suppose the $m \times n$ payoff matrix $A = \{a_{ij}\}$ has entries in $[0, 1]$ and player II is playing according to the Multiplicative Weights Algorithm. Let $\mathbf{x}_{\mathrm{emp}}^T \in \Delta_m$ be a row vector representing the empirical distribution of actions* *taken by player I in T steps; i.e., the ith coordinate of $\mathbf{x}_{\mathrm{emp}}^T$ is $\frac{|\{t \mid i_t=i\}|}{T}$. Then the total loss of player II satisfies*

$$\overline{L}_{\mathrm{MW}}^T \le T \min_{\mathbf{y}} \mathbf{x}_{\mathrm{emp}}^T A\mathbf{y} + \sqrt{\frac{T\log n}{2}}.$$

PROOF. By Theorem 18.3.7, player II's loss over the T rounds satisfies

$$\overline{L}_{\mathrm{MW}}^T \le L_j^T + \sqrt{\frac{T\log n}{2}}.$$

Since

$$L_j^T = T \sum_{t=1}^{T} \frac{a_{i_t,j}}{T} = T(\mathbf{x}_{\mathrm{emp}}^T A)_j,$$

we have

$$\min_j L_j^T = T \min_j (\mathbf{x}_{\mathrm{emp}}^T A)_j = T \min_{\mathbf{y}} \mathbf{x}_{\mathrm{emp}}^T A\mathbf{y},$$

and the proposition follows. □

REMARK 18.4.2. Suppose player I uses the mixed strategy ξ (a row vector) in all T rounds. If player II knows this, he can guarantee an expected loss of

$$\min_{\mathbf{y}} \xi A\mathbf{y},$$

which could be lower than V, the value of the game. In this case, $E(\mathbf{x}_{\mathrm{emp}}^T) = \xi$, so even if player II has no knowledge of ξ, the proposition bounds his expected loss by

$$T \min_{\mathbf{y}} \xi A\mathbf{y} + \sqrt{\frac{T\log n}{2}}.$$

Next, as promised, we rederive the Minimax Theorem as a corollary of Proposition 18.4.1.

THEOREM 18.4.3 (Minimax Theorem). *Let $A = \{a_{ij}\}$ be the payoff matrix of a zero-sum game. Let*

$$V_I = \max_{\mathbf{x}} \min_{\mathbf{y}} \mathbf{x}^T A\mathbf{y} = \max_{\mathbf{x}} \min_j (\mathbf{x}^T A)_j$$

and

$$V_{II} = \min_{\mathbf{y}} \max_{\mathbf{x}} \mathbf{x}^T A\mathbf{y} = \min_{\mathbf{y}} \max_i (A\mathbf{y})_i$$

be the safety values of the players. Then $v_I = v_{II}$.

PROOF. By adding a constant to all entries of the matrix and scaling, we may assume that all entries of A are in $[0, 1]$.

From Lemma 2.6.3, we have $V_I \le V_{II}$.

Suppose that in round t player II plays the mixed strategy $\mathbf{p}^t$ given by the Multiplicative Weights Algorithm and player I plays a best response; i.e.,

$$i_t = \mathrm{argmax}_i (A\mathbf{p}^t)_i.$$

Then

$$\overline{\ell}^t = \max_i (A\mathbf{p}^t)_i \ge \min_{\mathbf{y}} \max_i (A\mathbf{y})_i = V_{II},$$

from which

$$\overline{L}_{\mathrm{MW}}^T \ge TV_{II}. \tag{18.12}$$

(Note that the proof of (18.12) did not rely on any property of the Multiplicative Weights Algorithm.) On the other hand, from Proposition 18.4.1, we have

$$\overline{L}_{\mathrm{MW}}^T \leq T \min_{\mathbf{y}} \mathbf{x}_{\mathrm{emp}}^T A \mathbf{y} + \sqrt{\frac{T \log n}{2}},$$

and since $\min_{\mathbf{y}} \mathbf{x}_{\mathrm{emp}}^T A \mathbf{y} \leq V_{\mathrm{I}}$, we obtain

$$T V_{\mathrm{II}} \leq T V_{\mathrm{I}} + \sqrt{\frac{T \log n}{2}},$$

and hence $V_{\mathrm{II}} \leq V_I$. $\qquad\square$

18.5. Adaptive decision making as a zero-sum game*

Our goal in this section is to characterize the minimax regret in the setting of Definition 18.3.2, i.e.,

$$\min_{\mathcal{D}_{[0,1]}} \max_{\{\boldsymbol{\ell}^t\}} \mathcal{R}_T(\mathcal{D}_{[0,1]}, \{\boldsymbol{\ell}^t\}) \tag{18.13}$$

as the value of a finite zero-sum game between a decider and an adversary. In (18.13), the sequence of loss vectors $\{\boldsymbol{\ell}^t\}_{t=1}^T$ is in $[0,1]^{nT}$, and $\mathcal{D}_{[0,1]}$ is a sequence of functions $\{\mathbf{p}^t\}_{t=1}^T$, where

$$\mathbf{p}^t : [0,1]^{n(t-1)} \to \Delta_n$$

maps the losses from previous rounds to the decider's current mixed strategy over actions.

18.5.1. Minimax regret is attained in {0,1} losses. Given $\{\ell_i^t : 1 \leq i \leq n, 1 \leq t \leq T\}$, denote by $\{\hat{\ell}_i^t\}$ the sequence of independent $\{0,1\}$-valued random variables with

$$\mathbb{E}\left[\hat{\ell}_i^t\right] = \ell_i^t.$$

THEOREM 18.5.1 (**"Replacing losses by coin tosses"**). *For any decision strategy $\mathcal{D}$ that is defined only for $\{0,1\}$ losses, a corresponding decision strategy $\mathcal{D}_{[0,1]}$ is defined as follows: For each t, given $\{\boldsymbol{\ell}^j\}_{j=1}^{t-1}$, applying $\mathcal{D}$ to $\{\hat{\boldsymbol{\ell}}^j\}_{j=1}^{t-1}$ yields $\hat{\mathbf{p}}^t$. Use $\mathbf{p}^t = \mathbb{E}[\hat{\mathbf{p}}^t]$ at time t in $\mathcal{D}_{[0,1]}$. Then*

$$\mathbb{E}\left[\mathcal{R}_T(\mathcal{D}, \{\hat{\boldsymbol{\ell}}^t\})\right] \geq \mathcal{R}_T(\mathcal{D}_{[0,1]}, \{\boldsymbol{\ell}^t\}). \tag{18.14}$$

PROOF. Denote by $\mathbb{E}_t[\cdot]$ an expectation given the history prior to time t. We have

$$\mathbb{E}_t\left[\hat{\mathbf{p}}^t \cdot \hat{\boldsymbol{\ell}}^t\right] = \hat{\mathbf{p}}^t \cdot \mathbb{E}_t\left[\hat{\boldsymbol{\ell}}^t\right] = \hat{\mathbf{p}}^t \cdot \boldsymbol{\ell}^t$$

since $\hat{\mathbf{p}}^t$ is determined by $\{\hat{\boldsymbol{\ell}}^j\}_{j=1}^{t-1}$. Taking an expectation of both sides,

$$\mathbb{E}\left[\hat{\mathbf{p}}^t \cdot \hat{\boldsymbol{\ell}}^t\right] = \mathbf{p}^t \cdot \boldsymbol{\ell}^t,$$

i.e., the randomization does not change the expected loss of the decider. However, it may reduce the expected loss of the best expert since

$$\mathbb{E}\left[\min_i \hat{L}_i^T\right] \leq \min_i \mathbb{E}\left[\hat{L}_i^T\right] = \min_i L_i^T.$$

Thus,

$$\sum_t \mathbb{E}\left[\hat{\mathbf{p}}^t \cdot \hat{\boldsymbol{\ell}}^t\right] - \mathbb{E}\left[\min_i \hat{L}_i^T\right] \geq \sum_t \mathbf{p}^t \cdot \boldsymbol{\ell}^t - \min_i L_i^T,$$

yielding (18.14). □

REMARK 18.5.2. From an algorithmic perspective, there is no need to compute $\mathbf{p}^t = \mathbb{E}\left[\hat{\mathbf{p}}^t\right]$ in order to implement $\mathcal{D}_{[0,1]}$'s decision at time t. Rather, $\mathcal{D}_{[0,1]}$ can simply use $\hat{\mathbf{p}}^t$ at step t.

COROLLARY 18.5.3. *Minimax regret is attained in $\{0,1\}$ losses; i.e.,*

$$\min_{\mathcal{D}} \max_{\boldsymbol{\ell}^t \in \{0,1\}^{nT}} \mathcal{R}_T(\mathcal{D}, \{\boldsymbol{\ell}^t\}) = \min_{\mathcal{D}_{[0,1]}} \max_{\boldsymbol{\ell}^t \in [0,1]^{nT}} \mathcal{R}_T(\mathcal{D}_{[0,1]}, \{\boldsymbol{\ell}^t\}). \qquad (18.15)$$

PROOF. Given $\mathcal{D}$, construct $\mathcal{D}_{[0,1]}$ as in the lemma. Since

$$\max_{\boldsymbol{\ell}^t \in \{0,1\}^{nT}} \mathcal{R}_T(\mathcal{D}, \{\boldsymbol{\ell}^t\}) \geq \mathbb{E}\left[\mathcal{R}_T(\mathcal{D}, \{\hat{\boldsymbol{\ell}}^t\})\right] \geq \mathcal{R}_T(\mathcal{D}_{[0,1]}, \{\boldsymbol{\ell}^t\}) \quad \forall \boldsymbol{\ell}^t \in [0,1]^{nT},$$

we have

$$\max_{\boldsymbol{\ell}^t \in \{0,1\}^{nT}} \mathcal{R}_T(\mathcal{D}, \{\boldsymbol{\ell}^t\}) \geq \max_{\boldsymbol{\ell}^t \in [0,1]^{nT}} \mathcal{R}_T(\mathcal{D}_{[0,1]}, \{\boldsymbol{\ell}^t\}),$$

and (18.15) follows. □

18.5.2. Optimal adversary strategy. The adaptive decision-making problem can be seen as a finite two-player zero-sum game as follows. The pure strategies[2] of the adversary (player I) are the loss vectors $\{\boldsymbol{\ell}^t\} \in \{0,1\}^{nT}$. An adversary strategy $\mathcal{L}$ is a probability distribution over loss sequences $\underline{\boldsymbol{\ell}} := \{\boldsymbol{\ell}^t\}_{t=1}^T$. The pure strategies for the decider (player II) are $\underline{\mathbf{a}} := \{a_t\}_{t=1}^T$, where $a_t : \{0,1\}^{n(t-1)} \to [n]$ maps losses in the first $t-1$ steps to an action in the tth step. A decider strategy $\mathbb{D}$ is a probability distribution over pure strategies. Let

$$\mathcal{R}_T^*(\underline{\mathbf{a}}, \underline{\boldsymbol{\ell}}) := \sum_{t=1}^T \ell_{a_t}^t - \min_i \sum_t \ell_i^t.$$

By the Minimax Theorem

$$\min_{\mathbb{D}} \max_{\underline{\boldsymbol{\ell}}} \mathbb{E}\left[\mathcal{R}_T^*(\underline{\mathbf{a}}, \underline{\boldsymbol{\ell}})\right] = \max_{\mathcal{L}} \min_{\underline{\mathbf{a}}} \mathbb{E}\left[\mathcal{R}_T^*(\underline{\mathbf{a}}, \underline{\boldsymbol{\ell}})\right].$$

Observe that when taking an expectation over $\underline{\mathbf{a}} \sim \mathbb{D}$, with $\underline{\boldsymbol{\ell}}$ fixed, we get

$$\mathbb{E}\left[\mathcal{R}_T^*(\underline{\mathbf{a}}, \underline{\boldsymbol{\ell}})\right] = \sum_{t=1}^T \mathbf{p}^t \cdot \boldsymbol{\ell}^t - \min_i \sum_{t=1}^T \ell_i^t = \mathcal{R}_T(\mathcal{D}, \boldsymbol{\ell}),$$

where $\mathbf{p}^t = \mathbf{p}^t(\{\boldsymbol{\ell}^s\}_{s=1}^{t-1})$ is the marginal distribution over actions taken by the decider at time t and $\mathcal{D}$ is the sequence of functions $\{\mathbf{p}^t\}_{t=1}^T$.

Reducing to balanced adversary strategies

We say that an adversary strategy $\mathcal{L}$ is **balanced** if, for every time t, conditioned on the history of losses through time $t-1$, the expected loss of each expert is the same, i.e., for all pairs of actions i and j, we have $\mathbb{E}_t\left[\ell_i^t\right] = \mathbb{E}_t\left[\ell_j^t\right]$.

---

[2]These are *oblivious* strategies, which do not depend on previous decider actions. See the chapter notes for a discussion of *adaptive*, i.e., nonoblivious, adversary strategies.

For fixed $\mathbf{a}$ and $\boldsymbol{\ell} \sim \mathcal{L}$, write

$$\mathcal{R}_T^*(\mathbf{a}, \mathcal{L}) := \mathbb{E}\left[\mathcal{R}_T^*(\mathbf{a}, \boldsymbol{\ell})\right].$$

PROPOSITION 18.5.4. *Let* $\mathcal{R}_T^*(\mathcal{L}) := \min_{\mathbf{a}} \mathcal{R}_T^*(\mathbf{a}, \mathcal{L})$. *Then* $\max_{\mathcal{L}} \mathcal{R}_T^*(\mathcal{L})$ *is attained in* **balanced strategies**.

PROOF. Clearly $\min_{\mathbf{a}} \mathcal{R}_T^*(\mathbf{a}, \mathcal{L})$ is achieved by choosing, at each time step t, the action which has the smallest expected loss in that step, given the history of losses. We claim that for every $\mathcal{L}$ that is not balanced at time t for some history, there is an alternative strategy $\tilde{\mathcal{L}}$ that is balanced at t and has

$$\mathcal{R}_T^*(\tilde{\mathcal{L}}) \geq \mathcal{R}_T^*(\mathcal{L}).$$

Construct such a $\tilde{\mathcal{L}}$ as follows: Pick $\{\boldsymbol{\ell}^t\}$ according to $\mathcal{L}$. Let $\tilde{\boldsymbol{\ell}}^s = \boldsymbol{\ell}^s$ for all $s \neq t$. At time t, define for all i

$$\tilde{\ell}_i^t = \ell_i^t \theta_i$$

where θ_i is independent of the loss sequence,

$$\theta_i \in \{0, 1\}, \quad \text{and} \quad \mathbb{E}\left[\theta_i\right] = \frac{\min_j \mathbb{E}_t\left[\ell_j^t\right]}{\mathbb{E}_t\left[\ell_i^t\right]}.$$

(Recall that $\mathbb{E}_t\left[\cdot\right]$ denotes the expectation given the history prior to time t.) This change ensures that at time t, all experts have conditional expected loss equal to $\min_j \mathbb{E}_t\left[\ell_j^t\right]$. The best-response strategy of the decider at time t remains a best response. The same would hold at future times if the decider was given ℓ_i^t as well as $\tilde{\ell}_i^t$. Since he has less information, his expected loss at future times cannot decrease. Moreover, the benchmark loss $\min_j \mathbb{E}\left[L_j^T\right]$ is weakly reduced. □

Against a balanced adversary, the sequence of actions of the decider is irrelevant. Taking $\mathcal{D}$ to be the uniform distribution over actions (i.e., $p_i^t = 1/n$ for each i and t), we have

$$\mathcal{R}_T^*(\mathcal{L}) = \mathbb{E}\left[\frac{1}{n}\sum_{i=1}^n L_i^T - \min_i L_i^T\right], \tag{18.16}$$

where $\mathrm{L}_i^T = \sum_{t=1}^T \ell_i^t$.

18.5.3. The case of two actions. For $n = 2$, equation (18.16) reduces to

$$\mathcal{R}_T^*(\mathcal{L}) = \frac{1}{2}(L_1^T + L_2^T) - \min(L_1^T, L_2^T) = \frac{1}{2}|L_1^T - L_2^T|.$$

Write $X^t = \ell_1^t - \ell_2^t$ so that $X^t \in \{-1, 0, 1\}$ with $\mathbb{E}_t\left[X^t\right] = 0$. To maximize[3] $\mathbb{E}|\sum_{t=1}^T X_t| = 2\mathcal{R}_T^*(\mathcal{L})$, we let $X^t \in \{-1, 1\}$ so that $S_t := \sum_{k=1}^t X_k$ is a simple random walk. In other words, $\boldsymbol{\ell}^t$ is i.i.d., equally likely to be $(1, 0)$ or $(0, 1)$. Thus, by the Central Limit Theorem, we have

$$\mathcal{R}_T := \mathcal{R}_T^*(\mathcal{L}) = \frac{1}{2}\mathbb{E}|S_T| = \frac{1}{2}\sqrt{T}\,\mathbb{E}|Z|(1 + o(1))$$

with $Z \sim N(0, 1)$, so $\mathbb{E}|Z| = \sqrt{\frac{2}{\pi}}$.

---

[3]Let $\{S_k\}$ be a simple random walk on $\mathbb{Z}$. Condition on $|\{t \leq T : X_t \neq 0\}| = m$. Then it suffices to show that $\mathbb{E}\left[|S_m|\right] \leq \mathbb{E}\left[|S_T|\right]$, which holds since $\mathbb{E}\left[|S_m|\right] = \mathbb{E}\left[\sharp j \in \{0, \ldots, m-1\} \mid S_j = 0\right]$.

Optimal decider

To find the optimal $\mathcal{D}$, we consider an initial integer gap $h \geq 0$ between the actions and define

$$r_T(h) = \min_{\mathcal{D}} \max_{\mathcal{L}} \mathbb{E} \left[L_{\mathcal{D}}^T - \min\{L_1^T + h, L_2^T\} \right],$$

where $L_{\mathcal{D}}^T = \sum_{t=1}^T \boldsymbol{\ell}^t \cdot \mathbf{p}^t$. By the Minimax Theorem,

$$r_T(h) = \max_{\mathcal{L}} \min_{\mathcal{D}} \mathbb{E} \left[L_{\mathcal{D}}^T - \min\{L_1^T + h, L_2^T\} \right].$$

As in the discussion above, the optimal adversary is balanced, so we have

$$r_T(h) = \max_{\mathcal{L} \text{ balanced}} \mathbb{E} \left[\frac{1}{2}(L_1^T + L_2^T) - \min(L_1^T + h, L_2^T) \right]$$

$$= \max_{\mathcal{L} \text{ balanced}} \mathbb{E} \left[\frac{1}{2}\Big(|L_1^T + h - L_2^T| - h \Big) \right].$$

Again, the adversary's optimal strategy is to select $\boldsymbol{\ell}^t$ i.i.d., equally likely to be $(1,0)$ or $(0,1)$, so with $\{S_t\}$ denoting a simple random walk,

$$r_T(h) = \frac{1}{2} \mathbb{E} \left[|h + S_T| - h \right]. \tag{18.17}$$

Fix an optimal strategy $\mathcal{D}$ for the decider. To emphasize the dependence on T and h, write $\psi_T(h) = p_1^1$, the probability that $\mathcal{D}$ assigns to action 1 in the first step. If the adversary selects the loss vector $\boldsymbol{\ell}^1 = (1,0)$, then the gap between the losses of the actions becomes $h + 1$, the decider's expected loss in this step is $\psi_T(h)$, and

$$\min(L_1^T + h, L_2^T) = \min(\tilde{L}_1^{T-1} + h + 1, \tilde{L}_2^{T-1}),$$

where $\tilde{L}_i^{T-1}$ refers to losses in the time interval $[2, T]$. Thus, if the optimal adversary assigns $\boldsymbol{\ell}^1 = (1,0)$ positive probability, then

$$r_T(h) = r_{T-1}(h + 1) + \psi_T(h).$$

On the other hand, if the adversary selects $\boldsymbol{\ell}^1 = (0,1)$, then the gap between the losses becomes $h - 1$, the decider's expected loss is $1 - \psi_T(h)$, and

$$\min(L_1^T + h, L_2^T) = \min(\tilde{L}_1^{T-1} + h, \tilde{L}_2^{T-1} + 1) = 1 + \min(\tilde{L}_1^{T-1} + h - 1, \tilde{L}_2^{T-1}).$$

Thus, if the optimal adversary assigns $\boldsymbol{\ell}^1 = (0,1)$ positive probability, then

$$r_T(h) = r_{T-1}(h - 1) - 1 + 1 - \psi_T(h) = r_{T-1}(h - 1) - \psi_T(h).$$

To maximize regret, the adversary will select $\boldsymbol{\ell}^1$ so that

$$r_T(h) = \max\Big(r_{T-1}(h + 1) + \psi_T(h), \, r_{T-1}(h - 1) - \psi_T(h) \Big).$$

At optimality, the decider will choose $\psi_T(h)$ to equalize these costs[4], in which case

$$\psi_T(h) = \frac{r_{T-1}(h - 1) - r_{T-1}(h + 1)}{2}.$$

Thus by (18.17)

$$\psi_T(h) = \frac{1}{4} \mathbb{E}\Big[|h - 1 + S_{T-1}| - |h + 1 + S_{T-1}| + 2 \Big].$$

---

[4]This is possible since $0 \leq r_{T-1}(h - 1) - r_{T-1}(h + 1) \leq 2$ by definition.

Since for m integer

$$|m-1| - |m+1| = \begin{cases} -2 & \text{if } m > 0, \\ 0 & \text{if } m = 0, \\ 2 & \text{if } m < 0, \end{cases}$$

we conclude that

$$\psi_T(h) = \frac{\mathbb{E}\left[2 \cdot \mathbb{1}_{\{S_{T-1}+h=0\}} + 4 \cdot \mathbb{1}_{\{S_{T-1}+h<0\}}\right]}{4}$$

$$= \frac{1}{2}\mathbb{P}\left[S_{T-1} + h = 0\right] + \mathbb{P}\left[S_{T-1} + h < 0\right]. \tag{18.18}$$

In other words, $\psi_T(h)$ is the probability that the action currently lagging by h will be the leader in $T-1$ steps.

THEOREM 18.5.5. *For $n = 2$, with losses in $\{0,1\}$, the minimax optimal regret is*

$$\mathcal{R}_T = \sqrt{\frac{T}{2\pi}}\,(1 + o(1)).$$

The optimal adversary strategy is to take $\boldsymbol{\ell}^t$ i.i.d., equally likely to be $(1,0)$ or $(0,1)$.

The optimal decision strategy $\{\mathbf{p}^t\}_{t=1}^T$ is determined as follows: First, $\mathbf{p}^1 = (1/2, 1/2)$. For $t \in [1, T-1]$, let i_t be an action with the smaller cumulative loss at time t, so that $L_{i_t}^t = \min(L_1^t, L_2^t)$. Also, let $h_t = |L_1^t - L_2^t|$. At time $t+1$, take the action i_t with probability $p_{i_t}^{t+1} = 1 - \psi_{T-t}(h_t)$ and other action $3 - i_t$ with probability $p_{3-i_t}^{t+1} = \psi_{T-t}(h_t)$.

Let Φ denote the standard normal distribution function. By the Central Limit Theorem,

$$\psi_T(h) = \Phi(-h/\sqrt{T})(1 + o(1)) \qquad \text{as } T \to \infty,$$

so the optimal decision algorithm is easy to implement.

18.5.4. Adaptive versus oblivious adversaries. In the preceding, we assumed the adversary is oblivious, i.e., selects the loss vectors $\boldsymbol{\ell}^t = \boldsymbol{\ell}^t(\mathcal{D})$ independently of the actions of the decider. (He can still use the mixed strategy $\mathcal{D}$, but not the actual random choices.)

A more powerful adversary is *adaptive*, i.e., can select loss vectors $\boldsymbol{\ell}^t$ that depend on $\mathcal{D}$ and *also* on the previous actions $a_1, a_2, \ldots, a_{t-1}$. With the (standard) definition of regret that we used, for every $\mathcal{D}$, adaptive adversaries cause the same worst-case regret as oblivious ones; both regrets simply equal the maximum over individual loss sequences $\max_{\boldsymbol{\ell}} \mathcal{R}_T(\mathcal{D}, \boldsymbol{\ell})$. For this reason, it is often noted that low regret algorithms like the Multiplicative Weights Algorithm work against adaptive adversaries as well as against oblivious ones.

Against adaptive adversaries, the notion of regret we use here is

$$\mathcal{R}_T(\mathcal{D}, \boldsymbol{\ell}) = \mathbb{E}\left[L_{\mathcal{D}}^T - \min_i \sum_{t=1}^T \ell_i^t(a_1, \ldots, a_{t-1})\right]. \tag{18.19}$$

An alternative known as **policy regret** is

$$\hat{\mathcal{R}}_T(\mathcal{D}, \boldsymbol{\ell}) = \mathbb{E}\left[L_{\mathcal{D}}^T - \min_i \sum_{t=1}^T \ell_i^t(i, \dots i)\right]. \tag{18.20}$$

The notion of regret in (18.19) is useful in the setting of learning from expert advice where it measures the performance of the decider relative to the performance of the best expert. Next we give some examples where policy regret is more appropriate.

(1) Suppose the actions of the decider lead to big and equal losses to all the experts and hence also to him, while if he consistently followed the advice of expert 1, then he would have 0 loss. The usual regret for his destructive actions will be 0, but the policy regret will be large. Such a scenario could arise, for example, if the decider is a large investor in the stock market, whose actions affect the prices of various stocks. Another example is when the decider launches a military campaign causing huge losses all around.

(2) Let $\boldsymbol{\ell} = \{\boldsymbol{\ell}^t\}$ be any oblivious loss sequence. Imposing a switching cost can be modeled as an adaptive adversary $\tilde{\mathcal{L}}$ defined by

$$\tilde{\ell}_i^t = \ell_i^t + \mathbb{1}_{\{a_{t-1} \neq a_{t-2}\}} \quad \forall i, t.$$

The usual regret will ignore the switching cost, i.e.,

$$\mathcal{R}_T(\mathcal{D}, \mathcal{L}) = \mathcal{R}_T(\mathcal{D}, \tilde{\mathcal{L}}) \quad \forall \mathcal{D},$$

but policy regret will take it into account.

For example, if $\ell_i^t = \mathbb{1}_{\{i=t \mod 2\}}$ and there are two actions, then MW will eventually choose between the two actions with nearly equal probability, so its expected cost will be $T/2 + O(1)$ plus the expected number of switches, which is also $T/2$. The cost L_i^T will also be $T/2$ plus the expected number of switches that MW does. Thus $\mathcal{R}_T(\text{MW}, \tilde{\mathcal{L}}) = O(1)$. However, with policy regret, the benchmark $\min_i \sum_{t=1}^T \ell_i^t(i, \dots, i) = T/2$, so $\hat{\mathcal{R}}_T(\text{MW}, \tilde{\mathcal{L}}) = T/2 + O(1)$.

(3) Consider a decider playing repeated Prisoner's Dilemma (as discussed in Example 4.1.1 and §6.4) for T rounds.

player II

		cooperate	defect
player I	cooperate	$(-1, -1)$	$(-10, 0)$
	defect	$(0, -10)$	$(-8, -8)$

Suppose that the loss sequence $\boldsymbol{\ell}$ is defined by an opponent playing Tit-for-Tat.[5] In this case, defining $a_0 = C$, the loss of action i at time t

---

[5]Recall Definition 6.4.2: Tit-for-Tat is the strategy in which the player cooperates in round 1 and in every round thereafter plays the strategy his opponent played in the previous round.

is:

$$
\ell_i^t(a_{t-1}) = \begin{cases} 1 & (a_{t-1}, i) = (C, C), \\ 0 & (a_{t-1}, i) = (C, D), \\ 10 & (a_{t-1}, i) = (D, C), \\ 8 & (a_{t-1}, i) = (D, D). \end{cases}
$$

Since it is a dominant strategy to defect in Prisoner's Dilemma,

$$
L_{\text{defect}}^T > L_{\text{cooperate}}^T.
$$

(This holds for any opponent, not just Tit-for-Tat.) Thus, for any decider strategy,

$$
\mathbb{E}\left[\mathcal{R}_T(\mathcal{D}, \boldsymbol{\ell})\right] = \mathbb{E}\left[\sum_t \mathbb{1}_{\{a_t = C\}}\left(\mathbb{1}_{\{a_{t-1} = C\}} + 2\mathbb{1}_{\{a_{t-1} = D\}}\right)\right].
$$

Minimizing regret will lead the decider towards defecting every round and incurring a loss of $8(T-1)$. However, minimizing policy regret will lead the decider to cooperate for $T-1$ rounds, yielding a loss of $T-1$.

Notes

A pioneering paper [Han57] on adaptive decision-making was written by Hannan in 1957. (Algorithms with sublinear regret are also known as *Hannan consistent.*) Adaptive decision making received renewed attention starting in the 1990s due to its applications in machine learning and algorithms. For in-depth expositions of this topic, see the book by Cesa-Bianchi and Lugosi [CBL06] and the surveys by Blum and Mansour (Chapter 4 of [NRTV07]) and Arora, Hazan, and Kale [AHK12].

James Hannan

The binary prediction problem with a perfect expert is folklore and can be found, e.g., in [CBL06]; the sharp bound in Theorem 18.1.4 is new, to the best of our knowledge. The Weighted Majority Algorithm and the Multiplicative Weights Algorithm from §18.2.1 and §18.3.2 are due to Littlestone and Warmuth [LW89, LW94]. A suite of decision-making algorithms closely related to the Multiplicative Weights Algorithm [FS97, Vov90, CBFH+97] go under different names including Exponential Weights, Hedge, etc. The sharp analysis in §18.3.2 is due to Cesa-Bianchi et al. [CBFH+97]. The use of the Multiplicative Weights algorithm to play zero-sum games, discussed in §18.4, is due to Grigoriadis and Khachiyan [GK95] and Freund and Schapire [FS97]. Theorem 18.5.5 is due to [Cov65].

Our exposition follows [GPS14], where optimal strategies for three experts are also determined. The notion of policy regret discussed in §18.5.4 is due to Arora, Dekel, and Tewari [ADT12]. Example (3) of §18.5.4 is discussed in [CBL06].

There are several important extensions of the material in this chapter that we have not addressed: In the *multiarmed bandit problem*[6] [BCB12], the decider learns his loss in each round but does not learn the losses of actions he did not choose. Surprisingly, even with such limited feedback, the decider can achieve regret $O(\sqrt{Tn \log n})$ against adaptive adversaries [ACBFS02] and $O(\sqrt{Tn})$ against oblivious adversaries [AB10].

Instead of comparing his loss to that of the best action, the decider could examine how his loss would change had he swapped certain actions for others, say, each occurrence of action i by $f(i)$. Choosing the optimal $f(\cdot)$ in hindsight defines *swap regret*. Hart and Mas-Colell [HMC00b] showed how to achieve sublinear swap regret using Blackwell's Approachability Theorem [Bla56]. Blum and Mansour [BM05] showed how to achieve sublinear swap regret via a reduction to algorithms that achieve sublinear regret. When players playing a game use sublinear swap regret algorithms, the empirical distributions of play converge to a correlated equilibrium.

The adaptive learning problem of §18.3.2 can be viewed as online optimization of linear functions. An important generalization is online convex optimization. See, e.g., the survey by Shalev-Shwartz [SS11].

David Blackwell

Julia Robinson

In 1951, Brown [Bro51] proposed a simple strategy, known as *Fictitious Play* or *Follow the Leader*, for the repeated play of a two-person zero-sum game: At time t, each player plays a best response to the empirical distribution of play by the opponent in the first $t - 1$ rounds. Robinson [Rob51] showed that if both players use fictitious play, their empirical distributions converge to optimal strategies. However, as discussed in §18.1, this strategy does not achieve sublinear regret in the setting of binary prediction with expert advice. Hannan [Han57] analyzed a variant, now known as *Follow the Perturbed Leader*, in which each action/expert is initially given a random loss and then henceforth the leader is followed. Using a different perturbation, Kalai and Vempala [KV05] showed that this algorithm obtains essentially the same regret bounds as the Multiplicative Weights Algorithm. ([KV05] also handles switching costs.)

Exercise 18.4 is due to Avrim Blum.

Exercises

18.1.   Consider the setting of §18.2, and suppose that u_t (respectively d_t) is the number of leaders voting up (respectively down) at time t. Consider a trader algorithm $\mathcal{A}$ that decides up or down at time t with probability p_t,

---

[6]This problem is named after slot machines in Las Vegas known as one-armed bandits.

where $p_t = p_t(u_t, d_t)$. Then there is an adversary strategy that ensures that any such trader algorithm $\mathcal{A}$ incurs an expected loss of at least $\lfloor \log_4(n) \rfloor L_*^T$. Hint: Adapt the adversary strategy in Proposition 18.1.3, ensuring that no expert incurs more than one mistake during S_0. Repeat.

18.2.   In the setting of learning with n experts, at least one of them perfect, suppose each day there are $q > 2$, rather than two, possibilities to choose between. Observe that the adversary can still force at least $\lfloor \log_2(n) \rfloor / 2$ mistakes in expectation. Then show that the decider can still guarantee that the expected number of mistakes is at most $\log_4(n)$.

Hint: Follow the majority opinion among the ℓ current leaders (if such a majority exists) with probability $p(x)$ (where the majority has size $x\ell$, as in the binary case); with the remaining probability (which is 1 if there is no majority), follow a uniformly chosen random leader (not in the majority).

18.3.   Show that there are cases where any deterministic algorithm in the experts setting makes at least twice as many mistakes as the best expert; i.e., for some T, $L(T) \geq 2L_*^T$.

18.4.   Consider the following variation on the Weighted Majority Algorithm:
        On each day t, associate a weight w_i^t with each expert i.
        Initially, when $t = 1$, set $w_i^1 = 1$ for all i.

        Each day t, follow the **weighted majority** opinion: Let U_t be the set of experts predicting up on day t, and D_t the set predicting down. Predict "up" on day t if $W_U(t) = \sum_{i \in U_t} w_i^t \geq W_D(t) = \sum_{i \in D_t} w_i^t$ and "down" otherwise.
        On day $t+1$, for each i such that (a) expert i predicted incorrectly on day t and (b) $w_i^t \geq \frac{1}{4n} \sum_j w_j^t$, set

$$w_i^{t+1} = \frac{1}{2} w_i^t. \tag{18.21}$$

Show that the number of mistakes made by the algorithm during every contiguous subsequence of days, say $\tau, \tau + 1, \ldots, \tau + r$, is $O(m + \log n)$, where m is the fewest number of mistakes made by any expert on days $\tau, \tau + 1, \ldots, \tau + r$.

18.5.   Consider the sequential adaptive decision-making setting of §18.3 with unknown time horizon T. Adapt the Multiplicative Weights Algorithm by changing the parameter ϵ over time to a new value at $t = 2^j$ for $j = 0, 1, 2, \ldots$. (This is a "doubling trick".) Show that the sequence of ϵ values can be chosen so that for every action i,

$$\overline{L}^T \leq L_i^T + \frac{\sqrt{2}}{\sqrt{2} - 1} \sqrt{\frac{1}{2} T \log n}.$$

18.6.   Generalize the results of §18.5.4 for $n = 2$ to the case where the time horizon T is geometric with parameter δ; i.e., the process stops with probability δ in every round:
- Determine the minimax optimal adversary and the minimax regret.
- Determine the minimax optimal decision algorithm.

S   18.7.

(a) For Y a normal $N(0,1)$ random variable, show that

$$e^{-\frac{y^2}{2}(1+o(1))} \leq \mathbb{P}\left[Y > y\right] \leq e^{-\frac{y^2}{2}} \qquad \text{as} \qquad y \to \infty.$$

(b) Suppose that $Y_1, \ldots, Y_n$ are i.i.d. $N(0,1)$ random variables. Show that

$$\mathbb{E}\left[\max_{1 \leq i \leq n} Y_i\right] = \sqrt{2 \log n}\,(1 + o(1)) \qquad \text{as} \qquad n \to \infty. \tag{18.22}$$

18.8.   Consider an adaptive adversary with bounded memory; that is,

$$\ell_i^t = \ell_i^t(a_{t-m}, \ldots, a_{t-1})$$

for constant m. Suppose that a decider divides time into blocks of length b and uses a fixed action, determined by the Multiplicative Weights Algorithm in each block. Show that the policy regret of this decider strategy is $O(\sqrt{Tb \log n} + Tm/b)$. Then optimize over b.

Linear programming

A.1. The Minimax Theorem and linear programming

Suppose that we want to determine if player I in a two-person zero-sum game with $m \times n$ payoff matrix $A = (a_{ij})$ can guarantee an expected gain of at least V. It suffices for her to find a mixed strategy $\mathbf{x}$ which guarantees her an expected gain of at least V for each possible pure strategy j player II might play. These conditions are captured by the following system of inequalities:

$$x_1 a_{1j} + x_2 a_{2j} + \cdots + x_m a_{mj} \geq V \text{ for } 1 \leq j \leq n.$$

In matrix-vector notation, this system of inequalities becomes:

$$\mathbf{x}^T A \geq V \mathbf{e}^T,$$

where $\mathbf{e}$ is an all-1's vector. (Its length will be clear from context.)

Thus, to maximize her guaranteed expected gain, player I should find an $\mathbf{x} \in \mathbb{R}^m$ and a $V \in \mathbb{R}$ that

$$\text{maximize } V$$
$$\text{subject to} \quad x^T A \geq V \mathbf{e}^T, \tag{A.1}$$
$$\sum_{1 \leq i \leq m} x_i = 1,$$
$$x_i \geq 0 \text{ for all } 1 \leq i \leq m.$$

This is an example of a **linear programming problem (LP).** Linear programming is the process of minimizing or maximizing a linear function of a finite set of real-valued variables, subject to linear equality and inequality constraints on those variables. In the linear program (A.1), the variables are V and $x_1, \ldots, x_m$.

The problem of finding the optimal strategy for player II can similarly be formulated as a linear program (LP):

$$\text{minimize } V$$
$$\text{subject to} \quad A\mathbf{y} \leq V, \mathbf{e} \tag{A.2}$$
$$\sum_{1 \leq j \leq n} y_j = 1,$$
$$y_j \geq 0 \text{ for all } 1 \leq j \leq n.$$

Many fundamental optimization problems in engineering, economics, and transportation can be formulated as linear programs. A typical example is planning airline routes. Conveniently, there are well-known efficient (polynomial time) algorithms for solving linear programs (see notes) and, thus, we can use these algorithms to solve for optimal strategies in large zero-sum games. In the rest of this appendix, we give a brief introduction to the theory of linear programming.

A.2. Linear programming basics

EXAMPLE A.2.1. (**The protein problem**). Consider the dilemma faced by a student-athlete interested in maximizing her protein consumption, while consuming no more than 5 units of fat per day and spending no more than \$6 a day on protein. She considers two alternatives: steak, which costs \$4 per pound and contains 2 units of protein and 1 unit of fat per pound; and peanut butter, which costs \$1 per pound and contains 1 unit of protein and 2 units of fat per pound.

Let x_1 be the number of pounds of steak she buys per day, and let x_2 be the number of pounds of peanut butter she buys per day. Then her goal is to

$$\max \ 2x_1 + x_2$$
$$\text{subject to} \quad 4x_1 + x_2 \leq 6, \tag{A.3}$$
$$x_1 + 2x_2 \leq 5,$$
$$x_1, x_2 \geq 0.$$

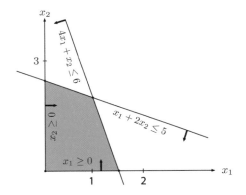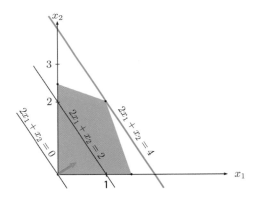

FIGURE A.1. The purple region in the above graphs is the feasible set for the linear program (A.3). The largest c for which the line $2x_1 + x_2 = c$ intersects the feasible set is $c = 4$. The red arrow from the origin on the right is perpendicular to all these lines.

The **objective function** of a linear program is the linear function being optimized, in this case $2x_1 + x_2$. The **feasible set** of a linear program is the set of **feasible** vectors that satisfy the constraints of the program, in this case, all nonnegative vectors (x_1, x_2) such that $4x_1 + x_2 \leq 6$ and $x_1 + 2x_2 \leq 5$.

The left-hand side of Figure A.1 shows this set. The question then becomes: which point in this feasible set maximizes $2x_1 + x_2$? For (A.3), this point is $(x_1, x_2) = (1, 2)$, and at this point $2x_1 + x_2 = 4$. Thus, the optimal solution to the linear program is 4.

A.2.1. Linear programming duality. The Minimax Theorem that we proved earlier shows that for any zero-sum game, the two linear programs (A.1) and (A.2) have the same optimal value V^*. This is a special case of the cornerstone of linear programming, **the Duality Theorem** (Theorem A.2.2 in the next section).

To motivate this theorem, let's consider the LP from the previous section more analytically. The first constraint of (A.3) immediately implies that the objective function is upper bounded by 6 on the feasible set. Doubling the second constraint gives a worse bound of 10. But combining them, we can do better.

Multiplying the first constraint by $y_1 \geq 0$, the second by $y_2 \geq 0$, and adding the results yields

$$y_1(4x_1 + x_2) + y_2(x_1 + 2x_2) \leq 6y_1 + 5y_2. \tag{A.4}$$

The left-hand side of (A.4) dominates the objective function $2x_1 + x_2$ as long as

$$4y_1 + y_2 \geq 2, \tag{A.5}$$
$$y_1 + 2y_2 \geq 1,$$
$$y_1, y_2 \geq 0.$$

Thus, for any (y_1, y_2) that is feasible for (A.5), we have $2x_1 + x_2 \leq 6y_1 + 5y_2$ for all feasible (x_1, x_2). The best upper bound we can obtain this way on the optimal value of (A.3) is the solution to the linear program

$$\min 6y_1 + 5y_2 \text{ subject to (A.5).} \tag{A.6}$$

This minimization problem is called the **dual** of LP (A.3). Observing that $(y_1, y_2) = (3/7, 2/7)$ is feasible for LP (A.6) with objective value 4, we can conclude that $(x_1, x_2) = (1, 2)$, which attains objective value 4 for the original problem, must be optimal.

A.2.2. Duality, more formally. We say that a maximization linear program is in standard form [1] if it can be written as

$$\left. \begin{array}{c} \max \mathbf{c}^T \mathbf{x} \\ \text{subject to} \\ A\mathbf{x} \leq \mathbf{b}, \\ \mathbf{x} \geq 0 \end{array} \right\}, \tag{P}$$

where $A \in \mathbb{R}^{m \times n}$, $\mathbf{x} \in \mathbb{R}^n$, $\mathbf{c} \in \mathbb{R}^n$, and $\mathbf{b} \in \mathbb{R}^m$. We will call such a linear program (P) the **primal LP**. We say the primal LP is **feasible** if the feasible set

$$\mathcal{F}(P) := \{\mathbf{x} \mid A\mathbf{x} \leq \mathbf{b}, \ \mathbf{x} \geq 0\}$$

is nonempty.

As in the example at the beginning of this section, if $\mathbf{y} \geq 0$ in $\mathbb{R}^m$ satisfies $\mathbf{y}^T A \geq \mathbf{c}^T$, then

$$\forall \mathbf{x} \in \mathcal{F}(P), \ \mathbf{y}^T \mathbf{b} \geq \mathbf{y}^T A \mathbf{x} \geq \mathbf{c}^T \mathbf{x}. \tag{A.7}$$

---

[1] It is a simple exercise to convert from nonstandard form (such as a game LP) to standard form. For example, an equality constraint such as $a_1x_1 + a_2x_2 + \cdots + a_nx_n = b$ can be converted to two inequalities: $a_1x_1 + a_2x_2 + \cdots + a_nx_n \geq b$ and $a_1x_1 + a_2x_2 + \cdots + a_nx_n \leq b$. A $\geq$ inequality can be converted to a $\leq$ inequality and vice versa by multiplying by -1. A variable x that is not constrained to be nonnegative, can be replaced by the difference $x' - x''$ of two nonnegative variables, and so on.

This motivates the general definition of the **dual LP**

$$
\left.
\begin{array}{c}
\min \mathbf{b}^T \mathbf{y} \\
\text{subject to} \\
\mathbf{y}^T A \geq \mathbf{c}^T, \\
\mathbf{y} \geq 0,
\end{array}
\right\}
\tag{D}
$$

where $\mathbf{y} \in \mathbb{R}^m$. As with the primal LP, we say the dual LP is feasible if the set

$$
\mathcal{F}(\mathrm{D}) := \{\mathbf{y} \mid \mathbf{y}^T A \geq \mathbf{c}^T;\ \mathbf{y} \geq 0\}
$$

is nonempty.

It is easy to check that the dual of the dual LP is the primal LP.[2]

THEOREM A.2.2 (**The Duality Theorem of Linear Programming**). *Suppose that $A \in \mathbb{R}^{m \times n}$, $\mathbf{x}, \mathbf{c} \in \mathbb{R}^n$, and $\mathbf{y}, \mathbf{b} \in \mathbb{R}^m$. If $\mathcal{F}(\mathrm{P})$ and $\mathcal{F}(\mathrm{D})$ are nonempty, then:*

- *$\mathbf{b}^T \mathbf{y} \geq \mathbf{c}^T \mathbf{x}$ for all $\mathbf{x} \in \mathcal{F}(\mathrm{P})$ and $\mathbf{y} \in \mathcal{F}(\mathrm{D})$. (This is called **weak duality**.)*
- *(P) has an optimal solution $\mathbf{x}^*$, (D) has an optimal solution $\mathbf{y}^*$, and $\mathbf{c}^T \mathbf{x}^* = \mathbf{b}^T \mathbf{y}^*$.*

REMARK A.2.3. The proof of the Duality Theorem is similar to the proof of the Minimax Theorem. This is not accidental; see the chapter notes.

COROLLARY A.2.4 (**Complementary Slackness**). *Let $\mathbf{x}^*$ be feasible for (P) and let $\mathbf{y}^*$ be feasible for (D). Then the following two statements are equivalent:*

(1) *$\mathbf{x}^*$ is optimal for (P) and $\mathbf{y}^*$ is optimal for (D).*
(2) *For each i such that $\sum_{1 \leq j \leq n} a_{ij} x_j^* < b_i$ we have $y_i^* = 0$, and for each j such that $c_j < \sum_{1 \leq i \leq m} y_i^* a_{ij}$ we have $x_j^* = 0$.*

PROOF. Feasibility of $\mathbf{y}^*$ and $\mathbf{x}^*$ implies that

$$
\sum_j c_j x_j^* \leq \sum_j x_j^* \sum_i y_i^* a_{ij} = \sum_i y_i^* \sum_j a_{ij} x_j^* \leq \sum_i b_i y_i^*.
\tag{A.8}
$$

By the Duality Theorem, optimality of $\mathbf{x}^*$ and $\mathbf{y}^*$ is equivalent to having equality hold throughout (A.8). Moreover, by feasibility, for each j we have $c_j x_j^* \leq x_j^* \sum_i y_i^* a_{ij}$, and for each i we have $y_i^* \sum_j a_{ij} x_j^* \leq b_i y_i^*$. Thus, equality holds in (A.8) if and only if (2) holds. □

A.2.3. An interpretation of a primal/dual pair. Consider an advertiser about to purchase advertising space in a set of n newspapers, and suppose that c_j is the price of placing an ad in newspaper j. The advertiser is targeting m different types of users, for example, based on geographic location, interests, age, and gender, and wants to ensure that, on average, b_i users of type i will see the ad. Denote by a_{ij} the number of type i users expected to see each ad in newspaper j. The advertiser must decide how many ads to place in each newspaper in order to meet his various demographic targets at minimum cost. To this end, the advertiser solves

---

[2]A standard form minimization LP can be converted to a maximization LP (and vice versa) by observing that minimizing $\mathbf{b}^T \mathbf{y}$ is the same as maximizing $-\mathbf{b}^T \mathbf{y}$, and $\geq$ inequalities can be converted to $\leq$ inequalities by multiplying the inequality by -1.

the following linear program, where x_j is the number of ad slots from newspaper j that she will purchase:

$$\min \sum_{1 \leq j \leq n} c_j x_j$$

$$\text{subject to} \quad \sum_{1 \leq j \leq n} a_{ij} x_j \geq b_i \quad \text{for all } 1 \leq i \leq m, \tag{A.9}$$

$$x_1, x_2 \ldots, x_n \geq 0.$$

The dual program is

$$\max \sum_{1 \leq i \leq m} b_i y_i$$

$$\text{subject to} \quad \sum_{1 \leq i \leq m} y_i a_{ij} \leq c_j \quad \text{for all } 1 \leq j \leq n, \tag{A.10}$$

$$y_1, y_2 \ldots, y_m \geq 0.$$

This dual program has a nice interpretation: Consider an online advertising exchange that matches advertisers with display ad slots. The exchange needs to determine y_i, how much to charge the advertiser for each impression (displayed ad) shown to a user of type i. Observing that $y_i a_{ij}$ is the expected cost of reaching the same number of type i users online that would be reached by placing a single ad in newspaper j, we see that if the prices y_i are set so that $\sum_{1 \leq i \leq m} y_i a_{ij} \leq c_j$, then the advertiser can switch from advertising in newspaper j to advertising online, reaching the same combination of user types without increasing her cost. If the advertiser switches entirely from advertising in newspapers to advertising online, the exchange's revenue will be

$$\sum_{1 \leq i \leq m} b_i y_i.$$

The Duality Theorem says that the exchange can price the impressions so as to satisfy (A.10) and incentivize the advertiser to switch, while still ensuring that its revenue $\sum_i b_i y_i$ (almost) matches the total revenue of the newspapers.

Moreover, Corollary A.2.4 implies that if inequality (A.9) is not tight for user type i in the optimal solution of the primal, i.e., $\sum_{1 \leq j \leq n} a_{ij} x_j > b_i$, then $y_i = 0$ in the optimal solution of the dual. In other words, if the optimal combination of ads the advertiser buys from the newspapers results in the advertisement being shown to more users of type i than necessary, then in the optimal pricing for the exchange, impressions shown to users of type i will be provided to the advertiser for free. This means that the exchange concentrates its fixed total charges on the user types which correspond to tight constraints in the primal. Thus, the advertiser can switch to advertising exclusively on the exchange without paying more and without sacrificing any of the "bonus" advertising the newspapers were providing.

(The fact that some impressions are free may seem counterintuitive since providing ads has a cost, but it is a consequence of the assumption that the exchange maximizes revenue from this advertiser. In reality, the exchange would maximize profit, and these goals are equivalent only when the cost of production is zero.)

Finally, the other consequence of Corollary A.2.4 is that if $x_j > 0$, i.e., some ads were purchased from newspaper j, then the corresponding dual constraint must be tight, i.e., $\sum_{1 \le i \le m} y_i a_{ij} = c_j$.

A.2.4. The proof of the Duality Theorem*. Weak duality follows from (A.7). We complete the proof of the Duality Theorem in two steps. First, we will use the Separating Hyperplane Theorem to show that $\sup_{\mathbf{x} \in \mathcal{F}(\mathrm{P})} \mathbf{c}^T \mathbf{x} = \inf_{\mathbf{y} \in \mathcal{F}(\mathrm{D})} \mathbf{b}^T \mathbf{y}$, and then we will show that the sup and inf above are attained. For the first step, we will need the following lemma.

LEMMA A.2.5. *Let $A \in \mathbb{R}^{m \times n}$, and let $S = \{A\mathbf{x} \mid \mathbf{x} \ge 0\}$. Then S is closed.*

PROOF. If the columns of A are linearly independent, then $A : \mathbb{R}^n \mapsto W = A(\mathbb{R}^n)$ is invertible, so there is a linear inverse $L : W \mapsto \mathbb{R}^n$, from which

$$\{A\mathbf{x} \mid \mathbf{x} \ge 0\} = L^{-1}\{\mathbf{x} \in \mathbb{R}^n \mid \mathbf{x} \ge 0\},$$

which is closed by continuity of L.

Otherwise, if the columns $A^{(j)}$ of A are dependent, then we claim that

$$\{A\mathbf{x} \mid \mathbf{x} \ge 0\} = \bigcup_{k=1}^{n} \Big\{ \sum_{j=1}^{n} z_j A^{(j)} \mid \mathbf{z} \ge 0,\ z_k = 0 \Big\}.$$

To see this, observe that there is $\boldsymbol{\lambda} \ne \mathbf{0}$ such that $A\boldsymbol{\lambda} = 0$. Without loss of generality, $\lambda_j < 0$ for some j; otherwise, negate $\boldsymbol{\lambda}$. Given $\mathbf{x} \in \mathbb{R}^n$, $\mathbf{x} \ge 0$, find the largest $t \ge 0$ such that $\mathbf{x} + t\boldsymbol{\lambda} \ge 0$. For this t, some $x_k + t\lambda_k = 0$. Thus, $A\mathbf{x} = A(\mathbf{x} + t\boldsymbol{\lambda}) \in \{\sum_{j=1}^{n} z_j A^{(j)} \mid \mathbf{z} \ge 0,\ z_k = 0\}$.

Using induction on n, we see that $\{A\mathbf{x} \mid \mathbf{x} \ge 0\}$ is the union of a finite number of closed sets, which is closed. □

Next, we establish the following "alternative" theorem known as **Farkas' Lemma**, from which the proof of duality will follow.

LEMMA A.2.6 (**Farkas' Lemma** – 2 versions). *Let $A \in \mathbb{R}^{m \times n}$ and $\mathbf{b} \in \mathbb{R}^m$. Then*

    (1) *Exactly one of the following holds:*
        (a) *There exists $\mathbf{x} \in R^n$ such that $A\mathbf{x} = \mathbf{b}$ and $\mathbf{x} \ge 0$ or*
        (b) *there exists $\mathbf{y} \in R^m$ such that $\mathbf{y}^T A \ge 0$ and $\mathbf{y}^T \mathbf{b} < 0$.*
    (2) *Exactly one of the following holds:*
        (a) *There exists $\mathbf{x} \in R^n$ such that $A\mathbf{x} \le \mathbf{b}$ and $\mathbf{x} \ge 0$ or*
        (b) *there exists $\mathbf{y} \in R^m$ such that $\mathbf{y}^T A \ge 0$, $\mathbf{y}^T \mathbf{b} < 0$ and $\mathbf{y} \ge 0$.*

PROOF. Part (1): (See Figure A.2 for a visualization of Part (1).) We first show by contradiction that (a) and (b) can't hold simultaneously: Suppose that $\mathbf{x}$ satisfies (a) and $\mathbf{y}$ satisfies (b). Then

$$0 > \mathbf{y}^T \mathbf{b} = \mathbf{y}^T A\mathbf{x} \ge 0,$$

a contradiction.

We next show that if (a) is infeasible, then (b) is feasible: Let $S = \{A\mathbf{x} \mid \mathbf{x} \ge 0\}$. Then S is convex, and by Lemma A.2.5, it is closed. In addition, $\mathbf{b} \notin S$ since (a) is infeasible. Therefore, by the Separating Hyperplane Theorem, there is a hyperplane that separates b from S; i.e., $\mathbf{y}^T \mathbf{b} < a$ and $\mathbf{y}^T \mathbf{z} \ge a$ for all $\mathbf{z} \in S$. Since 0 is in S, $a \le 0$ and therefore $\mathbf{y}^T \mathbf{b} < 0$. Moreover, all entries of $\mathbf{y}^T A$ are nonnegative: If not, say the kth entry is negative, then by taking x_k arbitrarily large and $x_i = 0$ for

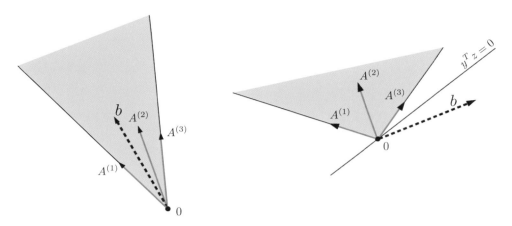

FIGURE A.2. The figure illustrates the two cases (1)(a) and (1)(b) of Farkas' Lemma. The shaded region represents all positive combinations of the columns of A.

$i \neq k$, the inequality $\mathbf{y}^T A \mathbf{x} \geq a$ would be violated for some $\mathbf{x} \geq 0$. Thus, it must be that $\mathbf{y}^T A \geq 0$.

Part (2): We apply Part (1) to an equivalent pair of systems. The existence of an $\mathbf{x} \in \mathbb{R}^n$ such that $A\mathbf{x} \leq \mathbf{b}$ and $\mathbf{x} \geq 0$ is equivalent to the existence of an $\mathbf{x} \geq 0$ in $\mathbb{R}^n$ and $\mathbf{v} \geq 0$ in $\mathbb{R}^m$ such that

$$A\mathbf{x} + I\mathbf{v} = \mathbf{b},$$

where I is the $m \times m$ identity matrix. Applying Part (1) to this system means that either it is feasible or there is a $\mathbf{y} \in \mathbb{R}^m$ such that

$$\begin{aligned} \mathbf{y}^T A &\geq 0, \\ I\mathbf{y} &\geq 0, \\ \mathbf{y}^T b &< 0, \end{aligned}$$

which is precisely equivalent to (b). □

COROLLARY A.2.7. *Under the assumptions of Theorem A.2.2,*

$$\sup_{\mathbf{x} \in \mathcal{F}(P)} \mathbf{c}^T \mathbf{x} = \inf_{\mathbf{y} \in \mathcal{F}(D)} \mathbf{b}^T \mathbf{y}.$$

PROOF. Suppose that $\sup_{\mathbf{x} \in \mathcal{F}(P)} \mathbf{c}^T \mathbf{x} < \gamma$. Then $\{A\mathbf{x} \leq \mathbf{b}; -\mathbf{c}^T \mathbf{x} \leq -\gamma; \mathbf{x} \geq 0\}$ is infeasible, and therefore by Part (2) of Farkas' Lemma, there is $(\mathbf{y}, \lambda) \geq 0$ in $\mathbb{R}^{m+1}$ such that $\mathbf{y}^T A - \lambda \mathbf{c}^T \geq 0$ and $\mathbf{y}^T \mathbf{b} - \lambda \gamma < 0$. Since there is an $\mathbf{x} \in \mathcal{F}(P)$, we have

$$0 \leq (\mathbf{y}^T A - \lambda \mathbf{c}^T)\mathbf{x} \leq \mathbf{y}^T \mathbf{b} - \lambda \mathbf{c}^T \mathbf{x}$$

and therefore $\lambda > 0$. We conclude that $\mathbf{y}/\lambda$ is feasible for (D), with objective value strictly less than γ. □

To complete the proof of the Duality Theorem, we need to show that the sup and inf in Corollary A.2.7 are attained. This will follow from the next theorem.

THEOREM A.2.8. *Let $A \in \mathbb{R}^{m \times n}$ and $\mathbf{b} \in \mathbb{R}^m$. Denote*

$$\mathcal{F}(\mathrm{P}_=) = \{\mathbf{x} \in \mathbb{R}^n \mid \mathbf{x} \geq 0 \text{ and } A\mathbf{x} = \mathbf{b}\}.$$

(i) *If $\mathcal{F}(\mathrm{P}_=) \neq \emptyset$ and $\sup\{\mathbf{c}^T\mathbf{x} \mid \mathbf{x} \in \mathcal{F}(\mathrm{P}_=)\} < \infty$, then this sup is attained.*
(ii) *If $\mathcal{F}(\mathrm{P}) \neq \emptyset$ and $\sup\{\mathbf{c}^T\mathbf{x} \mid \mathbf{x} \in \mathcal{F}(\mathrm{P})\} < \infty$, then this sup is attained.*

The proof of (i) will show that the sup is attained at one of a distinguished, finite set of points in $\mathcal{F}(\mathrm{P}_=)$ known as extreme points or vertices.

DEFINITION A.2.9.

(1) Let S be a convex set. A point $\mathbf{x} \in S$ is an **extreme point** of S if whenever $\mathbf{x} = \alpha\mathbf{u} + (1 - \alpha)\mathbf{v}$ with $\mathbf{u}, \mathbf{v} \in S$ and $0 < \alpha < 1$, we must have $\mathbf{x} = \mathbf{u} = \mathbf{v}$.
(2) If S is the feasible set of a linear program, then S is convex; an extreme point of S is called a **vertex**.

We will need the following lemma.

LEMMA A.2.10. *Let $\mathbf{x} \in \mathcal{F}(\mathrm{P}_=)$. Then $\mathbf{x}$ is a vertex of $\mathcal{F}(\mathrm{P}_=)$ if and only if the columns $\{A^{(j)} \mid x_j > 0\}$ are linearly independent.*

PROOF. Suppose $\mathbf{x}$ is not extreme; i.e., $\mathbf{x} = \alpha\mathbf{v} + (1 - \alpha)\mathbf{w}$, where $\mathbf{v} \neq \mathbf{w}$, $0 < \alpha < 1$, and $\mathbf{v}, \mathbf{w} \in \mathcal{F}(\mathrm{P}_=)$. Thus, $A(\mathbf{v} - \mathbf{w}) = 0$ and $\mathbf{v} - \mathbf{w} \neq \mathbf{0}$. Observe that $v_j = w_j = 0$ for all $j \notin S$, where $S = \{j \mid x_j > 0\}$, since $v_j, w_j \geq 0$. We conclude that the columns $\{A^{(j)} \mid x_j > 0\}$ are linearly dependent.

For the other direction, suppose that the vectors $\{A^{(j)} \mid x_j > 0\}$ are linearly dependent. Then there is $\mathbf{w} \neq 0$ such that $A\mathbf{w} = 0$ and $w_j = 0$ for all $j \notin S$. Then for ϵ sufficiently small $\mathbf{x} \pm \epsilon\mathbf{w} \in \mathcal{F}(\mathrm{P}_=)$, and therefore $\mathbf{x}$ is not extreme. $\square$

LEMMA A.2.11. *Suppose that $\sup\{\mathbf{c}^T\mathbf{x} \mid \mathbf{x} \in \mathcal{F}(\mathrm{P}_=)\} < \infty$. Then for any point $\mathbf{x} \in \mathcal{F}(\mathrm{P}_=)$, there is a vertex $\tilde{\mathbf{x}} \in \mathcal{F}(\mathrm{P}_=)$ with $\mathbf{c}^T\tilde{\mathbf{x}} \geq \mathbf{c}^T\mathbf{x}$.*

PROOF. We show that if $\mathbf{x}$ is not a vertex, then there exists $\mathbf{x}' \in \mathcal{F}(\mathrm{P}_=)$ with a strictly larger number of zero entries than $\mathbf{x}$, such that $\mathbf{c}^T\mathbf{x}' \geq \mathbf{c}^T\mathbf{x}$. This step can be applied only a finite number of times before we reach a vertex that satisfies the conditions of the lemma.

Let $S = \{j \mid x_j > 0\}$. If $\mathbf{x}$ is not a vertex, then the columns $\{A^{(j)} \mid j \in S\}$ are linearly dependent and there is a vector $\boldsymbol{\lambda} \neq \mathbf{0}$ such that $\sum_j \lambda_j A^{(j)} = A\boldsymbol{\lambda} = 0$ with $\lambda_j = 0$ for $j \notin S$.

Without loss of generality, $\mathbf{c}^T\boldsymbol{\lambda} \geq 0$ (if not, negate $\boldsymbol{\lambda}$). Consider the vector $\hat{\mathbf{x}}(t) = \mathbf{x} + t\boldsymbol{\lambda}$. For $t \geq 0$, we have $\mathbf{c}^T\hat{\mathbf{x}}(t) \geq \mathbf{c}^T\mathbf{x}$ and $A\hat{\mathbf{x}}(t) = \mathbf{b}$. For t sufficiently small, $\hat{\mathbf{x}}(t)$ is also nonnegative and thus feasible.

If there is $j \in S$ such that $\lambda_j < 0$, then there is a positive t such that $\hat{\mathbf{x}}(t)$ is feasible with strictly more zeros than $\mathbf{x}$, so we can take $\mathbf{x}' = \hat{\mathbf{x}}(t)$.

The same conclusion holds if $\lambda_j \geq 0$ for all j and $\mathbf{c}^T\boldsymbol{\lambda} = 0$; simply negate $\boldsymbol{\lambda}$ and apply the previous argument.

To complete the argument, we show that the previous two cases are exhaustive: if $\lambda_j \geq 0$ for all j and $\mathbf{c}^T\boldsymbol{\lambda} > 0$, then $\hat{\mathbf{x}}(t) \geq 0$ for all $t \geq 0$ and $\lim_{t \to \infty} \mathbf{c}^T\hat{\mathbf{x}}(t) = \infty$, contradicting the assumption that the objective value is bounded on $\mathcal{F}(\mathrm{P}_=)$. $\square$

PROOF OF THEOREM A.2.8. Part (i): Lemma A.2.11 shows that if the linear program

$$\text{maximize } \mathbf{c}^T\mathbf{x} \text{ subject to } \mathbf{x} \in \mathcal{F}(\mathrm{P_=})$$

is feasible and bounded, then for every feasible solution, there is a vertex with at least that objective value. Thus, we can search for the optimum of the linear program by considering only vertices of $\mathcal{F}(\mathrm{P_=})$. Since there are only finitely many, the optimum is attained.

Part (ii): We apply the reduction from Part (2) of the Farkas' Lemma to show that the linear program (P) is equivalent to a program of the type considered in Part (1) with a matrix $(A; I)$ in place of A. □

A.3. Notes

There are many books on linear programming; for an introduction, see [MG07]. As mentioned in the notes to Chapter 2, linear programming duality is due to von Neumann.

Linear programming problems were first formulated by Fourier in 1827. In 1939, Kantorovich [Kan39] published a book on applications of linear programming and suggested algorithms for their solution. He shared the 1975 Nobel Prize in Economics with T. Koopmans "for their contributions to the theory of optimum allocation of resources."

Dantzig's development of the Simplex Algorithm in 1947 (see [Dan51a]) was significant as, for many real-world linear programming problems, it was computationally efficient. Dantzig was largely motivated by the need to solve planning problems for the air force. See [Dan82] for the history of linear programming up to 1982. In 1970, Klee and Minty [KM72] showed that the Simplex Algorithm could require exponential time in the worst case. Leonid Khachiyan [Kha80] developed the first polynomial time algorithm in 1979; *The New York Times* referred to this paper as "the mathematical Sputnik." In 1984, Narendra Karmarkar [Kar84] introduced a more efficient algorithm using interior point methods. These are just a few of the high points in the extensive research literature on linear programming.

We showed that the Minimax Theorem follows from linear programming duality. In fact, the converse also holds; see, e.g., [Dan51b, Adl13].

Exercises

A.1.   Prove that linear programs (A.1) and (A.2) are dual to each other.

A.2.   Prove Theorem 17.1.1 using linear programming duality and the Birkhoff – von Neumann Theorem (Exercise 3.4).

APPENDIX B

Some useful probability tools

B.1. The second moment method

LEMMA B.1.1. *Let X be a nonnegative random variable. Then*

$$\mathbb{P}(X > 0) \geq \frac{(\mathbb{E}[X])^2}{\mathbb{E}[X^2]}.$$

PROOF. The lemma follows from this version of the Cauchy-Schwarz inequality:

$$(\mathbb{E}[XY])^2 \leq \mathbb{E}[X^2]\,\mathbb{E}[Y^2]. \tag{B.1}$$

Applying (B.1) to X and $Y = \mathbb{1}_{X>0}$, we obtain

$$(\mathbb{E}[X])^2 \leq \mathbb{E}[X^2]\,\mathbb{E}[Y^2] = \mathbb{E}[X^2]\,\mathbb{P}(X > 0).$$

Finally, we prove (B.1). Without loss of generality $\mathbb{E}[X^2]$ and $\mathbb{E}[Y^2]$ are both positive. Letting $U = X/\sqrt{\mathbb{E}[X^2]}$ and $V = Y/\sqrt{\mathbb{E}[Y^2]}$ and using the fact that $2UV \leq U^2 + V^2$, we obtain

$$2\mathbb{E}[UV] \leq \mathbb{E}[U^2] + \mathbb{E}[V^2] = 2.$$

Therefore,

$$(\mathbb{E}[UV])^2 \leq 1,$$

which is equivalent to (B.1). $\qquad\square$

B.2. The Hoeffding-Azuma Inequality

LEMMA B.2.1 (Hoeffding Lemma [Hoe63]). *Suppose that X is a random variable with distribution F such that $a \leq X \leq a + 1$ for some $a \leq 0$ and $\mathbb{E}[X] = 0$. Then for any $\lambda \in \mathbb{R}$,*

$$\mathbb{E}\left[e^{\lambda X}\right] \leq e^{\lambda^2/8}.$$

PROOF. This proof is from [BLM13]. Let

$$\Psi(\lambda) = \log \mathbb{E}\left[e^{\lambda X}\right].$$

Observe that

$$\Psi'(\lambda) = \frac{\mathbb{E}\left[Xe^{\lambda X}\right]}{\mathbb{E}\left[e^{\lambda X}\right]} = \int x\,dF_\lambda,$$

where

$$F_\lambda(u) = \frac{\int_{-\infty}^u e^{\lambda x}\,dF}{\int_{-\infty}^\infty e^{\lambda x}\,dF}.$$

Also,

$$\Psi''(\lambda) = \frac{\mathbb{E}\left[e^{\lambda X}\right]\mathbb{E}\left[X^2 e^{\lambda X}\right] - \left(\mathbb{E}\left[Xe^{\lambda X}\right]\right)^2}{\left(\mathbb{E}\left[e^{\lambda X}\right]\right)^2}$$

$$= \int x^2\,dF_\lambda - \left(\int x\,dF_\lambda\right)^2 = \mathrm{Var}(X_\lambda),$$

where X_λ has law F_λ.

For any random variable Y with $a \leq Y \leq a + 1$, we have

$$\mathrm{Var}(Y) \leq \mathbb{E}\left[\left(Y - a - \frac{1}{2}\right)^2\right] \leq \frac{1}{4}.$$

In particular,

$$|\Psi''(\lambda)| \leq 1/4$$

for all λ. Since $\Psi(0) = \Psi'(0) = 0$, it follows that $|\Psi'(\lambda)| \leq \frac{|\lambda|}{4}$, and thus

$$\Psi(\lambda) \leq \left|\int_0^\lambda \frac{\theta}{4} d\theta\right| = \frac{\lambda^2}{8}$$

for all λ. $\square$

THEOREM B.2.2 (Hoeffding-Azuma Inequality [Hoe63]). *Let $S_t = \sum_{i=1}^t X_i$ be a martingale; i.e., $\mathbb{E}[S_{t+1}|H_t] = S_t$ where $H_t = (X_1, X_2, \ldots, X_t)$ represents the history. If all $|X_t| \leq 1$, then*

$$\mathbb{P}[S_t \geq R] \leq e^{-R^2/2t}.$$

PROOF. Since $-\frac{1}{2} \leq \frac{X_{t+1}}{2} \leq \frac{1}{2}$, the previous lemma gives

$$\mathbb{E}\left[e^{\lambda X_{t+1}}|H_t\right] \leq e^{\frac{(2\lambda)^2}{8}} = e^{\lambda^2/2},$$

so

$$\mathbb{E}\left[e^{\lambda S_{t+1}}|H_t\right] = e^{\lambda S_t}\mathbb{E}\left[e^{\lambda X_{t+1}}|H_t\right] \leq e^{\lambda^2/2}e^{\lambda S_t}.$$

Taking expectations,

$$\mathbb{E}\left[e^{\lambda S_{t+1}}\right] \leq e^{\lambda^2/2}\mathbb{E}\left[e^{\lambda S_t}\right],$$

so by induction on t

$$\mathbb{E}\left[e^{\lambda S_t}\right] \leq e^{t\lambda^2/2}.$$

Finally, by Markov's Inequality,

$$\mathbb{P}[S_t \geq R] = \mathbb{P}\left[e^{\lambda S_t} \geq e^{\lambda R}\right] \leq e^{-\lambda R}e^{t\lambda^2/2}.$$

Optimizing, we choose $\lambda = R/t$, so

$$\mathbb{P}[S_t \geq R] \leq e^{-R^2/2t}.$$ $\square$

APPENDIX C

Convex functions

We review some basic facts about convex functions:

(1) A function $f : [a, b] \to \mathbb{R}$ is convex if for all $x, z \in [a, b]$ and $\alpha \in (0, 1)$ we have

$$f(\alpha x + (1 - \alpha)z) \leq \alpha f(x) + (1 - \alpha)f(z). \tag{C.1}$$

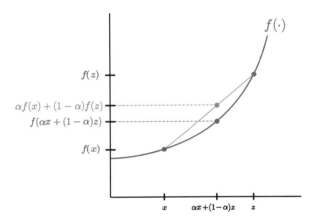

FIGURE C.1. A convex function f.

(2) The definition implies that the supremum of any family of convex functions is convex.

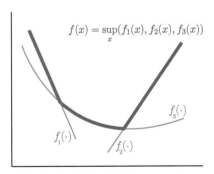

FIGURE C.2. The supremum of three convex functions.

(3) For $x < y$ in $[a, b]$ denote by $S(x, y) = \frac{f(y) - f(x)}{y - x}$ the slope of f on $[x, y]$. Convexity of f is equivalent to the inequality

$$S(x, y) \leq S(y, z)$$

holding for all $x < y < z$ in $[a, b]$.

(4) For $x < y < z$, the inequality in (3) is equivalent to $S(x, y) \leq S(x, z)$ and to $S(x, z) \leq S(y, z)$. Thus, for f convex in $[a, b]$, the slope $S(x, y)$ is (weakly) monotone increasing in x and in y as long as x, y are in $[a, b]$. This implies continuity of f in (a, b).

(5) It follows from (3) and the Mean Value Theorem that if f is continuous in $[a, b]$ and has a (weakly) increasing derivative in (a, b), then f is convex in $[a, b]$.

(6) The monotonicity in (4) implies that a convex function f in $[a, b]$ has an increasing right derivative f'_+ in $[a, b)$ and an increasing left derivative f'_- in $(a, b]$. Since $f'_+(x) \leq f'_-(y)$ for any $x < y$, we infer that f is differentiable at every point of continuity in (a, b) of f'_+.

(7) Since increasing functions can have only countably many discontinuities, a convex function is differentiable with at most countably many exceptions. The convex function $f(x) = \sum_{n \geq 1} |x - 1/n|/n^2$ indeed has countably many points of nondifferentiability.

(8) *Definition:* We say that $s \in \mathbb{R}$ is a **subgradient** of f at x if

$$f(y) \geq f(x) + s \cdot (y - x) \qquad \forall y \in [a, b]. \tag{C.2}$$

The right-hand side as a function of y is called a **supporting line** of f at x. See Figure C.3 and Figure C.4

(9) If $s(t)$ is a subgradient of f at t for each $t \in [a, b]$, then

$$s(y)(x - y) \leq f(x) - f(y) \leq s(x)(x - y) \quad \forall x, y \in [a, b]. \tag{C.3}$$

These inequalities imply that $s(\cdot)$ is weakly increasing and $f(\cdot)$ is continuous on $[a, b]$.

(10) *Fact:* Let $f : [a, b] \to \mathbb{R}$ be any function. Then f has a subgradient for all $x \in [a, b]$ if and only if f is convex and continuous on $[a, b]$.
Proof:
$\Longrightarrow$: f is the supremum of affine functions (the supporting lines). Continuity at the endpoints follows from the existence of a subgradient at these points.
$\Longleftarrow$: By (4), any $s \in [f'_-(x), f'_+(x)]$ is a subgradient.

(11) *Proposition:* If $s(x)$ is a subgradient of f at x for every $x \in [a, b]$, then

$$f(t) = f(a) + \int_a^t s(x) \, dx \qquad \forall t \in [a, b].$$

Proof: By translation, we may assume that $a = 0$. Fix $t \in (0, b]$ and $n > 1$. Define

$$t_k := \frac{kt}{n}.$$

For $x \in [t_{k-1}, t_k)$, define

$$g_n(x) = s(t_{k-1}) \quad \text{and} \quad h_n(x) = s(t_k).$$

Then $g_n(\cdot) \leq s(\cdot) \leq h_n(\cdot)$ in $[0, t)$, so

$$\int_0^t g_n(x) \, dx \leq \int_0^t s(x) \, dx \leq \int_0^t h_n(x) \, dx. \tag{C.4}$$

By (C.3),

$$\frac{t}{n} s(t_{k-1}) \leq f(t_k) - f(t_{k-1}) \leq \frac{t}{n} s(t_k) \quad \forall k \in [1, n].$$

Summing over $k \in [1, n]$ yields

$$\int_0^t g_n(x)\, dx \leq f(t) - f(0) \leq \int_0^t h_n(x)\, dx. \qquad \text{(C.5)}$$

Direct calculation gives that

$$\int_0^t h_n(x)\, dx - \int_0^t g_n(x)\, dx = [s(t) - s(0)]\frac{t}{n},$$

so by (C.4) and (C.5), we deduce that

$$\left| f(t) - f(0) - \int_0^t s(x)\, dx \right| \leq [s(t) - s(0)]\frac{t}{n}.$$

Taking $n \to \infty$ completes the proof.

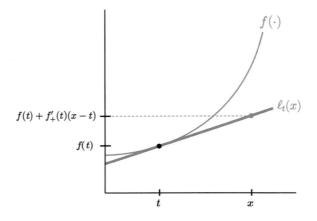

FIGURE C.3. The line $\ell_t(\cdot)$ is a supporting line at t and $f'_+(t)$ is a subgradient of f at t.

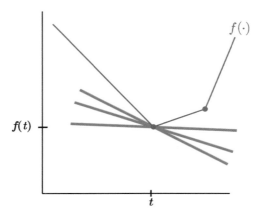

FIGURE C.4. A collection of supporting lines at t.

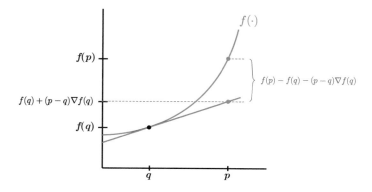

FIGURE C.5

(12) **Jensen's inequality:** If $f : [a, b] \to \mathbb{R}$ is convex and X is a random variable taking values in $[a, b]$, then $f(\mathbb{E}\,[X]) \leq \mathbb{E}\,[f(X)]$. (Note that for X taking just two values, this is the definition of convexity.)

Proof: Let $\ell(\cdot)$ be a supporting line for f at $\mathbb{E}\,[X]$. Then by linearity of expectation,
$$f(\mathbb{E}\,[X]) = \ell(\mathbb{E}\,[X]) = \mathbb{E}\,[\ell(X)] \leq \mathbb{E}\,[f(X)].$$

(13) The definition (C.1) of convex functions extends naturally to any function defined on a convex set K in a vector space. Observe that the function $f : K \to R$ is convex if and only if for any $\mathbf{x}, \mathbf{y} \in K$, the function
$$\Psi(t) = f(t\mathbf{x} + (1 - t)\mathbf{y})$$
is convex on $[0, 1]$. It follows that
$$\Psi(1) \geq \Psi(0) + \Psi'_+(0);$$
i.e., for all $\mathbf{x}, \mathbf{y} \in K$,
$$f(\mathbf{x}) \geq f(\mathbf{y}) + \nabla f(\mathbf{y}) \cdot (\mathbf{x} - \mathbf{y}). \tag{C.6}$$

A vector $\mathbf{v} \in \mathbb{R}^n$ is a **subgradient** of a convex function $f : \mathbb{R}^n \to \mathbb{R}$ at $\mathbf{y}$ if for all $\mathbf{x}$
$$f(\mathbf{x}) \geq f(\mathbf{y}) + \mathbf{v} \cdot (\mathbf{x} - \mathbf{y}).$$
If f is differentiable at $\mathbf{y}$, then the only subgradient is $\nabla f(\mathbf{y})$.

APPENDIX D

Solution sketches for selected exercises

Chapter 2

2.a.   Consider the betting game with the following payoff matrix:

player II

	L	R
T	0	2
B	5	1

player I

Draw graphs for this game analogous to those shown in Figure 2.2, and determine the value of the game.

Solution sketch.

Suppose that player I plays T with probability x_1 and B with probability $1 - x_1$, and suppose that player II plays L with probability y_1 and R with probability $1 - y_1$. (We note that in this game there is no saddle point.)

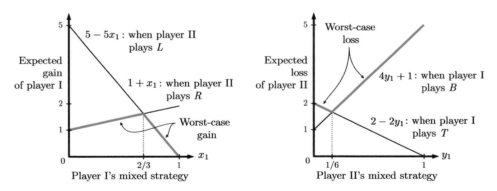

FIGURE D.1. The left side of the figure shows the worst-case expected gain of player I as a function of her mixed strategy (where she plays T with probability x_1 and B with probability $1 - x_1$). This worst-case expected gain is maximized when she plays T with probability $2/3$ and B with probability $1/3$. The right side of the figure shows the worst-case expected loss of player II as a function of his mixed strategy when he plays L with probability y_1 and R with probability $1 - y_1$. The worst-case expected loss is minimized when he plays L with probability $1/6$ and R with probability $5/6$.

Reasoning from player I's perspective, her expected gain is $2(1 - y_2)$ for playing the pure strategy T, and $4y_2 + 1$ for playing the pure strategy B. Thus, if she knows y_2, she will pick the strategy corresponding to the maximum of $2(1 - y_2)$ and $4y_2 + 1$. Player II can choose $y_2 = 1/6$ so as to minimize this maximum, and the expected amount player II will pay player I is $5/3$. This is the player II strategy that minimizes his worst-case loss. See Figure D.1 for an illustration.

From player II's perspective, his expected loss is $5(1 - x_1)$ if he plays the pure strategy L and $1 + x_1$ if he plays the pure strategy R, and he will aim to minimize this expected payout. In order to maximize this minimum, player I will choose $x_1 = 2/3$, which again yields an expected gain of $5/3$.

2.c.   Prove that if equation (2.9) holds, then player I can safely ignore row i.

Solution sketch. Consider any mixed strategy $\mathbf{x}$ for player I, and use it to construct a new strategy $\mathbf{z}$ in which $z_i = 0$, $z_\ell = x_\ell + \beta_\ell x_i$ for $\ell \in I$, and $z_k = x_k$ for $k \notin I \cup \{i\}$. Then, against player II's j-th strategy

$$(\mathbf{z}^T A - \mathbf{x}^T A)_j = \sum_{\ell \in I} (x_\ell + \beta_\ell x_i - x_\ell) a_{\ell j} - x_i a_{ij} \geq 0.$$

2.e.   Two players each choose a number in $[0, 1]$. If they choose the same number, the payoff is 0. Otherwise, the player that chose the lower number pays \$1 to the player who chose the higher number, unless the higher number is 1, in which case the payment is reversed. Show that this game has no mixed Nash equilibrium. Show that the safety values for players I and II are -1 and 1 respectively.

Solution sketch. Given a mixed strategy $F \in \Delta$ for player I and given $\epsilon > 0$, find a point α such that $F(\alpha) > 1 - \epsilon$. Then taking G supported in $[\alpha, 1)$ yields a payoff of at least $1 - \epsilon$ to player II.

2.6.   Given a 5×5 zero-sum game, such as the following, how would you quickly determine by hand if it has a saddle point:

$$\begin{pmatrix} 20 & 1 & 4 & 3 & 1 \\ 2 & 3 & 8 & 4 & 4 \\ 10 & 8 & 7 & 6 & 9 \\ 5 & 6 & 1 & 2 & 2 \\ 3 & 7 & 9 & 1 & 5 \end{pmatrix} ?$$

Solution sketch. A simple approach is to "star" the maximum element in each column and underline the minimum element in each row. (If there is more than one, star/underline all of them.) Any element in the matrix that is both starred and underlined is a saddle point. In the example below, the 6 at position $(3, 4)$ is a saddle point.

$$\begin{pmatrix} 20^* & \underline{1} & 4 & 3 & \underline{1} \\ \underline{2} & 3 & 8 & 4 & 4 \\ 10 & 8^* & 7 & \underline{6}^* & 9^* \\ 5 & 6 & \underline{1} & 2 & 2 \\ 3 & 7 & 9^* & \underline{1} & 5 \end{pmatrix}$$

2.17.   Consider a directed graph $G = (V, E)$ with nonnegative weights w_{ij} on each edge (i, j). Let $W_i = \sum_j w_{ij}$. Each player chooses a vertex, say i for player I and j for player II. Player I receives a payoff of w_{ij} if $i \neq j$ and loses $W_i - w_{ii}$ if $i = j$. Thus, the payoff matrix A has entries $a_{ij} = w_{ij} - 1_{\{i=j\}} W_i$. If $n = 2$ and the w_{ij}'s are all 1, this game is called Matching Pennies.

- Show that the game has value 0.
  Solution sketch. $\sum_j a_{ij} = 0$ for all i, so by giving all vertices equal weight, player II can ensure a loss of at most 0.
  Conversely, for any strategy $\mathbf{y} \in \Delta_n$ for player II, player I can select action i with $y_i = \min_k y_k$, yielding a payoff of
  $$\sum_j a_{ij} y_j = \sum_j w_{ij} (y_j - y_i) \geq 0.$$

- Deduce that for some $x \in \Delta_n$, $\mathbf{x}^T A = 0$.
  Solution sketch. By the Minimax Theorem, $\exists \mathbf{x} \in \Delta_n$ with $\mathbf{x} A \geq 0$. Since $\mathbf{x} A 1 = 0$, we must have $\mathbf{x} A = 0$.

2.19.   Prove that if set $G \subseteq \mathbb{R}^d$ is compact and $H \subseteq \mathbb{R}^d$ is closed, then $G + H$ is closed. (This fact is used in the proof of the Minimax Theorem to show that the set K is closed.)
Solution sketch. Suppose that $x_n + y_n \to z$, where $x_n \in G$ and $y_n \in H$ for all n. Then there is a subsequence $x_{n_k} \to x \in G$; we infer that $y_{n_k} \to z - x$, whence $z - x \in H$.

2.20.   Find two sets $F_1, F_2 \subset \mathbb{R}^2$ that are closed such that $F_1 - F_2$ is not closed.

Solution sketch. $F_1 = \{xy \geq 1\}$, $F_2 = \{x = 0\}$, $F_1 + F_2 = \{x > 0\}$.

2.21.   Consider a zero-sum game A and suppose that π and σ are permutations of player I's strategies $\{1, \dots, m\}$ and player II's strategies $\{1, \dots, n\}$, respectively, such that
$$a_{\pi(i)\sigma(j)} = a_{ij} \tag{D.1}$$
for all i and j. Show that there exist optimal strategies $\mathbf{x}^*$ and $\mathbf{y}^*$ such that $x_i^* = x_{\pi(i)}^*$ for all i and $y_j^* = y_{\sigma(j)}^*$ for all j.

Solution sketch. First, observe that there is an ℓ such that π^ℓ is the identity permutation (since there must be $k > r$ with $\pi^k = \pi^r$, in which case $\ell = k - r$). Let $(\pi \mathbf{x})_i = x_{\pi(i)}$ and $(\sigma \mathbf{y})_j = y_{\sigma(j)}$.

Let $\Psi(\mathbf{x}) = \min_{\mathbf{y}} \mathbf{x}^T A \mathbf{y}$. Since $(\pi \mathbf{x})^T A(\sigma \mathbf{y}) = \mathbf{x}^T A \mathbf{y}$, we have $\Psi(\mathbf{x}) = \Psi(\pi \mathbf{x})$ for all $\mathbf{x} \in \Delta_m$. Therefore, for all $\mathbf{y} \in \Delta_n$

$$\left(\frac{1}{\ell} \sum_{k=0}^{\ell-1} \pi^k \mathbf{x} \right)^T A \mathbf{y} \geq \frac{1}{\ell} \sum_{k=0}^{\ell-1} \Psi(\pi^k \mathbf{x}) = \Psi(\mathbf{x}).$$

Thus, if $\mathbf{x}$ is optimal, so is $\mathbf{x}^* = \frac{1}{\ell} \sum_{k=0}^{\ell-1} \pi^k \mathbf{x}$. Clearly $\pi \mathbf{x}^* = \mathbf{x}^*$.

2.22.   Player I chooses a positive integer $x > 0$ and player II chooses a positive integer $y > 0$. The player with the lower number pays a dollar to the player with the higher number unless the higher number is more than twice larger in which case the payments are reversed.

$$A(x, y) = \begin{cases} 1 & \text{if } y < x \leq 2y \text{ or } x < y/2, \\ -1 & \text{if } x < y \leq 2x \text{ or } y < x/2, \\ 0 & \text{if } x = y. \end{cases}$$

Find the unique optimal strategy in this game.

Solution sketch. 1 strictly dominates any $x > 4$, and 4 strictly dominates 3. Restricting to 1, 2, and 4, we get Rock-Paper-Scissors.

Chapter 4

4.6.   Show that there is no pure Nash equilibrium, only a unique mixed one, and both commitment strategy pairs have the property that the player who did not make the commitment still gets the Nash equilibrium payoff.

player II

		C	D
player I	A	$(6, -10)$	$(0, 10)$
	B	$(4, 1)$	$(1, 0)$

Solution sketch. In this game, there is no pure Nash equilibrium (one of the players always prefers another strategy, in a cyclic fashion). For mixed strategies, if player I plays (A, B) with probabilities $(p, 1 - p)$ and player II plays (C, D) with probabilities $(q, 1 - q)$, then the expected payoffs are $1 + 3q - p + 3pq$ for player I and $10p + q - 21pq$ for player II. We easily get that the unique mixed equilibrium is $p = 1/21$ and $q = 1/3$, with payoffs 2 for player I and $10/21$ for player II.

If player I can make a commitment, then by choosing $p = 1/21 - \varepsilon$ for some small $\varepsilon > 0$, she will make player II choose $q = 1$, and the payoffs will be $4 + 2/21 - 2\varepsilon$ for player I and $10/21 + 11\varepsilon$ for player II. If player II can make a commitment, then by choosing $q = 1/3 + \varepsilon$, he will make player I choose $p = 1$, and the payoffs will be $2 + 6\varepsilon$ for player I and $10/3 - 11\varepsilon$ for player II. Notice that in both of these commitment strategy pairs, the player who did not make the commitment gets a larger payoff than he does in the symmetric Nash equilibrium.

Chapter 5

5.6.   Show that any d-simplex in $\mathbb{R}^d$ contains a ball.

Solution sketch. The d-simplex Δ_0 with vertices the origin and the standard basis $\mathbf{e}_1, \ldots, \mathbf{e}_d$ in $\mathbb{R}^d$ contains the ball $B(\mathbf{y}, \frac{1}{2d})$, where $\mathbf{y} := \frac{1}{2d}(\mathbf{e}_1 + \cdots + \mathbf{e}_d)$. Given an arbitrary d-simplex Δ, by translation we may assume its vertices are $0, \mathbf{v}_1, \ldots, \mathbf{v}_d$. Let A be the square matrix with columns $\mathbf{v}_i$ for $i \le d$. Since these columns are linearly independent, A is invertible. Then Δ contains $B(A\mathbf{y}, \epsilon)$, where $\epsilon := \min\{\|Ax\| \text{ such that } \|x\| = 1/d\} > 0$.

5.7.   Let $K \subset \mathbb{R}^d$ be a compact convex set which contains a d-simplex. Show that K is homeomorphic to a closed ball.

Solution sketch. Suggested steps:

(i) By Exercise 5.6, K contains a ball $B(z, \epsilon)$. By translation, assume without loss of generality that $B(0, \epsilon) \subset K$.

(ii) Show that $\rho : \mathbb{R}^d \to \mathbb{R}$ defined by

$$\rho(x) = \inf\{r > 0 \ : \ \frac{x}{r} \in K\}$$

is subadditive (i.e., $\rho(x+y) \le \rho(x) + \rho(y)$) and satisfies

$$\frac{\|x\|}{\text{diam}(K)} \le \rho(x) \le \frac{\|x\|}{\epsilon}$$

for all $\mathbf{x}$. Deduce that ρ is continuous.

Solution of step (ii): Suppose $\frac{x}{r} \in K$ and $\frac{y}{s} \in K$, where $r, s \ge 0$. Then

$$\frac{x+y}{r+s} = \frac{r}{r+s} \cdot \frac{x}{r} + \frac{s}{r+s} \cdot \frac{y}{s} \in K,$$

so $\rho(x+y) \le r + s$. Therefore $\rho(x+y) \le \rho(x) + \rho(y)$, from which

$$\rho(x+y) - \rho(x) \le \frac{\|y\|}{\epsilon}.$$

Similarly,

$$\rho(x) - \rho(x+y) \le \rho(x+y-y) - \rho(x+y) \le \frac{\|y\|}{\epsilon}.$$

(iii) Define

$$h(x) = \frac{\rho(x)}{\|x\|} x$$

for $\mathbf{x} \ne 0$ and $h(0) = 0$ and show that $h : K \to \overline{B(0,1)}$ is a homeomorphism.

Chapter 6

6.3.   Consider the zero-sum two-player game in which the game to be played is randomized by a fair coin toss. (This example was discussed in §2.5.1.) If the toss comes up heads, the payoff matrix is given by A^H, and if tails, it is given by A^T:

$$A^H =$$

player II

	L	R
U	8	2
D	6	0

player I

and

$$A^T =$$

player II

	L	R
U	2	6
D	4	10

player I

For each of the settings below, draw the Bayesian game tree, convert to normal form, and find the value of the game.

(a) Suppose that player I is told the result of the coin toss and both players play simultaneously.

(b) Suppose that player I is told the result of the coin toss, but she must reveal her move first.

Solution sketch.

(a) In what follows U_H (respectively, U_T) means that player I plays U if the coin toss is heads (respectively, tails), and D_H (respectively, D_T) means that the play I plays D if the coin toss is heads (respectively, tails).

player II

	L	R
U_H, U_T	5	4
U_H, D_T	6	6
D_H, U_T	4	3
D_H, D_T	3	5

player I

The value of the game is 6, since row 2 dominates all other rows.

(b) A column strategy such as L_U, L_D means that that player II plays L regardless of what move player I reveals, whereas strategy L_U, R_D means that player II plays L if player I reveals U, but plays R if player I reveals D.

player II

	L_U, L_D	L_U, R_D	R_U, R_D	R_U, L_D
U_H, U_T	5	5	4	4
U_H, D_T	6	9	6	3
D_H, U_T	4	1	3	6
D_H, D_T	5	5	5	5

player I

The value of the game is 5. Clearly the value of the game is at least 5, since player I can play the pure strategy D_H, D_T. To see that it is at most 5, observe that row 1 is dominated by row 4 for player I and column 1 is dominated by column 3 for player II. By playing L_U, R_D with probability 0.5 and R_U, R_D with probability 0.5, player II can ensure that player I's payoff is at most 5.

Chapter 7

7.2.   Consider the following symmetric game as played by two drivers, both trying to get from Here to There (or two computers routing messages along cables of different bandwidths). There are two routes from Here to There; one is wider, and therefore faster, but congestion will slow them down if both take the same route. Denote the wide route W and the narrower route N. The payoff matrix is

player II (yellow)

		W	N
player I (red)	W	$(3,3)$	$(5,4)$
	N	$(4,5)$	$(2,2)$

FIGURE D.2. The left-most image shows the payoffs when both drivers drive on the narrower route, the middle image shows the payoffs when both drivers drive on the wider route, and the right-most image shows what happens when the red driver (player I) chooses the wide route and the yellow driver (player II) chooses the narrow route.

Find all Nash equilibria and determine which ones are evolutionarily stable.

Solution sketch. There are two pure Nash equilibria: (W, N) and (N, W).

If player I chooses W with probability x, player II's payoff for choosing W is $3x + 5(1 - x)$, and for choosing N is $4x + 2(1 - x)$. Equating these, we get that the symmetric Nash equilibrium is when both players take the wide route with probability $x = 0.75$, resulting in an expected payoff of 3.5 for both players.

Is this an evolutionarily stable equilibrium? Let $\mathbf{x} = (.75, .25)$ be our equilibrium strategy. We already checked that $\mathbf{x}^T A \mathbf{x} = \mathbf{z}^T A \mathbf{x}$ for all pure strategies $\mathbf{z}$;

we need only check that $\mathbf{z}^T A \mathbf{z} < \mathbf{x}^T A \mathbf{z}$. For $\mathbf{z} = (1,0)$, $\mathbf{x}^T A \mathbf{z} = 3.25 > \mathbf{z}^T A \mathbf{z} = 3$ and for $\mathbf{z} = (0,1)$, $\mathbf{x}^T A \mathbf{z} = 4.25 > \mathbf{z}^T A \mathbf{z} = 2$, implying that $\mathbf{x}$ is evolutionarily stable.

Chapter 8

8.7.   Show that the price of anarchy bound for the market sharing game from §8.3 can be improved to $2 - 1/k$ when there are k teams. Show that this bound is tight.

Solution sketch. We know that S includes some top j cities and that S^* covers the top k. For each i, we have $u_i(c_i, \mathbf{c}_{-i}) \geq \frac{V(S^*) - V(S)}{k-j}$. So $\sum_i (k-j) u_i(c_i, \mathbf{c}_{-i}) \geq k(V(S^*) - V(S))$ or $kV(S^*) \leq (2k-j)V(S)$, so $V(S^*) \leq (2 - 1/k)V(S)$ since $j \geq 1$.

8.8.   Consider an auctioneer selling a single item via a first-price auction: Each of the n bidders submits a bid, say b_i for the ith bidder, and, given the bid vector $\mathbf{b} = (b_1, \ldots, b_n)$, the auctioneer allocates the item to the highest bidder at a price equal to her bid. (The auctioneer employs some deterministic tie-breaking rule.) Each bidder has a **value** v_i for the item. A bidder's **utility** from the auction when the bid vector is $\mathbf{b}$ and her value is v_i is

$$u_i[\mathbf{b}|v_i] := \begin{cases} v_i - b_i, & i \text{ wins the auction,} \\ 0, & \text{otherwise.} \end{cases}$$

Each bidder will bid in the auction so as to maximize her (expected) utility. The expectation here is over any randomness in the bidder strategies. The **social surplus** $V(\mathbf{b})$ of the auction is the sum of the utilities of the bidders and the auctioneer revenue. Since the auctioneer revenue equals the winning bid, we have

$$V(\mathbf{b}) := \text{value of winning bidder.}$$

Show that the price of anarchy is at most $1 - 1/e$; that is, for $\mathbf{b}$ a Nash equilibrium,

$$\mathbb{E}[V(\mathbf{b})] \geq \left(1 - \frac{1}{e}\right) \max_i v_i.$$

Hint: Consider what happens when bidder i deviates from b_i to the distribution with density $f(x) = 1/(v_i - x)$, with support $[0, (1 - 1/e)v_i]$.

Solution sketch. We first show that the price of anarchy is at most $1/2$. Suppose the n bidders have values $v_1, \ldots, v_n$ and their bids, in Nash equilibrium, are $\mathbf{b} = (b_1, \ldots, b_n)$. Without loss of generality, suppose that $v_1 \geq v_i$ for all i. Consider what happens when bidder 1 deviates from b_1 to $b_1^* := v_1/2$. We have

$$u_1[\mathbf{b}|v_1] \geq u_1[b_1^*, \mathbf{b}_{-i}|v_1] \geq \frac{v_1}{2} - \max_i b_i, \tag{D.2}$$

and since $u_i[\mathbf{b}|v_i] \geq 0$ for all $i \neq 1$, we have

$$\sum_i u_i[\mathbf{b}|v_i] \geq \frac{v_1}{2} - \max_i b_i.$$

On the other hand,

$$\sum_i u_i[\mathbf{b}|v_i] = v_{i^*} - \max_i b_i,$$

where i^* is the winning bidder, so

$$v_{i^*} \geq \frac{v_1}{2}.$$

To extend this to $1 - 1/e$, consider instead what happens when bidder 1 deviates from b_1 to the distribution with density $f(x) = 1/(v_1 - x)$, with support $[0, (1 - 1/e)v_1]$.

Let $p := \max_{i>1} b_i$. Then instead of (D.2), we get

$$u_1[b_1^*, \mathbf{b}_{-i}|v_1] \geq \int_p^{(1-1/e)v_1} (v_1 - x)f(x)\,dx \geq \left(1 - \frac{1}{e}\right)v_1 - \max_i b_i.$$

As above, from this we conclude that the value of the winning bidder v_{i^*} satisfies

$$v_{i^*} \geq \left(1 - \frac{1}{e}\right)v_1.$$

Chapter 11

11.1.   Show that the Talmud rule is monotone in A for all n and coincides with the garment rule for $n = 2$.

Solution sketch. The monotonicity of the Talmud rule follows from the monotonicity of CEA and CEL and the observation that if $A = C/2$, then $a_i = c_i/2$ for all i. Thus, if A is increased from a value that is at most $C/2$ to a value that is above $C/2$, then the allocation to each claimant i also increases from below $c_i/2$ to above $c_i/2$.

Suppose that $n = 2$, and let $c_1 \leq c_2$. We consider the two cases:

- $A \leq C/2$: Then $c_2 \geq A$.
    - If also $c_1 \geq A$, then $(a_1, a_2) = (A/2, A/2)$.
    - If $c_1 < A$, then $a_1 = \frac{c_1}{2}$ and $a_2 = A - \frac{c_1}{2} \leq \frac{c_2}{2}$.
  In both cases, this is the CEA rule with claims $(\frac{c_1}{2}, \frac{c_2}{2})$.
- $A > C/2$: Then $c_1 < A$.
    - If $c_2 \geq A$, then $a_1 = c_1/2$ and $a_2 = A - \frac{c_1}{2}$, so $\ell_1 = \frac{c_1}{2}$ and $\ell_2 = c_2 - A + \frac{c_1}{2} \leq \frac{c_2}{2}$.
    - If $c_2 < A$, then $a_1 = \frac{c_1}{2} + \frac{1}{2}(A - c_2)$ and $a_2 = \frac{c_2}{2} + \frac{1}{2}(A - c_1)$, so $\ell_1 = \ell_2 = \frac{C-A}{2}$.
  In both cases, this is the CEL rule with claims $(\frac{c_1}{2}, \frac{c_2}{2})$.

Chapter 13

13.4.   Show that Instant Runoff violates the Condorcet winner criterion, IIA with preference strengths, and cancellation of ranking cycles.

Solution sketch. If 40% of the population has preference order $A \succ B \succ C$, 40% of the population has $C \succ B \succ A$, and 20% of the population has $B \succ A \succ C$, then B is a Condorcet winner but loses an IRV election. To see that IRV violates IIA with preference strength, consider what happens when C is moved to the bottom for the second 40% group. To see that IRV violates cancellation of ranking cycles, suppose that 40% of the population has preference order $A \succ B \succ C$, 35% of the population has $B \succ C \succ A$, and 25% of the population has $C \succ A \succ B$. Then A is the winner of IRV, but eliminating the 75% of the population that are in a ranking cycle will change the winner to B.

Chapter 14

14.5.   Find a symmetric equilibrium in the war of attrition auction discussed in §14.4.3, under the assumption that bids are committed to up-front, rather than in the more natural setting where a player's bid (the decision as to how long to stay in) can be adjusted over the course of the auction.

Solution sketch. Let β be a symmetric strictly increasing equilibrium strategy. The expected payment $p(v)$ of an agent in a war-of-attrition auction in which all bidders use β is

$$p(v) = F(v)^{n-1}\mathbb{E}\left[\max_{i\leq n-1}\beta(V_i) \mid \max_{i\leq n-1}V_i \leq v\right] + (1 - F(v)^{n-1})\beta(v).$$

Equating this with $p(v)$ from (14.9), we have

$$\int_0^v (F(v)^{n-1} - F(w)^{n-1})dw = \int_0^v \beta(w)(n-1)F(w)^{n-2}f(w)dw + (1 - F(v)^{n-1})\beta(v).$$

Differentiating both sides with respect to v, cancelling common terms, and simplifying yields

$$\beta'(v) = \frac{(n-1)vF(v)^{n-2}f(v)}{1 - F(v)^{n-1}},$$

and hence

$$\beta(v) = \int_0^v \frac{(n-1)wF(w)^{n-2}f(w)}{1 - F(w)^{n-1}}dw.$$

For two players with F uniform on $[0,1]$ this yields

$$\beta(v) = \int_0^v \frac{w}{1-w}dw = -v - \log(1-v).$$

14.13.   Determine the explicit payment rule for the three tie-breaking rules just discussed.

Solution sketch. Fix a bidder, say 1. We consider all the possibilities for when bidder 1 might win and what his payment is in each case. Suppose that

$$\varphi = \max_{i\geq 2}\psi_i(b_i)$$

is attained k times by bidders $i \geq 2$. Let

$$[v_-(\varphi), v_+(\varphi)] = \{b \;:\; \psi_1(b) = \varphi\}$$

and

$$b_* = \max\{b_i \;:\; \psi_i(b_i) = \varphi,\; i \geq 2\}.$$

- Tie-breaking by bid:
    - If $\psi_1(b_1) > \varphi$, then bidder 1 wins and pays $\max\{b_*, v_-(\varphi)\}$.
    - If $\psi_1(b_1) = \varphi$ and b_1 is largest among those with virtual valuation at least φ, then bidder 1 wins and pays $\max\{b_*, v_-(\varphi)\}$.
- Tie-breaking according to a fixed ordering of bidders:
    - If $\psi_1(b_1) = \varphi$ and bidder 1 wins (has the highest rank), then his payment is $v_-(\varphi)$.
    - If $\psi_1(b_1) > \varphi$, then his payment is $v_-(\varphi)$ if he has the highest rank and $v_+(\varphi)$ otherwise.
- Random tie-breaking:
    - If $\psi_1(b_1) = \varphi$, then bidder 1 wins with probability $\frac{1}{k+1}$, and if bidder 1 wins, he is charged $v_-(\varphi)$.
    - If $\psi_1(b_1) > \varphi$, then bidder 1 wins, and he is charged

$$\frac{1}{k+1}v_-(\varphi) + \frac{k}{k+1}v_+(\varphi),$$

because in $\frac{1}{k+1}$ of the permutations he will be ranked above the other k bidders with virtual value φ.

14.14.   Consider two bidders where bidder 1's value is drawn from an exponential dis-
          tribution with parameter 1 and bidder 2's value is drawn independently from
          Unif$[0, 1]$. What is the Myerson optimal auction in this case? Show that if
          $(v_1, v_2) = (1.5, 0.8)$, then bidder 2 wins.

          Solution sketch. We first compute the virtual value functions:

          $$\psi_1(v_1) = v_1 - 1 \quad \text{and} \quad \psi_2(v_2) = v_2 - \frac{1 - v_2}{1} = 2v_2 - 1.$$

          Thus, bidder 1 wins when $v_1 - 1 \geq \max(0, 2v_2 - 1)$, whereas bidder 2 wins when
          $2v_2 - 1 \geq \max(0, v_1 - 1)$. If $(v_1, v_2) = (1.5, 0.8)$, then bidder 2 wins and pays
          $\psi_2^{-1}(\psi_1(1.5)) = 0.75$. This shows that in the optimal auction with non-i.i.d. bid-
          ders, the bidder with the highest value might not win.

Chapter 18

18.7.   (a) For Y a normal random variable, $N(0, 1)$, show that

          $$e^{-\frac{y^2}{2}(1+o(1))} \leq \mathbb{P}\left[Y > y\right] \leq e^{-\frac{y^2}{2}} \quad \text{as} \quad y \to \infty.$$

          (b) Suppose that $Y_1, \ldots, Y_n$ are i.i.d. $N(0, 1)$ random variables. Show that

          $$\mathbb{E}\left[\max_{1 \leq i \leq n} Y_i\right] = \sqrt{2 \log n}\,(1 + o(1)) \quad \text{as} \quad n \to \infty. \tag{D.3}$$

          Solution sketch.
          (a)
          $$\int_y^\infty e^{-x^2/2}\,dx \leq \frac{1}{y}\int_y^\infty x e^{-x^2/2}\,dx = \frac{1}{y}e^{-y^2/2}$$
          $$\text{and} \quad \int_y^{y+1} e^{-x^2/2}\,dx \geq e^{-\frac{(y+1)^2}{2}}.$$

          (b) Let $M_n = \mathbb{E}\left[\max_{1 \leq i \leq n} Y_i\right]$. Then by a union bound

          $$\mathbb{P}\left[M_n \geq \sqrt{2 \log n} + \frac{x}{\sqrt{2 \log n}}\right] \leq n e^{-(\log n + x)} = e^{-x}.$$

          On the other hand,

          $$\mathbb{P}\left[Y_i > \sqrt{2\alpha \log n}\right] = n^{-\alpha + o(1)},$$

          so $\mathbb{P}\left[M_n \geq \sqrt{2\alpha \log n}\right] = \left(1 - n^{-\alpha + o(1)}\right)^n \to 0 \quad \text{for } \alpha < 1.$

Bibliography

The bold number(s) at the end of a citation identify the page(s) on which the reference was cited.

[AABR09]   Jacob Abernethy, Alekh Agarwal, Peter L. Bartlett, and Alexander Rakhlin. A stochastic view of optimal regret through minimax duality. In *COLT 2009 — The 22nd Conference on Learning Theory,* 2009.

[AAE13]   Baruch Awerbuch, Yossi Azar, and Amir Epstein. The price of routing unsplittable flow. *SIAM J. Comput.,* 42(1):160–177, 2013. **155**

[AB10]   Jean-Yves Audibert and Sébastien Bubeck. Regret bounds and minimax policies under partial monitoring. *J. Mach. Learn. Res.,* 11:2785–2836, 2010. **320**

[ABC+09]   Yossi Azar, Benjamin Birnbaum, L. Elisa Celis, Nikhil R. Devanur, and Yuval Peres. Convergence of local dynamics to balanced outcomes in exchange networks. In *2009 50th Annual IEEE Symposium on Foundations of Computer Science (FOCS 2009),* pages 293–302. IEEE Computer Soc., Los Alamitos, CA, 2009. **300**

[ACBFS02] P. Auer, N. Cesa-Bianchi, Y. Freund, and R. Schapire. The nonstochastic multi-armed bandit problem. *SIAM Journal on Computing,* 32(1):48–77, 2002. **320**

[ADG91]   Ahmet Alkan, Gabrielle Demange, and David Gale. Fair allocation of indivisible goods and criteria of justice. *Econometrica: Journal of the Econometric Society,* pages 1023–1039, 1991. **300**

[ADK+08]   Elliot Anshelevich, Anirban Dasgupta, Jon Kleinberg, Eva Tardos, Tom Wexler, and Tim Roughgarden. The price of stability for network design with fair cost allocation. *SIAM Journal on Computing,* 38(4):1602–1623, 2008. **155**

[Adl13]   Ilan Adler. The equivalence of linear programs and zero-sum games. *International Journal of Game Theory,* 42(1):165–177, 2013. **331**

[ADT12]   Raman Arora, Ofer Dekel, and Ambuj Tewari. Online bandit learning against an adaptive adversary: From regret to policy regret. In John Langford and Joelle Pineau, editors, *Proceedings of the 29th International Conference on Machine Learning,* ICML '12, pages 1503–1510, New York, NY, USA, July 2012. Omnipress. **320**

[AFG+11]   Gagan Aggarwal, Amos Fiat, Andrew V. Goldberg, Jason D. Hartline, Nicole Immorlica, and Madhu Sudan. Derandomization of auctions. *Games Econom. Behav.,* 72(1):1–11, 2011. **274**

[AG03]   Steve Alpern and Shmuel Gal. *The theory of search games and rendezvous.* International Series in Operations Research & Management Science, 55. Kluwer Academic Publishers, Boston, MA, 2003. **60**

[AH81]   Robert Axelrod and William Donald Hamilton. The evolution of cooperation. *Science,* 211(4489):1390–1396, 1981. **125**

[AH92]   Robert J. Aumann and Sergiu Hart, editors. *Handbook of game theory with economic applications. Vol. I.* North-Holland Publishing Co., Amsterdam, 1992. **xxv**

[AH94]   Robert J. Aumann and Sergiu Hart, editors. *Handbook of game theory with economic applications. Vol. II.* North-Holland Publishing Co., Amsterdam, 1994. **218, 220**

[AHK12]   Sanjeev Arora, Elad Hazan, and Satyen Kale. The multiplicative weights update method: A meta-algorithm and applications. *Theory Comput.,* 8:121–164, 2012. **319**

[AK08]   Aaron Archer and Robert Kleinberg. Characterizing truthful mechanisms with
           convex type spaces. *ACM SIGecom Exchanges*, 7(3):5, 2008. **286**

[Ake70]   George A. Akerlof. The market for "lemons": Quality uncertainty and the market
           mechanism. *The Quarterly Journal of Economics*, pages 488–500, 1970. **83**

[Alo87]   Noga Alon. Splitting necklaces. *Adv. in Math.*, 63(3):247–253, 1987. **192**

[AM85]   Robert J. Aumann and Michael Maschler. Game theoretic analysis of a bankruptcy
           problem from the Talmud. *J. Econom. Theory*, 36(2):195–213, 1985. **193, 204**

[AM95]   Robert J. Aumann and Michael B. Maschler. *Repeated games with incomplete
           information*. MIT Press, Cambridge, MA, 1995. With the collaboration of Richard
           E. Stearns. **125**

[AM06]   Lawrence M. Ausubel and Paul Milgrom. The lovely but lonely Vickrey auction.
           In *Combinatorial Auctions, Chapter 1*, pages 22–26, 2006. **286**

[AM16]   Haris Aziz and Simon Mackenzie. A discrete and bounded envy-free cake cutting
           protocol for any number of agents. In FOCS 2016 — *Proceedings of the 7th Annual
           IEEE Symposium on Foundations of Computer Science*, IEEE Computer Soc.,
           2016. **193**

[AMS09]   Saeed Alaei, Azarakhsh Malekian, and Aravind Srinivasan. On random sampling
           auctions for digital goods. In *Proceedings of the 10th ACM Conference on Elec-
           tronic Commerce*, pages 187–196. ACM, 2009. **274**

[Ansa]   Hex information. http://www.cs.unimaas.nl/icga/games/hex/.

[Ansb]   V. V. Anshelevich. The game of Hex: An automatic theorem proving approach
           to game programming. http://home.earthlink.net/~vanshel/VAnshelevich-01.
           pdf.

[Arr51]   Kenneth J. Arrow. *Social Choice and Individual Values*. Cowles Commission Mono-
           graph, No. 12. John Wiley & Sons Inc., New York, N. Y., 1951. **220**

[Arr02]   A. Arratia. On the descriptive complexity of a simplified game of Hex. *Log. J.
           IGPL*, 10:105–122, 2002.

[ARS+03]   Micah Adler, Harald Räcke, Naveen Sivadasan, Christian Sohler, and Berthold
           Vöcking. Randomized pursuit-evasion in graphs. *Combin. Probab. Comput.*,
           12(3):225–244, 2003. **52, 60**

[AS94]   Robert J. Aumann and Lloyd S. Shapley. Long-term competition—a game-
           theoretic analysis. In *Essays in Game Theory (Stony Brook, NY, 1992)*, pages
           1–15. Springer, New York, 1994. **125**

[ASS02]   Kenneth J. Arrow, Amartya K. Sen, and Kotaro Suzumura, editors. *Handbook
           of social choice and welfare. Vol. 1*. Handbooks in Economics. North-Holland,
           Amsterdam, 2002. **218, 220**

[ASS11]   Kenneth J. Arrow, Amartya Sen, and Kotaro Suzumura, editors. *Handbook of
           social choice and welfare. Vol. 2*. Handbooks in Economics. North-Holland, Ams-
           terdam, 2011. **218, 220**

[ASÜ04]   Atila Abdulkadiroğlu, Tayfun Sönmez, and M. Utku Ünver. Room assignment-rent
           division: A market approach. *Social Choice and Welfare*, 22(3):515–538, 2004. **300**

[AT01]   Aaron Archer and Éva Tardos. Truthful mechanisms for one-parameter agents.
           In FOCS 2001 — *Proceedings of the 42nd IEEE Symposium on Foundations of
           Computer Science*, pages 482–491. IEEE Computer Soc., 2001.

[Aum74]   Robert J. Aumann. Subjectivity and correlation in randomized strategies. *J. Math.
           Econom.*, 1(1):67–96, 1974. **136**

[Aum81]   Robert J. Aumann. Survey of repeated games. *Essays in game theory and mathe-
           matical economics in honor of Oskar Morgenstern*, 1981. **125**

[Aum85]   Robert Aumann. Repeated games. *Issues in contemporary microeconomics and
           welfare*, pages 209–242, 1985. **125**

[Aum87]   Robert J. Aumann. Correlated equilibrium as an expression of Bayesian rationality.
           Econometrica, 55(1):1–18, Jan. 1987. **136**

[Axe84]   Robert Axelrod. *The Evolution of Cooperation*. Basic Books, 387 Park Avenue So.,
           New York, NY 10016, 1984. **83, 125**

[Axe97]   Robert Axelrod. The evolution of strategies in the iterated prisoner's dilemma. In
           The dynamics of norms, Cambridge Stud. Probab. Induc. Decis. Theory, pages
           1–16. Cambridge Univ. Press, Cambridge, 1997.

[Ban51]   Thøger Bang. A solution of the "plank problem". *Proc. Amer. Math. Soc.*, 2:990–993, 1951.

[BBC69]   A. Beck, M. Bleicher, and J. Crow. *Excursions into Mathematics*. Worth, 1969.

[BCB12]   Sébastien Bubeck and Nicolo Cesa-Bianchi. Regret analysis of stochastic and non-stochastic multi-armed bandit problems. *Foundations and Trends® in Machine Learning*, 5(1):1–122, 2012. **320**

[BCG82a]   E. R. Berlekamp, J. H. Conway, and R. K. Guy. *Winning Ways for Your Mathematical Plays*, volume 1. Academic Press, 1982. **21, 22**

[BCG82b]   E. R. Berlekamp, J. H. Conway, and R. K. Guy. *Winning Ways for Your Mathematical Plays*, volume 2. Academic Press, 1982. **22**

[BCG04]   Elwyn R. Berlekamp, John H. Conway, and Richard K. Guy. *Winning ways for your mathematical plays. Vols. 1-4*. A K Peters, Ltd., Wellesley, MA, second edition, 2004. **22**

[BdVSV11]   Sushil Bikhchandani, Sven de Vries, James Schummer, and Rakesh V. Vohra. An ascending Vickrey auction for selling bases of a matroid. *Operations Research*, 59(2):400–413, 2011. **273**

[BEDL10]   Avrim Blum, Eyal Even-Dar, and Katrina Ligett. Routing without regret: On convergence to Nash equilibria of regret-minimizing algorithms in routing games. *Theory Comput.*, 6:179–199, 2010. **155**

[Ber06]   Carl Bergstrom. Honest signalling theory. http://octavia.zoology.washington.edu/handicap/, 2006. **124**

[BHLR08]   Avrim Blum, MohammadTaghi Hajiaghayi, Katrina Ligett, and Aaron Roth. Regret minimization and the price of total anarchy. In *Proceedings of the Fortieth Annual ACM Symposium on Theory of Computing*, pages 373–382, 2008. **155**

[BI64]   John F. Banzhaf III. Weighted voting doesn't work: A mathematical analysis. *Rutgers Law Review*, 19:317, 1964. **203**

[Bik99]   Sushil Bikhchandani. Auctions of heterogeneous objects. *Games and Economic Behavior*, 26(2):193–220, 1999.

[BILW14]   Moshe Babaioff, Nicole Immorlica, Brendan Lucier, and S. Matthew Weinberg. A simple and approximately optimal mechanism for an additive buyer. In *55th Annual IEEE Symposium on Foundations of Computer Science — (FOCS) 2014*, pages 21–30. IEEE Computer Soc., Los Alamitos, CA, 2014. **252**

[BK94]   Jeremy Bulow and Paul Klemperer. Auctions vs. negotiations. Technical report, National Bureau of Economic Research, 1994. **252**

[BK99]   Jeremy Bulow and Paul Klemperer. The generalized war of attrition. *American Economic Review*, pages 175–189, 1999. **253**

[Bla56]   David Blackwell. An analog of the minimax theorem for vector payoffs. *Pacific J. Math.*, 6:1–8, 1956. **320**

[BLM13]   Stéphane Boucheron, Gábor Lugosi, and Pascal Massart. *Concentration inequalities: A nonasymptotic theory of independence*. Oxford University Press, Oxford, 2013. **332**

[BM05]   Avrim Blum and Yishay Mansour. From external to internal regret. In *Learning Theory*, pages 621–636. Springer, 2005. **320**

[BMW56]   Martin Beckmann, C. B. McGuire, and Christopher B. Winsten. *Studies in the economics of transportation*. Yale University Press, 1956. **155**

[Bor85]   Kim C. Border. *Fixed point theorems with applications to economics and game theory*. Cambridge University Press, Cambridge, 1985. **102**

[Bou02]   Charles L. Bouton. Nim, a game with a complete mathematical theory. *Ann. of Math. (2)*, 3(1-4):35–39, 1901/1902. **21**

[BPP+14]   Yakov Babichenko, Yuval Peres, Ron Peretz, Perla Sousi, and Peter Winkler. Hunter, Cauchy rabbit, and optimal Kakeya sets. *Trans. Amer. Math. Soc.*, 366(10):5567–5586, 2014. **52, 60**

[BR89]   Jeremy Bulow and John Roberts. The simple economics of optimal auctions. *The Journal of Political Economy*, pages 1060–1090, 1989. **252**

[Bra68]   D. Braess. Über ein paradoxon aus der verkehrsplanung. *Unternehmensforschung*, 12(1):258–268, 1968. **155**

[Bra12]   Steven J. Brams. Fair division. In *Computational complexity. Vols. 1-6*, pages 1073–1079. Springer, New York, 2012.

[Bri50]   Glenn W. Brier. Verification of forecasts expressed in terms of probability. *Monthly weather review*, 78(1):1–3, 1950. **286**

[Bro11]   L. E. J. Brouwer. Über Abbildung von Mannigfaltigkeiten. *Math. Ann.*, 71(1):97–115, 1911. **102**

[Bro51]   George W. Brown. Iterative solution of games by fictitious play. *Activity analysis of production and allocation*, 13(1):374–376, 1951. **320**

[Bro00]   C. Browne. *Hex Strategy: Making the Right Connections*. A K Peters, CRC Press, 2000.

[BSW05]   Itai Benjamini, Oded Schramm, and David B. Wilson. Balanced Boolean functions that can be evaluated so that every input bit is unlikely to be read. In *Proc. 37th Symposium on the Theory of Computing (STOC)*, 2005.

[BT95]   Steven J. Brams and Alan D. Taylor. An envy-free cake division protocol. *Amer. Math. Monthly*, 102(1):9–18, 1995. **193**

[BT96]   Steven J. Brams and Alan D. Taylor. *Fair division*. From cake-cutting to dispute resolution. Cambridge University Press, Cambridge, 1996. **193**

[BTT89]   J. Bartholdi, C. A. Tovey, and M. A. Trick. Voting schemes for which it can be difficult to tell who won the election. *Social Choice and Welfare*, 6(2):157–165, 1989. **219**

[CBFH+97]   Nicolò Cesa-Bianchi, Yoav Freund, David Haussler, David P. Helmbold, Robert E. Schapire, and Manfred K. Warmuth. How to use expert advice. *J. ACM*, 44(3):427–485, May 1997. **319**

[CBL06]   Nicolò Cesa-Bianchi and Gabor Lugosi. *Prediction, Learning, and Games*. Cambridge University Press, New York, NY, 2006. **319, 320**

[CD06]   Xi Chen and Xiaotie Deng. Settling the complexity of two-player Nash equilibrium. In *FOCS 2006 — In Proceedings of the 47th Annual IEEE Symposium on Foundations of Computer Science*, 2006. **84**

[CDT06]   Xi Chen, Xiaotie Deng, and Shang-Hua Teng. Computing Nash equilibria: Approximation and smoothed complexity. In *FOCS 2006 — Proceedings of the 47th Annual IEEE Symposium on Foundations of Computer Science*, pages 603–612. IEEE, 2006. **84**

[CDW16]   Yang Cai, Nikhil R Devanur, and S Matthew Weinberg. A duality based unified approach to Bayesian mechanism design. In *Proceedings of the 48th Annual ACM Symposium on Theory of Computing (STOC)*, 2016. **253**

[CGL14]   Ning Chen, Nick Gravin, and Pinyan Lu. Optimal competitive auctions. In *Proceedings of the 46th Annual ACM Symposium on Theory of Computing (STOC)*, pages 253–262, 2014. **274**

[CH13]   Shuchi Chawla and Jason D. Hartline. Auctions with unique equilibria. In *ACM Conference on Electronic Commerce*, pages 181–196, 2013.

[CK81]   Vincent P. Crawford and Elsie Marie Knoer. Job matching with heterogeneous firms and workers. *Econometrica: Journal of the Econometric Society*, pages 437–450, 1981. **300**

[CK05a]   George Christodoulou and Elias Koutsoupias. On the price of anarchy and stability of correlated equilibria of linear congestion games. In *Algorithms–ESA 2005*, pages 59–70. Springer, 2005. **155**

[CK05b]   George Christodoulou and Elias Koutsoupias. The price of anarchy of finite congestion games. In *Proceedings of the Thirty-Seventh Annual ACM Symposium on Theory of Computing*, pages 67–73. ACM, 2005. **155**

[CKKK11]   Ioannis Caragiannis, Christos Kaklamanis, Panagiotis Kanellopoulos, and Maria Kyropoulou. On the efficiency of equilibria in generalized second price auctions. In *Proceedings of the 12th ACM Conference on Electronic Commerce*, pages 81–90. ACM, 2011.

[CKS08]   George Christodoulou, Annamária Kovács, and Michael Schapira. Bayesian combinatorial auctions. In *Automata, Languages and Programming*, pages 820–832. Springer, 2008.

[Cla71]   Edward H. Clarke. Multipart pricing of public goods. *Public choice*, 11(1):17–33, 1971. **273, 286**

[CM88]   Jacques Cremer and Richard P. McLean. Full extraction of the surplus in Bayesian
              and dominant strategy auctions. *Econometrica: Journal of the Econometric Soci-
              ety*, pages 1247–1257, 1988. **253**

[Cov65]   Thomas M. Cover. Behavior of sequential predictors of binary sequences. In *Pro-
              ceedings of the 4th Prague Conference on Information Theory, Statistical Decision
              Functions, Random Processes*, pages 263–272, 1965. **319**

[CP10]   Yiling Chen and David M. Pennock. Designing markets for prediction. *AI Maga-
              zine*, 31(4):42–52, 2010. **287**

[CS14]   Shuchi Chawla and Balasubramanian Sivan. Bayesian algorithmic mechanism de-
              sign. *ACM SIGecom Exchanges*, 13(1):5–49, 2014. **253**

[CSSM04]   José R. Correa, Andreas S. Schulz, and Nicolás E. Stier-Moses. Selfish routing in
              capacitated networks. *Mathematics of Operations Research*, 29(4):961–976, 2004.
              155

[Dan51a]   George B. Dantzig. Maximization of a linear function of variables subject to linear
              inequalities. In *Activity Analysis of Production and Allocation*, Cowles Commission
              Monograph No. 13, pages 339–347. John Wiley & Sons, Inc., New York, N. Y.;
              Chapman & Hall, Ltd., London, 1951. **331**

[Dan51b]   George B. Dantzig. A proof of the equivalence of the programming problem and
              the game problem. In *Activity analysis of production and allocation*, Cowles Com-
              mission Monograph No. 13. edition, TC Koopmans. New York: John Wiley, 1951.
              331

[Dan82]   George B. Dantzig. Reminiscences about the origins of linear programming. *Oper.
              Res. Lett.*, 1(2):43–48, April 1982. **39, 331**

[Das13]   Constantinos Daskalakis. On the complexity of approximating a Nash equilibrium.
              ACM Transactions on Algorithms (TALG), 9(3):23, 2013. **84**

[Das15]   Constantinos Daskalakis. Multi-item auctions defying intuition? *ACM SIGecom
              Exchanges*, 14(1):41–75, 2015. **253**

[Daw06]   Richard Dawkins. *The Selfish Gene*. Number 199. Oxford University Press, 2006.
              253

[dC85]   Marquis de Condorcet. Essai sur l'application de l'analyse à la probabilité des
              décisions rendues à la pluralité des voix. In *The French Revolution Research Col-
              lection*. Pergamon Press, 1785. **219**

[DD92]   Robert W. Dimand and Mary Ann Dimand. The early history of the theory
              of strategic games from Waldegrave to Borel. *History of Political Economy*,
              24(Supplement):15–27, 1992. **xxvi**

[DDT14]   Constantinos Daskalakis, Alan Deckelbaum, and Christos Tzamos. The complexity
              of optimal mechanism design. In *SODA 2014 — Proceedings of the Twenty-Fifth
              Annual ACM-SIAM Symposium on Discrete Algorithms*. ACM-SIAM, 2014. **252**

[Dem82]   Gabrielle Demange. Strategyproofness in the assignment market game. *Laboratoire
              d'Économétrie de l'École Polytechnique, Paris*, 1982. **300**

[DF81]   L. E. Dubins and D. A. Freedman. Machiavelli and the Gale-Shapley algorithm.
              Amer. Math. Monthly, 88(7):485–494, 1981. **177**

[DG85]   Gabrielle Demange and David Gale. The strategy structure of two-sided matching
              markets. *Econometrica: Journal of the Econometric Society*, pages 873–888, 1985.

[DGP09]   Constantinos Daskalakis, Paul W Goldberg, and Christos H. Papadimitriou. The
              complexity of computing a Nash equilibrium. *SIAM Journal on Computing*,
              39(1):195–259, 2009. **84**

[DGS86]   Gabrielle Demange, David Gale, and Marilda Sotomayor. Multi-item auctions. *The
              Journal of Political Economy*, pages 863–872, 1986. **300**

[DHM79]   Partha Dasgupta, Peter Hammond, and Eric Maskin. The implementation of social
              choice rules: Some general results on incentive compatibility. *Rev. Econom. Stud.*,
              46(2):185–216, 1979. **286**

[DN08]   Avinash K. Dixit and Barry Nalebuff. *The art of strategy: A game theorist's guide
              to success in business & life*. W. W. Norton & Company, 2008. **124**

[DPS15]   Nikhil R. Devanur, Yuval Peres, and Balasubramanian Sivan. Perfect Bayesian
              equilibria in repeated sales. In *Proceedings of the Twenty-Sixth Annual ACM-
              SIAM Symposium on Discrete Algorithms (SODA)*, pages 983–1002. SIAM, 2015.
              253

[DS61]   L. E. Dubins and E. H. Spanier. How to cut a cake fairly. *Amer. Math. Monthly*, 68:1–17, 1961. **192**

[DS84]   Peter G. Doyle and J. Laurie Snell. *Random walks and electric networks*, volume 22 of *Carus Mathematical Monographs*. Mathematical Association of America, Washington, DC, 1984. **60**

[Dub57]   L. E. Dubins. *A discrete evasion game*. Princeton University Press, Princeton, N.J., 1957. **60**

[Ede62]   M. Edelstein. On fixed and periodic points under contractive mappings. *J. London Math. Soc.*, 37:74–79, 1962. **102**

[Ege31]   J Egerváry. Matrixok kombinatorius tulajdonsgairl [on combinatorial properties of matrices]. *Matematikai Zs Fizikai Lapok*, 38, 1931. See *Logistic Papers* (George Washington University), **11**, 1955 for a translation by Harold Kuhn. **299**

[EOS07]   Benjamin Edelman, Michael Ostrovsky, and Michael Schwarz. Internet advertising and the generalized second-price auction: Selling billions of dollars worth of keywords. *American Economic Review*, 97(1):242–259, 2007. **274**

[ET76]   S. Even and R. E. Tarjan. A combinatorial problem which is complete in polynomial space. *J. Assoc. Comput. Mach.*, 23(4):710–719, 1976. **21**

[Eve57]   H. Everett. Recursive games. In *Contributions to the theory of games, vol. 3*, Annals of Mathematics Studies, no. 39, pages 47–78. Princeton University Press, Princeton, N. J., 1957. **39**

[Fer08]   Thomas Ferguson. *Game Theory*. 2008. `http://www.math.ucla.edu/~tom/Game_Theory/Contents.html`. **xxv, 39**

[FGHK02]   Amos Fiat, Andrew V. Goldberg, Jason D. Hartline, and Anna R. Karlin. Competitive generalized auctions. In *Proceedings of the Thiry-Fourth Annual ACM Symposium on Theory of Computing (STOC)*, pages 72–81. ACM, 2002. **274**

[FJM04]   Lisa Fleischer, Kamal Jain, and Mohammad Mahdian. Tolls for heterogeneous selfish users in multicommodity networks and generalized congestion games. In *Foundations of Computer Science, 2004. Proceedings. 45th Annual IEEE Symposium*, pages 277–285. IEEE, 2004. **155**

[FLM+03]   Alex Fabrikant, Ankur Luthra, Elitza Maneva, Christos H. Papadimitriou, and Scott Shenker. On a network creation game. In *Proceedings of the Twenty-Second Annual Symposium on Principles of Distributed Computing (PODC)*, pages 347–351. ACM, 2003. **155**

[FPT04]   Alex Fabrikant, Christos Papadimitriou, and Kunal Talwar. The complexity of pure Nash equilibria. In *Proceedings of the 36th Annual ACM Symposium on Theory of Computing (STOC)*, pages 604–612. ACM, New York, 2004. **83**

[Fri71]   James W. Friedman. A non-cooperative equilibrium for supergames. *The Review of Economic Studies*, pages 1–12, 1971. **125**

[Fro17]   Georg Frobenius. Über zerlegbare determinanten. *Sitzungsberichte Der Königlich Preussischen Akademie Der Wissenschaften Zu Berlin*, 1917. **60**

[FS97]   Yoav Freund and Robert E. Schapire. A decision-theoretic generalization of online learning and an application to boosting. *J. Comput. Syst. Sci.*, 55(1):119–139, August 1997. **319**

[FT91]   Drew Fudenberg and Jean Tirole. *Game theory*. MIT Press, Cambridge, MA, 1991. **125**

[FV93]   Dean P. Foster and Rakesh V. Vohra. A randomization rule for selecting forecasts. *Operations Research*, 41(4):704–709, July 1993. `http://www.jstor.org/pss/171965`.

[FV97]   Dean P. Foster and Rakesh V. Vohra. Calibrated learning and correlated equilibrium. *Games Econom. Behav.*, 21(1-2):40–55, 1997.

[FV99]   Dean P. Foster and Rakesh V. Vohra. Calibration, expected utility and local optimality. Discussion Papers 1254, Northwestern University, Center for Mathematical Studies in Economics and Management Science, March 1999. `http://ideas.repec.org/p/nwu/cmsems/1254.html`. **287**

[Gal79]   D. Gale. The game of Hex and the Brouwer fixed-point theorem. *Amer. Math. Monthly*, 86:818–827, 1979. **21, 102**

[Gar88]   Martin Gardner. *Hexaflexagons and other mathematical diversions: The first Sci-entific American book of puzzles and games*. University of Chicago Press, 1988. **21**

[Gar00]   Andrey Garnaev. *Search games and other applications of game theory*, volume 485 of *Lecture Notes in Economics and Mathematical Systems*. Springer-Verlag, Berlin, 2000. **38**

[GBOW88]   S. Goldwasser, M. Ben-Or, and A. Wigderson. Completeness theorems for non-cryptographic fault-tolerant distributed computing. In *Proc. of the 20th STOC*, pages 1–10, 1988.

[GDGJFJ10]   Julio González-Díaz, Ignacio García-Jurado, and M Gloria Fiestras-Janeiro. *An introductory course on mathematical game theory*, volume 115. American Mathematical Society Providence, 2010. **xxv**

[Gea04]   J. Geanakoplos. *Three Brief Proofs of Arrow's Impossibility Theorem*. Cowles Commission Monograph No. 1123R4. Cowles Foundation for Research in Economics, Yale University, Box 208281, New Haven, Connecticut 06520-8281, 1996 (updated 2004). `http://cowles.econ.yale.edu/`. **220**

[GHK+06]   Andrew V. Goldberg, Jason D. Hartline, Anna R. Karlin, Michael Saks, and Andrew Wright. Competitive auctions. *Games and Economic Behavior*, 55(2):242–269, 2006. **274**

[GHKS04]   Andrew V. Goldberg, Jason D. Hartline, Anna R. Karlin, and Michael Saks. A lower bound on the competitive ratio of truthful auctions. In *STACS 2004 — Symposium on Theoretical Aspects of Computer Science*, pages 644–655. Springer, 2004. **274**

[GHS09]   A. Gierzynski, W. Hamilton, and W. D. Smith. Burlington, Vermont 2009 IRV mayor election: Thwarted-majority, non-monotonicity and other failures (oops). http://rangevoting.org/Burlington.html, 2009. **220**

[GHW01]   Andrew V. Goldberg, Jason D. Hartline, and Andrew Wright. Competitive auctions and digital goods. In *Proceedings of the Twelfth Annual ACM-SIAM Symposium on Discrete Algorithms (SODA)*, pages 735–744. Society for Industrial and Applied Mathematics, 2001. **274**

[GI89]   Dan Gusfield and Robert W. Irving. *The stable marriage problem: Structure and algorithms*. MIT Press, 1989. **177**

[Gib73]   Allan Gibbard. Manipulation of voting schemes: A general result. *Econometrica*, 41:587–601, 1973. **220**

[Gil59]   Donald B. Gillies. Solutions to general non-zero-sum games. *Contributions to the Theory of Games*, 4(40):47–85, 1959. **203**

[Gin00]   H. Gintis. *Game Theory Evolving; A problem-centered introduction to modeling strategic interaction*. Princeton University Press, Princeton, New Jersey, 2000. **xxv, 83, 129**

[GK95]   Michael D. Grigoriadis and Leonid G. Khachiyan. A sublinear-time randomized approximation algorithm for matrix games. *Oper. Res. Lett.*, 18(2):53–58, 1995. **319**

[GL77]   Jerry Green and Jean-Jacques Laffont. Characterization of satisfactory mechanisms for the revelation of preferences for public goods. *Econometrica: Journal of the Econometric Society*, pages 427–438, 1977. **273, 274, 286**

[Gli52]   I. L. Glicksberg. A further generalization of the Kakutani fixed theorem, with application to Nash equilibrium points. *Proc. Amer. Math. Soc.*, 3:170–174, 1952. **38**

[GMV04]   Hongwei Gui, Rudolf Müller, and Rakesh V. Vohra. Dominant strategy mechanisms with multidimensional types. Discussion paper, Center for Mathematical Studies in Economics and Management Science, 2004. **286**

[GMV05]   Michel Goemans, Vahab Mirrokni, and Adrian Vetta. Sink equilibria and convergence. In *Foundations of Computer Science (FOCS)*, pages 142–151. IEEE, 2005. **155**

[GO80]   E. Goles and J. Olivos. Periodic behaviour of generalized threshold functions. *Discrete Math.*, 30(2):187–189, 1980. **83**

[GPS14]   Nick Gravin, Yuval Peres, and Balasubramanian Sivan. Towards optimal algorithms for prediction with expert advice. In *SODA 2016 — Proceedings of the*

Twenty-Seventh Annual ACM-SIAM Symposium on Discrete Algorithms, 2014. **320**

[GR07]   Tilmann Gneiting and Adrian E. Raftery. Strictly proper scoring rules, prediction, and estimation. *J. Amer. Statist. Assoc.*, 102(477):359–378, 2007. **287**

[Gra99]   Peter Grassberger. Pair connectedness and shortest-path scaling in critical percolation. *J. Phys. A*, 32(35):6233–6238, 1999.

[Gro79]   Theodore Groves. Efficient collective choice when compensation is possible. *The Review of Economic Studies*, 46(2):227–241, 1979. **273, 286**

[Gru39]   Patrick M. Grundy. Mathematics and games. *Eureka*, 2(5):6–8, 1939. **21**

[GS62]   D. Gale and L. S. Shapley. College Admissions and the Stability of Marriage. *Amer. Math. Monthly*, 69(1):9–15, 1962. **170, 176**

[GS85]   David Gale and Marilda Sotomayor. Some remarks on the stable matching problem. *Discrete Appl. Math.*, 11(3):223–232, 1985. **177**

[Hal35]   Philip Hall. On representatives of subsets. *J. London Math. Soc*, 10(1):26–30, 1935. **60**

[Han57]   James Hannan. Approximation to Bayes risk in repeated play. *Contributions to the Theory of Games*, 3:97–139, 1957. **319, 320**

[Han03]   Robin Hanson. Combinatorial information market design. *Information Systems Frontiers*, 5(1):107–119, 2003. **287**

[Han12]   Robin Hanson. Logarithmic market scoring rules for modular combinatorial information aggregation. *The Journal of Prediction Markets*, 1(1):3–15, 2012. **287**

[Har67]   John C. Harsanyi. Games with incomplete information played by Bayesian players, I–III: Part I. The basic model. *Management Science*, 50(12-supplement):1804–1817, 1967. **124**

[Har68a]   Garrett Hardin. The tragedy of the commons. *Science*, 162(3859):1243–1248, 1968. **83**

[Har68b]   John C. Harsanyi. Games with incomplete information played by Bayesian players, Part II. Bayesian equilibrium points. *Management Science*, 14(5):320–334, 1968. **124**

[Har68c]   John C. Harsanyi. Games with incomplete information played by Bayesian players, Part III. The basic probability distribution of the game. *Management Science*, 14(7):486–502, 1968. **124**

[Har17]   Jason Hartline. *Mechanism Design and Approximation*. Cambridge University Press. Forthcoming. **155, 252, 253, 274, 286**

[HH07]   Vincent F. Hendricks and Pelle Guldborg Hansen. *Game Theory: 5 Questions*. Automatic Press, 2007. **135**

[HHP06]   Christopher Hoffman, Alexander E. Holroyd, and Yuval Peres. A stable marriage of Poisson and Lebesgue. *Ann. Probab.*, 34(4):1241–1272, 2006. **178**

[HHT14]   Jason Hartline, Darrell Hoy, and Sam Taggart. Price of anarchy for auction revenue. In *Proceedings of the Fifteenth ACM Conference on Economics and Computation*, pages 693–710. ACM, 2014. **155, 253**

[HMC00a]   Sergiu Hart and Andreu Mas-Colell. A simple adaptive procedure leading to correlated equilibrium. *Econometrica*, 68(5):1127–1150, Sept. 2000. http://www.jstor.org/pss/2999445. **287**

[HMC00b]   Sergiu Hart and Andreu Mas-Colell. A simple adaptive procedure leading to correlated equilibrium. *Econometrica*, 68(5):1127–1150, 2000. **320**

[HN12]   Sergiu Hart and Noam Nisan. Approximate revenue maximization with multiple items. *ACM Conference on Electronic Commerce*, 2012. **252**

[Hoe63]   Wassily Hoeffding. Probability inequalities for sums of bounded random variables. *J. Amer. Statist. Assoc.*, 58:13–30, 1963. **332, 333**

[HPPS09]   Alexander E. Holroyd, Robin Pemantle, Yuval Peres, and Oded Schramm. Poisson matching. *Ann. Inst. Henri Poincaré Probab. Stat.*, 45(1):266–287, 2009. **178**

[HS89]   Sergiu Hart and David Schmeidler. Existence of correlated equilibria. *Math. Oper. Res.*, 14(1):18–25, 1989. **39, 136**

[HS98]   Josef Hofbauer and Karl Sigmund. *Evolutionary games and population dynamics*. Cambridge University Press, Cambridge, 1998. **135**

[HT88]   Oliver D. Hart and Jean Tirole. Contract renegotiation and Coasian dynamics. *The Review of Economic Studies*, pages 509–540, 1988. **253**

[Isa55]   Rufus Isaacs. The problem of aiming and evasion. *Naval Res. Logist. Quart.*, 2:47–67, 1955. **60**

[Isa65]   Rufus Isaacs. *Differential games. A mathematical theory with applications to warfare and pursuit, control and optimization.* John Wiley & Sons, Inc., New York-London-Sydney, 1965. **60**

[Jac90a]   Carl Gustav Jacob Jacobi. De aequationum differentialium systemate non normali ad formam normalem revocando (translation: The reduction to normal form of a non-normal system of differential equations). *Borchardt Journal für die reine und angewandte Mathematik*, 1890.

[Jac90b]   Carl Gustav Jacob Jacobi. De investigando ordine systematis aequationum differentialium vulgarium cujuscunque (translation: About the research of the order of a system of arbitrary ordinary differential equations). *Borchardt Journal für die reine und angewandte Mathematik*, 64, 1890. **299**

[Jac05]   Matthew O. Jackson. A survey of network formation models: Stability and efficiency. *Group Formation in Economics: Networks, Clubs, and Coalitions*, pages 11–49, 2005. **155**

[JM08]   Bernard J. Jansen and Tracy Mullen. Sponsored search: An overview of the concept, history, and technology. *International Journal of Electronic Business*, 6(2):114–131, 2008. **274**

[Kal10]   Gil Kalai. Noise sensitivity and chaos in social choice theory. In *Fete of combinatorics and computer science*, volume 20 of *Bolyai Soc. Math. Stud.*, pages 173–212. János Bolyai Math. Soc., Budapest, 2010. **220**

[Kan39]   L. V. Kantorovich. Mathematical methods of organizing and planning production (translation published in 1960). *Management Sci.*, 6:366–422, 1939. **331**

[Kar57]   Samuel Karlin. An infinite move game with a lag. In *Contributions to the theory of games, vol. 3*, Annals of Mathematics Studies, no. 39, pages 257–272. Princeton University Press, Princeton, N. J., 1957. **60**

[Kar59]   S. Karlin. *Mathematical methods and theory in games, programming and economics*, volume 2. Addison-Wesley, 1959. **39**

[Kar84]   Narendra Karmarkar. A new polynomial-time algorithm for linear programming. In *Proceedings of the Sixteenth Annual ACM Symposium on Theory of Computing (STOC)*, pages 302–311. ACM, 1984. **331**

[KBB+11]   Yashodhan Kanoria, Mohsen Bayati, Christian Borgs, Jennifer Chayes, and Andrea Montanari. Fast convergence of natural bargaining dynamics in exchange networks. In *Proceedings of the Twenty-Second Annual ACM-SIAM Symposium on Discrete Algorithms*, pages 1518–1537. SIAM, 2011. **300**

[Kha80]   Leonid G Khachiyan. Polynomial algorithms in linear programming. *USSR Computational Mathematics and Mathematical Physics*, 20(1):53–72, 1980. **331**

[Kir06]   René Kirkegaard. A short proof of the Bulow-Klemperer auctions vs. negotiations result. *Economic Theory*, 28(2):449–452, 2006. **252**

[KJC82]   Alexander S. Kelso Jr. and Vincent P. Crawford. Job matching, coalition formation, and gross substitutes. *Econometrica: Journal of the Econometric Society*, pages 1483–1504, 1982.

[Kle98]   Paul Klemperer. Auctions with almost common values: The wallet game and its applications. *European Economic Review*, 42(3):757–769, 1998. **253**

[Kle99]   Paul Klemperer. Auction theory: A guide to the literature. *Journal of Economic Surveys*, 13(3):227–286, 1999. **253**

[KM72]   Victor Klee and George J. Minty. How good is the simplex algorithm? In *Inequalities, III (Proc. Third Sympos., Univ. California, Los Angeles, Calif., 1969; dedicated to the memory of Theodore S. Motzkin)*, pages 159–175. Academic Press, New York, 1972. **331**

[KM15]   Gil Kalai and Elchanan Mossel. Sharp thresholds for monotone non-Boolean functions and social choice theory. *Math. Oper. Res.*, 40(4):915–925, 2015. **220**

[KN02]   H. W. Kuhn and S. Nasar, editors. *The Essential John Nash*. Princeton University Press, 2002.

[Knu97]   Donald Ervin Knuth. *Stable marriage and its relation to other combinatorial problems: An introduction to the mathematical analysis of algorithms*, volume 10. American Mathematical Soc., 1997. **177**

[Kön31]   Dénes König. Graphen und matrizen. *Mat. Fiz. Lapok*, 38:116–119, 1931. **60, 299**

[KP99]   Elias Koutsoupias and Christos Papadimitriou. Worst-case equilibria. In *STACS 99 — Symposium on Theoretical Aspects of Computer Science*, pages 404–413. Springer, 1999. **154**

[KP09]   Elias Koutsoupias and Christos Papadimitriou. Worst-case equilibria. *Computer Science Review*, 3(2):65–69, 2009. **154, 155**

[Kri09]   Vijay Krishna. *Auction Theory*. Academic Press, 2009. **252, 253, 286**

[KS75]   Ehud Kalai and Meir Smorodinsky. Other solutions to Nash's bargaining problem. *Econometrica*, 43:513–518, 1975. **204**

[KT06]   Jon Kleinberg and Éva Tardos. *Algorithm design*. Pearson, 2006. **60**

[KT08]   Jon Kleinberg and Éva Tardos. Balanced outcomes in social exchange networks. In *STOC 2008 — Proceedings of the 40th Annual ACM Symposium on Theory of Computing*. ACM, 2008. **300**

[Kuh53]   Harold W. Kuhn. Extensive games and the problem of information. *Contributions to the Theory of Games*, 2(28):193–216, 1953. **124**

[Kuh55]   Harold W. Kuhn. The Hungarian method for the assignment problem. *Naval Research Logistics Quarterly*, 2(1-2):83–97, 1955. **299**

[KV05]   Adam Kalai and Santosh Vempala. Efficient algorithms for online decision problems. *J. Comput. System Sci.*, 71(3):291–307, 2005. **320**

[Las31]   E. Lasker. *Brettspiele der Völker, Rätsel und mathematische Spiele*. Berlin, 1931.

[Leh64]   Alfred Lehman. A solution of the Shannon switching game. *J. Soc. Indust. Appl. Math.*, 12:687–725, 1964. **21**

[Leo83]   Herman B. Leonard. Elicitation of honest preferences for the assignment of individuals to positions. *The Journal of Political Economy*, pages 461–479, 1983. **300**

[Leo10]   Robert Leonard. *Von Neumann, Morgenstern, and the creation of game theory: From chess to social science, 1900–1960*. Cambridge University Press, 2010. **205**

[Lin66]   Joram Lindenstrauss. A short proof of Liapounoff's convexity theorem. *J. Math. Mech.*, 15:971–972, 1966. **192**

[LLP+99]   Andrew J. Lazarus, Daniel E. Loeb, James G. Propp, Walter R. Stromquist, and Daniel H. Ullman. Combinatorial games under auction play. *Games Econom. Behav.*, 27(2):229–264, 1999.

[LLPU96]   Andrew J. Lazarus, Daniel E. Loeb, James G. Propp, and Daniel Ullman. Richman games. In Richard J. Nowakowski, editor, *Games of No Chance*, volume 29 of *MSRI Publications*, pages 439–449. Cambridge Univ. Press, Cambridge, 1996.

[LOS02]   Daniel Lehmann, Liadan Ita Oćallaghan, and Yoav Shoham. Truth revelation in approximately efficient combinatorial auctions. *Journal of the ACM (JACM)*, 49(5):577–602, 2002. **286**

[LP09]   László Lovász and Michael D. Plummer. *Matching theory*. AMS Chelsea Publishing, Providence, RI, 2009. Corrected reprint of the 1986 original [MR0859549]. **60, 299**

[LPL11]   Brendan Lucier and Renato Paes Leme. GSP auctions with correlated types. In *Proceedings of the 12th ACM Conference on Electronic Commerce*, pages 71–80. ACM, 2011. **253**

[LPW09]   David A. Levin, Yuval Peres, and Elizabeth L. Wilmer. *Markov chains and mixing times*. With a chapter by James G. Propp and David B. Wilson. American Mathematical Society, Providence, RI, 2009. **60**

[LR57]   R. Duncan Luce and Howard Raiffa. *Games and decisions: Introduction and critical survey*. John Wiley & Sons, Inc., New York, N.Y., 1957. **xxv**

[LW89]   Nick Littlestone and M. K. Warmuth. The weighted majority algorithm. In *Proceedings of the 30th Annual IEEE Symposium on Foundations of Computer Science*. IEEE, 1989. **319**

[LW94]   Nick Littlestone and Manfred K. Warmuth. The weighted majority algorithm. *Information and Computation*, 108(2):212–261, February 1994. **319**

[LY94]   Richard J. Lipton and Neal E. Young. Simple strategies for large zero-sum games with applications to complexity theory. In *Proceedings of the Twenty-Sixth Annual ACM Symposium on Theory of Computing*, pages 734–740. ACM, 1994. **39**

[Lya40]   A. Lyapunov. Sur les fonctions-vecteurs complètement additives. *Bull. Acad. Sci. USSR*, 4:465–478, 1940. **192**

[Man96]   Richard Mansfield. Strategies for the Shannon switching game. *Amer. Math. Monthly*, 103(3):250–252, 1996. **22**

[Mar98]   Donald A. Martin. The determinacy of Blackwell games. *J. Symbolic Logic*, 63(4):1565–1581, 1998.

[MBC+16]   Hervé Moulin, Felix Brandt, Vincent Conitzer, Ulle Endriss, Jérôme Lang, and Ariel D. Procaccia. *Handbook of computational social choice.* Cambridge University Press, 2016. **220**

[MCWJ95]   Andreu Mas-Collel, W. Whinston, and J. Green. Microeconomic theory. Oxford University Press, 1995. **286**

[Met15]   Cade Metz. Facebook doesn't make as much money as it could—on purpose. Wired. http://www.wired.com/2015/09/facebook-doesnt-make-much-money-couldon-purpose/, 2015. **274**

[MG07]   Jiri Matousek and Bernd Gärtner. *Understanding and using linear programming.* Springer Science & Business Media, 2007. **39, 136, 331**

[Mil95]   John Milnor. A Nobel Prize for John Nash. *Math. Intelligencer*, 17(3):11–17, 1995. **21**

[Mil04]   Paul Robert Milgrom. *Putting auction theory to work.* Cambridge University Press, 2004. **252, 286**

[MM05]   Flávio Marques Menezes and Paulo Klinger Monteiro. *An introduction to auction theory.* Oxford University Press, 2005. **252**

[Moo10]   E. H. Moore. A generalization of the game called NIM. *Ann. of Math.* (Ser. 2), 11:93–94, 1909–1910.

[Mou88a]   Hervé Moulin. *Axioms of cooperative decision making*, volume 15 of *Econometric Society Monographs.* With a foreword by Amartya Sen. Cambridge University Press, Cambridge, 1988.

[Mou88b]   Hervé Moulin. Condorcet's principle implies the no show paradox. *J. Econom. Theory*, 45(1):53–64, 1988. **219**

[MP92]   Richard D. McKelvey and Thomas R. Palfrey. An experimental study of the centipede game. *Econometrica: Journal of the Econometric Society*, pages 803–836, 1992. **124**

[MS83]   Roger B. Myerson and Mark A. Satterthwaite. Efficient mechanisms for bilateral trading. *J. Econom. Theory*, 29(2):265–281, 1983. **274**

[MS96]   Dov Monderer and Lloyd S. Shapley. Potential games. *Games Econom. Behav.*, 14(1):124–143, 1996. **83**

[MS01]   Hervé Moulin and Scott Shenker. Strategyproof sharing of submodular costs: Budget balance versus efficiency. *Economic Theory*, 18(3):511–533, 2001. **274**

[MSZ13]   Michael Maschler, Eilon Solan, and Shmuel Zamir. *Game theory.* Translated from Hebrew by Ziv Hellman and edited by Mike Borns. Cambridge University Press, Cambridge, 2013. **xxv, 84, 124, 125, 136, 204**

[MSZ15]   Jean-Francois Mertens, Sylvain Sorin, and Shmuel Zamir. *Repeated Games.* Part of Econometric Society Monographs. Cambridge University Press, 2015. **125**

[Mun57]   James Munkres. Algorithms for the assignment and transportation problems. *J. Soc. Indust. Appl. Math.*, 5:32–38, 1957. **299**

[MV04]   Vahab S. Mirrokni and Adrian Vetta. Convergence issues in competitive games. In *Approximation, Randomization, and Combinatorial Optimization. Algorithms and Techniques*, pages 183–194. Springer, 2004. **155**

[Mye81]   Roger B. Myerson. Optimal auction design. *Mathematics of operations research*, 6(1):58–73, 1981. **252, 253, 274, 286**

[Mye12]   Roger Myerson. Perspectives on mechanism design in economic theory. Nobel Prize lecture. 2012. **252**

[Nas50a]   John F. Nash Jr. Equilibrium points in *n*-person games. *Proc. Nat. Acad. Sci. U.S.A.*, 36:48–49, 1950. **83, 102**

[Nas50b]   John F. Nash Jr. *Non-cooperative games.* ProQuest LLC, Ann Arbor, MI, 1950. Thesis (Ph.D.)–Princeton University. **135**

[Nis99]   Noam Nisan. Algorithms for selfish agents. In *STACS 1999 — Symposium on Theoretical Aspects of Computer Science.* Springer, 1999. **274**

[NJ50]   John F. Nash Jr. The bargaining problem. *Econometrica: Journal of the Econometric Society*, pages 155–162, 1950. **204**

[Now96]   Richard J. Nowakowski, editor. *Games of no chance*, volume 29 of *Mathematical Sciences Research Institute Publications*. Cambridge University Press, Cambridge, 1996. **22**

[Now02]   Richard Nowakowski, editor. *More games of no chance*, volume 42 of *Mathematical Sciences Research Institute Publications*. Cambridge University Press, Cambridge, 2002. **22**

[NR01]   Noam Nisan and Amir Ronen. Algorithmic mechanism design. Economics and artificial intelligence. *Games Econom. Behav.*, 35(1-2):166–196, 2001. **274, 286**

[NR07]   Noam Nisan and Amir Ronen. Computationally feasible VCG mechanisms. *J. Artif. Intell. Res.(JAIR)*, 29:19–47, 2007. **286**

[NRTV07]   Noam Nisan, Tim Roughgarden, Éva Tardos and Vijay Vazirani (editors). *Algorithmic game theory*. Cambridge University Press, 2007. **xxv, 83, 154, 155, 178, 252, 274, 286, 319**

[NS03]   Abraham Neyman and Sylvain Sorin, editors. *Stochastic games and applications*, volume 570 of *NATO Science Series C: Mathematical and Physical Sciences*, Dordrecht, Kluwer Academic Publishers, 2003.

[O'N82]   Barry O'Neill. A problem of rights arbitration from the Talmud. *Math. Social Sci.*, 2(4):345–371, 1982. **193, 204**

[O'N94]   Barry O'Neill. Game theory models of peace and war. *Handbook of game theory with economic applications, Vol. II*, volume 11 of *Handbooks in Economics*, pages 995–1053. North-Holland, Amsterdam, 1994. **83**

[OR94]   Martin J. Osborne and Ariel Rubinstein. *A course in game theory*. MIT Press, 1994. **xxv**

[OR14]   J. J. O'Connor and E. F. Robertson. Emanuel Sperner. http://www-history.mcs.st-andrews.ac.uk/history/Biographies/Sperner.html 2014. **102**

[OS07]   Ryan O'Donnell and Rocco Servedio. Learning monotone decision trees in polynomial time. *SIAM Journal on Computing* 37:3, pages 827-844, 2007. **166**

[Owe67]   Guillermo Owen. An elementary proof of the minimax theorem. *Management Sci.*, 13:765, 1967. **38**

[Owe95]   Guillermo Owen. *Game theory*. Academic Press, Inc., San Diego, CA, third edition, 1995. **xxv, 39, 203**

[Pac15]   Eric Pacuit. Voting methods. https://plato.stanford.edu/entries/voting-methods/. 2015. **220**

[Pap01]   Christos Papadimitriou. Algorithms, games, and the internet. In *Proceedings of the Thirty-Third Annual ACM Symposium on Theory of Computing*, pages 749–753. ACM, 2001. **154, 286**

[PH03]   Ignacio Palacios-Huerta. Professionals play minimax. *The Review of Economic Studies*, 70(2):395–415, 2003. **xxvi**

[PHV09]   Ignacio Palacios-Huerta and Oscar Volij. Field centipedes. *The American Economic Review*, 99(4):1619–1635, 2009. **124**

[Pig20]   Arthur Cecil Pigou. *The economics of welfare*. Palgrave Macmillan, 1920. **154**

[Pou11]   William Poundstone. *Prisoner's dilemma*. Anchor, 2011. **83**

[PPS15]   David C. Parkes, Ariel D. Procaccia, and Nisarg Shah. Beyond dominant resource fairness: Extensions, limitations, and indivisibilities. *ACM Transactions on Economics and Computation*, 3(1):3, 2015.

[Pro11]   Ariel D. Procaccia. Computational social choice: The first four centuries. *XRDS: Crossroads, The ACM Magazine for Students*, 18(2):31–34, 2011. **218, 220**

[Pro13]   Ariel D. Procaccia. Cake cutting: Not just child's play. *Communications of the ACM*, 56(7):78–87, 2013. **193**

[PS12a]   Panagiota Panagopoulou and Paul Spirakis. Playing a game to bound the chromatic number. *Amer. Math. Monthly*, 114(5):373–387, 2012. http://arxiv.org/math/0508580.

[PS12b]   Panagiota N. Panagopoulou and Paul G. Spirakis. Playing a game to bound the chromatic number. *Amer. Math. Monthly*, 119(9):771–778, 2012. **83**

[PSSW07]   Yuval Peres, Oded Schramm, Scott Sheffield, and David B. Wilson. Random-turn Hex and other selection games. *Amer. Math. Monthly*, 114(5):373–387, 2007. `http://arxiv.org/math/0508580`. **xi, 166**

[Ras07]   Eric Rasmusen. *Games and information.* An introduction to game theory. Black-well Publishing, Malden, MA, fourth edition, 2007. **125**

[Rei81]   S. Reisch. Hex ist PSPACE-vollständig. *Acta Inform.*, 15:167–191, 1981. **21**

[Rob51]   Julia Robinson. An iterative method of solving a game. *Ann. of Math. (2)*, 54:296–301, 1951. **320**

[Rob79]   Kevin Roberts. The characterization of implementable choice rules. *Aggregation and revelation of preferences*, 12(2):321–348, 1979. **286**

[Roc84]   Sharon C. Rochford. Symmetrically pairwise-bargained allocations in an assign-ment market. *Journal of Economic Theory*, 34(2):262–281, 1984. **300**

[Ron01]   Amir Ronen. On approximating optimal auctions. In *Proceedings of the 3rd ACM Conference on Electronic Commerce*, pages 11–17. ACM, 2001. **252**

[Ros73]   R. W. Rosenthal. A class of games possessing pure-strategy Nash equilibria. *International Journal of Game Theory*, 2:65–67, 1973. **83**

[Ros81]   Robert W. Rosenthal. Games of perfect information, predatory pricing and the chain-store paradox. *Journal of Economic Theory*, 25(1):92–100, 1981. **124**

[Rot82]   Alvin E. Roth. The economics of matching: Stability and incentives. *Math. Oper. Res.*, 7(4):617–628, 1982. **177**

[Rot86]   Alvin E. Roth. On the allocation of residents to rural hospitals: A general property of two-sided matching markets. *Econometrica: Journal of the Econometric Society*, pages 425–427, 1986. **178**

[Rot88]   Alvin E. Roth. *The Shapley value: Essays in honor of Lloyd S. Shapley.* Cambridge University Press, 1988. **203**

[Rot15]   Alvin E. Roth. *Who Gets What and Why: The New Economics of Matchmaking and Market Design.* Houghton Mifflin Harcourt, 2015. **177**

[Rou03]   Tim Roughgarden. The price of anarchy is independent of the network topology. *Journal of Computer and System Sciences*, 67(2):341–364, 2003. **155**

[Rou05]   Tim Roughgarden. *Selfish routing and the price of anarchy*, volume 174. MIT Press, Cambridge, 2005. **154**

[Rou09]   Tim Roughgarden. Intrinsic robustness of the price of anarchy. In *Proceedings of the Forty-First Annual ACM Symposium on Theory of Computing (STOC)*, pages 513–522. ACM, 2009. **155**

[Rou12]   Tim Roughgarden. The price of anarchy in games of incomplete information. In *Proceedings of the 13th ACM Conference on Electronic Commerce*, pages 862–879. ACM, 2012. **155, 253**

[Rou13]   Tim Roughgarden. *Algorithmic Mechanism Design, CSE 364A Lecture Notes.* 2013. **155, 286**

[Rou14]   Tim Roughgarden. *Frontiers in Mechanism Design, CSE 364B Lecture Notes.* 2014. **155, 252, 253, 286**

[RRVV93]   Alvin E. Roth, Uriel G. Rothblum, and John H. Vande Vate. Stable matchings, optimal assignments, and linear programming. *Math. Oper. Res.*, 18(4):803–828, 1993. **178**

[RS81]   John G. Riley and William F. Samuelson. Optimal auctions. *The American Economic Review*, pages 381–392, 1981. **252**

[RS92]   Alvin E. Roth and Marilda A. Oliveira Sotomayor. *Two-sided matching: A study in game-theoretic modeling and analysis.* Cambridge University Press, 1992. **177, 300**

[RT02]   Tim Roughgarden and Éva Tardos. How bad is selfish routing? *Journal of the ACM (JACM)*, 49(2):236–259, 2002. **154, 155**

[Rub79]   Ariel Rubinstein. Equilibrium in supergames with the overtaking criterion. *J. Econom. Theory*, 21(1):1–9, 1979. **125**

[Rub82]   Ariel Rubinstein. Perfect equilibrium in a bargaining model. *Econometrica*, 50(1):97–109, 1982. **204**

[Rub98]   Ariel Rubinstein. *Modeling bounded rationality*, volume 1. MIT Press, 1998. **83**

[Rud76]   Walter Rudin. *Principles of mathematical analysis*. International Series in
             Pure and Applied Mathematics. McGraw-Hill Book Co., New York-Auckland-
             Düsseldorf, third edition, 1976. **xxiv, 102**

[RW98]   Jack Robertson and William Webb. *Cake-cutting algorithms. Be fair if you can.*
             A K Peters, Ltd., Natick, MA, 1998. **193**

[RW15]   Aviad Rubinstein and S. Matthew Weinberg. Simple mechanisms for a subadditive
             buyer and applications to revenue monotonicity. In *Proceedings of the Sixteenth
             ACM Conference on Economics and Computation*, pages 377–394. ACM, 2015.
             253

[Saa90]   D. G. Saari. The Borda dictionary. *Social Choice and Welfare*, 7(4):279–317, 1990.

[Saa95]   Donald G. Saari. *Basic geometry of voting*. Springer-Verlag, Berlin, 1995. **219,
             220**

[Saa01a]   Donald Saari. *Chaotic elections!: A mathematician looks at voting*. American
             Mathematical Soc., 2001. **220**

[Saa01b]   Donald G. Saari. *Decisions and elections*. Explaining the unexpected. Cambridge
             University Press, Cambridge, 2001. **220**

[Saa06]   Donald G. Saari. Which is better: The Condorcet or Borda winner? *Social Choice
             and Welfare*, 26(1):107–129, 2006.

[Sat75]   Mark Allen Satterthwaite. Strategy-proofness and Arrow's conditions: Existence
             and correspondence theorems for voting procedures and social welfare functions.
             J. Econom. Theory, 10(2):187–217, 1975. **220**

[Sch69]   David Schmeidler. The nucleolus of a characteristic function game. *SIAM J. Appl.
             Math.*, 17:1163–1170, 1969. **203**

[Sch93]   Klaus M. Schmidt. Commitment through incomplete information in a simple re-
             peated bargaining game. *Journal of Economic Theory*, 60(1):114–139, 1993. **253**

[Sch03]   Alexander Schrijver. *Combinatorial optimization: Polyhedra and efficiency*, vol-
             ume 24. Springer Science & Business Media, 2003. **60, 299**

[Sel65]   Reinhard Selten. Spieltheoretische behandlung eines oligopolmodells mit nach-
             frageträgheit: Teil i: Bestimmung des dynamischen preisgleichgewichts. *Zeitschrift
             für die gesamte Staatswissenschaft/Journal of Institutional and Theoretical Eco-
             nomics*, pages 301–324, 1965. **124**

[Sel75]   Reinhard Selten. Reexamination of the perfectness concept for equilibrium points
             in extensive games. *International Journal of Game Theory*, 4(1):25–55, 1975. **84**

[Sel98]   Reinhard Selten. Axiomatic characterization of the quadratic scoring rule. *Exper-
             imental Economics*, 1(1):43–62, 1998. **286**

[Sha53a]   C. E. Shannon. Computers and automata. *Proc. Inst. Radio Eng.*, 41:1234–1241,
             1953.

[Sha53b]   L. S. Shapley. A value for *n*-person games. In *Contributions to the theory of games,
             vol. 2*, Annals of Mathematics Studies, no. 28, pages 307–317. Princeton University
             Press, Princeton, N.J., 1953. **203**

[Sha79]   Adi Shamir. How to share a secret. *Comm. ACM*, 22(11):612–613, 1979.

[Sha17]   Haim Shapira. *Gladiators, Pirates and Games of Trust*. Watkins Publishing, Lon-
             don, 2017. **83**

[Sie13]   Aaron N. Siegel. *Combinatorial game theory*, American Mathematical Society,
             Providence, RI, 2013. **22**

[Sig05]   Karl Sigmund. John Maynard Smith and evolutionary game theory. *Theoretical
             population biology*, 68(1):7–10, 2005. **130**

[Sio58]   Maurice Sion. On general minimax theorems. *Pacific J. Math.*, 8:171–176, 1958.
             38

[Sky04]   Brian Skyrms. *The stag hunt and the evolution of social structure*. Cambridge
             University Press, 2004. **83**

[SL96]   B. Sinervo and C. M. Lively. The rock-paper-scissors game and the evolution of
             alternative male strategies. *Nature*, 380:240–243, March 1996. **136**

[Smi74]   J. Maynard Smith. The theory of games and the evolution of animal conflicts.
             Journal of Theoretical Biology, 47(1):209–221, 1974. **253**

[Smi82]   John Maynard Smith. *Evolution and the Theory of Games*. Cambridge University
             Press, 1982. **135**

[Sot96]   Marilda Sotomayor. A non-constructive elementary proof of the existence of stable marriages. *Games Econom. Behav.*, 13(1):135–137, 1996. **177**

[SP73]   John Maynard Smith and George. R. Price. The logic of animal conflict. *Nature*, 246:15, 1973. **135**

[Spe28]   E. Sperner. Neuer beweis für die invarianz der dimensionszahl und des gebietes. *Abh. Math. Sem. Univ. Hamburg*, 6(1):265–272, 1928. **102**

[Spe74]   Andrew Michael Spence. *Market signaling: Informational transfer in hiring and related screening processes.* Harvard Univ Press, 1974. **124**

[Spr36]   R. Sprague. Über mathematische kampfspiele. *Tôhoku Math. J.*, 41:438–444, 1935–36. **21**

[Spr37]   R. Sprague. Über zwei abarten von Nim. *Tôhoku Math. J.*, 43:351–359, 1937. **21**

[SS71]   Lloyd S. Shapley and Martin Shubik. The assignment game i: The core. *International Journal of Game Theory*, 1(1):111–130, 1971. **300**

[SS74]   Lloyd Shapley and Herbert Scarf. On cores and indivisibility. *Journal of Mathematical Economics*, 1(1):23–37, 1974. **178**

[SS11]   Shai Shalev-Shwartz. Online learning and online convex optimization. *Foundations and Trends in Machine Learning*, 4(2):107–194, 2011. **320**

[ST13]   Vasilis Syrgkanis and Éva Tardos. Composable and efficient mechanisms. In *Proceedings of the Forty-Fifth Annual ACM Symposium on Theory of Computing*, pages 211–220. ACM, 2013. **155, 253**

[Sta53]   John M. Stalnaker. The matching program for intern placement. *Academic Medicine*, 28(11):13–19, 1953. **176**

[Ste48]   Hugo Steinhaus. The problem of fair division. *Econometrica*, 16(1), 1948. **192**

[Ste00]   I. Stewart. Hex marks the spot. *Sci. Amer.*, 283:100–103, Sept. 2000.

[Stu02]   Bernd Sturmfels. *Solving systems of polynomial equations*, volume 97 of *CBMS Regional Conference Series in Mathematics*. American Mathematical Society, Providence, RI, 2002. **83**

[STZ07]   Subhash Suri, Csaba D. Tóth, and Yunhong Zhou. Selfish load balancing and atomic congestion games. *Algorithmica*, 47(1):79–96, 2007.

[Su99]   Francis Edward Su. Rental harmony: Sperner's lemma in fair division. *Amer. Math. Monthly*, 106(10):930–942, 1999. **192, 300**

[SW57]   Maurice Sion and Philip Wolfe. On a game without a value. In *Contributions to the theory of games, vol. 3*, Annals of Mathematics Studies, no. 39, pages 299–306. Princeton University Press, Princeton, N. J., 1957. **38**

[SW01]   Ulrich Schwalbe and Paul Walker. Zermelo and the early history of game theory. *Games and Economic Behavior*, 34(1):123–137, 2001. **124**

[SY05]   Michael Saks and Lan Yu. Weak monotonicity suffices for truthfulness on convex domains. In *Proceedings of the 6th ACM Conference on Electronic Commerce*, pages 286–293. ACM, 2005. **286**

[Syr12]   Vasilis Syrgkanis. Bayesian games and the smoothness framework. *arXiv preprint arXiv:1203.5155*, 2012. **253**

[Syr14]   Vasilis Syrgkanis. *Efficiency of Mechanisms in Complex Markets.* PhD thesis, Cornell University, 2014. **155, 253**

[Sch60]   Thomas Schelling. *The strategy of conflict.* Harvard University Press, 1960. **84**

[Tod63]   Masanao Toda. Measurement of subjective probability distributions. Technical report, DTIC Document, 1963. **287**

[Tra14]   Kevin Trahan. Purple pricing: Northwestern finds a solution to the flawed way college football teams sell tickets. *Forbes Magazine*, 2014. `http://www.forbes.com/sites/kevintrahan/2014/10/21/a-solution-to-the-flawed-way-college-football-teams-sell-tickets/`. **253**

[TvS02]   Theodore L. Turocy and Bernhard von Stengel. Game theory. In Encyclopedia of Information Systems. Technical report, CDAM Research Report LSE-CDAM-2001-09, 2002. **83, 124**

[Var07]   Hal R. Varian. Position auctions. *International Journal of Industrial Organization*, 25(6):1163–1178, 2007. **274**

[Vat89]   John H. Vande Vate. Linear programming brings marital bliss. *Operations Research Letters*, 8(3):147–153, 1989.

[Vet02]   Adrian Vetta. Nash equilibria in competitive societies, with applications to facility location, traffic routing and auctions. In *FOCS 2002 — Proceedings of the 43rd Annual IEEE Symposium on Foundations of Computer Science*. IEEE, 2002. **155**

[Vic61]   William Vickrey. Counterspeculation, auctions, and competitive sealed tenders. *The Journal of Finance*, 16(1):8–37, 1961. **252, 273, 286**

[vLW01]   J. H. van Lint and R. M. Wilson. *A course in combinatorics*. Cambridge University Press, Cambridge, second edition, 2001. **60**

[vN28]   J. v. Neumann. Zur Theorie der Gesellschaftsspiele. *Math. Ann.*, 100(1):295–320, 1928. **24, 38**

[vN53]   John von Neumann. A certain zero-sum two-person game equivalent to the optimal assignment problem. In *Contributions to the theory of games, vol. 2*, Annals of Mathematics Studies, no. 28, pages 5–12. Princeton University Press, Princeton, N.J., 1953. **60, 299**

[vNM53]   J. von Neumann and O. Morgenstern. *Theory of Games and Economic Behaviour*. Princeton University Press, Princeton, NJ., third edition, 1953. **38, 203**

[Voh11]   Rakesh V. Vohra. *Mechanism design: A linear programming approach*, volume 47. Cambridge University Press, 2011.

[Vov90]   Volodimir G. Vovk. Aggregating strategies. In *Proceedings of the Third Annual Workshop on Computational Learning Theory*, pages 371–386. Morgan Kaufmann Publishers Inc., 1990. **319**

[vR00]   J. van Rijswijck. Computer Hex: Are bees better than fruitflies? Master's thesis, University of Alberta, 2000.

[vR02]   J. van Rijswijck. Search and evaluation in Hex, 2002. http://www.cse.iitb.ac.in/~nikhilh/NASH/y-hex.pdf.

[War52]   John Glen Wardrop. Some theoretical aspects of road traffic research. In *Proceedings of Institute of Civil Engineers: Engineering Divisions*, volume 1, pages 325–362, 1952. **155**

[Wey50]   Hermann Weyl. Elementary proof of a minimax theorem due to von Neumann. In *Contributions to the Theory of Games*, Annals of Mathematics Studies, no. 24, pages 19–25. Princeton University Press, Princeton, N.J., 1950. **38**

[Wil86]   J. D. Williams. *The Compleat Strategyst: Being a primer on the theory of games of strategy*. Dover Publications Inc., New York, second edition, 1986.

[Win69]   Robert L. Winkler. Scoring rules and the evaluation of probability assessors. *Journal of the American Statistical Association*, 64(327):1073–1078, 1969. **287**

[WM68]   Robert L. Winkler and Allan H. Murphy. "Good" probability assessors. *Journal of Applied Meteorology*, 7(5):751–758, 1968. **287**

[Wyt07]   W. A. Wythoff. A modification of the game of Nim. *Nieuw Arch. Wisk.*, 7:199–202, 1907.

[Yan]   Jing Yang. Hex solutions. http://www.ee.umanitoba.ca/~jingyang/. **21**

[Yao15]   Andrew Chi-Chih Yao. An n-to-1 bidder reduction for multi-item auctions and its applications. In *Proceedings of the Twenty-Sixth Annual ACM-SIAM Symposium on Discrete Algorithms*, pages 92–109. SIAM, 2015. **253**

[YLP03]   J. Yang, S. Liao, and M. Pawlak. New winning and losing positions for Hex. In J. Schaeffer, M. Muller, and Y. Bjornsson, editors, *Computers and Games: Third International Conference*, pages 230–248. Springer-Verlag, 2003.

[YZ15]   Peyton Young and Shmuel Zamir. *Handbook of game theory*. Elsevier, 2015. **22, 135, 136, 286**

[Zah75]   Amotz Zahavi. Mate selection—a selection for a handicap. *Journal of Theoretical Biology*, 53(1):205–214, 1975. **124**

[Zer13]   Ernst Zermelo. Über eine anwendung der mengenlehre auf die theorie des schachspiels. In *Proceedings of the Fifth International Congress of Mathematicians*, volume 2, pages 501–504. II, Cambridge UP, Cambridge, 1913. **124**

[Zif99]   Robert M. Ziff. Exact critical exponent for the shortest-path scaling function in percolation. *J. Phys. A*, 32(43):L457–L459, 1999.

Index